MW00337104

BEACH AND DUNE RESTORATION

This new edition – now with Nancy L. Jackson as a coauthor – continues the themes of the first edition: the need to restore the biodiversity, ecosystem health, and ecosystem services provided by coastal landforms and habitats, especially in the light of climate change. The second edition reports on progress made on practices identified in the first edition, presents additional case studies, and addresses new and emerging issues. It analyzes the trade-offs involved in restoring beaches and dunes – especially on developed coasts – the most effective approaches to use, and how stakeholders can play an active role. The concept of restoration is broad, and includes physical, ecological, economic, social, and ethical principles and ideals. The book will be valuable for coastal scientists, engineers, planners, and managers, as well as shorefront residents. It will also serve as a useful supplementary reference textbook in courses dealing with issues of coastal management and ecology.

KARL F. NORDSTROM is Distinguished Professor Emeritus in the Department of Marine and Coastal Sciences at Rutgers University. He has 45 years of experience in conducting coastal research. He is a fellow of the American Association for the Advancement of Science and the Geological Society of America. His books include *Estuarine Beaches* (1992), *Beaches and Dunes of Developed Coasts* (2000), and *Beach and Dune Restoration* (2008). He has published more than 160 scholarly articles.

NANCY L. JACKSON is Professor Emerita in the Department of Chemistry and Environmental Science at New Jersey Institute of Technology. She has 30 years of research experience on beach and dune systems. She is a fellow of the American Association for the Advancement of Science and the Geological Society of America and was Fulbright Distinguished Chair and Scholar. She has published more than 100 scholarly articles.

From reviews of the first edition:

'... informs and educates stakeholders about potential viable alternative methods of managing developing landforms with the view to maintaining their function in line with stakeholder interests, while allowing natural processes to progress, further improving stability and diversity in beach and dune systems.'

– Environmental Conservation

"...an excellent, well-written resource ... recommended."

– CHOICE

Praise for the second edition:

'Read this book for a thorough and up-to-date account of the methods currently used in dune and beach restoration. Nordstrom and Jackson are world leaders in this field and they use a multitude of real-life case studies to illustrate the methods described. The work is contextualized in the framework of international agreements on biodiversity and habitat preservation, that are tempered by local demands and actions. Importantly, the authors talk about various categories and goals of restoration that allow the reader to differentiate the various paradigms within which restoration is undertaken. This will appeal to those involved in coastal conservation, engineering and management.'

– Andrew Cooper, Ulster University

'Nordstrom and Jackson deliver fundamental insights into the complex dynamics of the world's human-altered coastlines. Essential reading for understanding the enigmatic ways in which humans change the physical coastal systems in which we live ... a masterwork on the geomorphic interventions that typify human-dominated coastlines. Anyone thinking about future coastal change needs this book.'

– Eli Lazarus, University of Southampton

BEACH AND DUNE RESTORATION

Second Edition

KARL F. NORDSTROM
Rutgers University

NANCY L. JACKSON
New Jersey Institute of Technology

CAMBRIDGE
UNIVERSITY PRESS

University Printing House, Cambridge CB2 8BS, United Kingdom

One Liberty Plaza, 20th Floor, New York, NY 10006, USA

477 Williamstown Road, Port Melbourne, VIC 3207, Australia

314–321, 3rd Floor, Plot 3, Splendor Forum, Jasola District Centre, New Delhi – 110025, India

103 Penang Road, #05–06/07, Visioncrest Commercial, Singapore 238467

Cambridge University Press is part of the University of Cambridge.

It furthers the University's mission by disseminating knowledge in the pursuit of education, learning, and research at the highest international levels of excellence.

www.cambridge.org
Information on this title: www.cambridge.org/9781316516157
DOI: 10.1017/9781108866453

© Cambridge University Press 2022

First published 2022

A catalogue record for this publication is available from the British Library.

Library of Congress Cataloging-in-Publication Data
Names: Nordstrom, Karl F., author. | Jackson, Nancy L., author.
Title: Beach and dune restoration / Karl F. Nordstrom, Rutgers University, Nancy L. Jackson, New Jersey Institute of Technology.
Description: Cambridge, United Kingdom ; New York, NY, USA : Cambridge University Press, 2022. | Includes bibliographical references and index.
Identifiers: LCCN 2021026917 (print) | LCCN 2021026918 (ebook) | ISBN 9781316516157 (hardback) | ISBN 9781108791687 (paperback) | ISBN 9781108866453 (epub)
Subjects: LCSH: Beach nourishment. | Sand dune conservation. | Shore protection. | Beach erosion.
Classification: LCC TC332 .N67 2021 (print) | LCC TC332 (ebook) | DDC 627/.58–dc23
LC record available at https://lccn.loc.gov/2021026917
LC ebook record available at https://lccn.loc.gov/2021026918

ISBN 978-1-316-51615-7 Hardback

Contents

Contributors

Edward Anthony, Aix-Marseille University, France

Bas Arens, Bureau for Beach and Dune Research, Soest, The Netherlands

Deon van Eeden, Vula Environmental Services, Cape Town, South Africa

Juan Gallego-Fernandez, University of Seville, Spain

Tim Kana, Coastal Science and Engineering, Inc., Columbia, SC, USA

Teresa Konlechner, University of Melbourne, Australia

Roy Lubke, Rhodes University, Grahamstown, South Africa

Marisa Martínez, Institute of Ecology A.C., Xalapa, Mexico

Luana Portz, The Coast University, Barranquilla, Colombia

Enzo Pranzini, University of Florence, Italy

Ken Pye, Kenneth Pye Associates, Crowthorne, Berkshire, United Kingdom

Preface

This book is an update of *Beach and Dune Restoration* (Nordstrom 2008). The aims of this second edition are to report on new research in the application of coastal geomorphology, ecology, and management to restoration; present results of progress on practices identified in the first edition; and address issues that have increased in popularity recently. These new issues include finding ways to address increases in rates of sea level rise as well as increases in the number of extreme events, adapting to change through managed realignment (retreat) on exposed coasts, altering shore protection structures to make beaches and dunes more dynamic, implementing hybrid projects and living shorelines that combine soft and hard solutions, rejuvenating dune landscapes by removing vegetation, and conducting mega nourishments (e.g., the "sand motor"). Attention has also increased on developing strategies to incorporate participation of local stake-holders and evaluating shorelines as a coupled natural–human system, with humans as intrinsic agents of landscape evolution. These new developments reflect an emerging shift toward thinking about coastlines as a product of physical, ecological, and human processes and the need to integrate these processes in restoration practice. The book is intended to offer readers an understanding of how basic and applied research findings can inform beach and dune restoration efforts at local and regional scales.

Many past transformations of the coastal landscape, even those involving construction of new landforms, were done with little thought given to the accompanying environmental losses and the potential for achieving new environmental gains. Traditional beach- and dune-building practices emphasized the use of landforms for protection and recreation, but that does not preclude adding new natural resource values compatible with those uses. In many cases, traditional shore protection projects can be modified to achieve nature goals with little change in design or cost. We acknowledge that human-use functions will be the driving force for managing beaches in developed areas, so a return to a condition of pristine nature is not an option. Restored landforms and habitats will be subject to direct human use or indirect effects resulting from land uses in adjacent areas, so restored landforms may require periodic human adjustments to survive. The impossibility of returning to pristine nature should not deter efforts to regain elements of the natural environment and reverse the trend toward environmental loss.

The great competition for space near the land–ocean interface and the increasing demands of different interest groups require evaluation of beaches and dunes in a framework that considers physical, ecological, and social goals and objectives and the trade-offs and compromises involved. This focus on compromise and the need to accommodate different user groups is a distinguishing characteristic of this book. Another difference between this book and other books on restoration and management is the insight provided about restoration efforts at the local (municipal and property owner) scale. To many people, coastal restoration implies nourishing beaches, building dunes, and eliminating exotic species in dune preserves. We feel that restoration actions can extend beyond the ways they have traditionally been applied. Our working assumption is that some nature is better than none, even if it is imperfect, providing that no better option is available given the economic or political climate at the time. We consider this assumption valid if the restored environments are considered interim states that will be improved as natural features become more acceptable to stakeholders and greater resources are devoted to sustaining them.

Specific examples are used in many parts of the book to illustrate management practices. Several countries are highlighted, reflecting the greater number and scale of restoration activities there and the number of publications generated. The case studies mentioned may be in a specific location, but results are framed in terms of generic needs and capabilities and include citations to studies from other countries that support the findings. Species names may be different in different parts of the world, but pioneer and dune-building species and exotic species, for example, can play similar roles throughout the world, regardless of their species name. Many information sheets produced by government departments and environmental commissions for management of beaches and dunes are specific to their regions and readily available online. These information sheets and numerous technical reports provide practical guidelines for activities such as emplacing sediment, installing sand-trapping fences, and planting vegetation on dunes. Our intent is to provide a companion volume to design manuals rather than a substitute for them by presenting the broader rationale for restoration and introducing practitioners to approaches that may be unfamiliar to them but can be tailored to enhance local projects. Shore protection and restoration projects are inherently interdisciplinary, with geomorphologic, sedimentologic, biologic, economic, engineering, and regulatory inputs, requiring a synthesis of these interrelated themes.

Readers familiar with the first edition will note the inclusion of a new chapter (Chapter 5) that addresses the rationale and results of changing the effects of hard structures to make them more compatible with restoration goals, and a change in the order of Chapter 8 that is placed after the chapter that preceded it in the first edition. Many advances have been made in beach and dune restoration since 2008 as reflected in the addition of more than 500 references to this new edition. The recent literature on beaches and dunes is vast and we had to be somewhat selective. We retained the earlier citations to give credit to the people who originated many of the ideas that are supported in subsequent studies. We selected new studies that specifically identify the goals and implications of restoration projects or address landscape alterations that have implications for restoration, even if restoration was not an original goal. Regional remote sensing datasets are now publically available in many countries as well as many different techniques to survey beach/dune systems, including airborne LiDAR, ARGUS cameras, drones, and terrestrial laser scanners. Model studies of potential effects of landscape modifications and scientific studies of beaches and dunes provide insight to beach and dune change. Studies that incorporate these techniques are only included here if they are accompanied by specific recommendations for changes in restoration practice. Similarly, we do not include studies that concentrate on single species unless those studies provide information that can inform restoration of more inclusive sub-environments or unless those efforts appear to potentially cause degradation of other natural functions or alter evolutionary trajectories in an undesirable way.

Our emphasis is on trying to find ways to modify beaches and dunes to enhance natural processes and make natural habitats as dynamic and resilient as possible while maintaining their value for human use. This is a difficult goal to accomplish. We hope that identifying the many ways restoration can be envisioned and practiced will encourage managers to try new ways to enhance coastal landscapes within their jurisdictions and contribute to local and broader sustainability goals.

Acknowledgments

Financial support for the many projects that led to results published in this book was provided by the US Fulbright Commission, National Geographic Society, Interdisciplinary Global Joint Research Grant of Nihon University for 2001 for the Study on Erosion Control of National Land, US National Science Foundation, US National Park Service, and US National Oceanic and Atmospheric Administration Office of Sea Grant.

We are grateful to the following people for contributing to the first and second editions of this book, either by contributing information and ideas or for help gathering information in the field: Pierluigi Aminti, Bas Arens, Amanda Babson, Derry Bennett, Peter Best, John van Boxel, Alan Brampton, Harry de Butts, Dave Carter, Laura Caruso, Massimo Coli, Christopher Constantino, Skip Davis, Ian Eliot, Lucia Fanini, Giorgio Fontolan, Amy Freestone, Ulrike Gamper, Emir Garilao, Jeff Gebert, Gregorio Gómez-Pina, Rosana Grafals-Soto, D'Arcy Greene, Steven Handel, Jean Marie Hartman, Patrick Hesp, Woody Hobbs, Jacobus Hofstede, Shintaro Hotta, David Jenkins, Jim Johannessen, Marcha Johnson, Kayla Kaplan, Kate Korotky, Reinhard Lampe, Sonja Leipe, Robert Martucci, Brooke Maslo, Mark Mauriello, Anton McLachlan, Frank van der Meulen, Chris Miller, Julian Orford, Orrin Pilkey, Luana Portz, Enzo Pranzini, Patricia Rafferty, Nicole Raineault, Tracy Rice, Charles Roman, Helene Ruz, Sherestha Saini, Felicita Scapini, Douglas Sherman, Hugh Shipman, William Skaradek, David Smith, Horst Sterr, Thomas Terich, Kim Tripp, Lisa Vandemark, Allan Williams, Eric Wojciechowski, and Kit Wright.

We are especially grateful to the following people who provided text and photos about case studies in their countries: Edward Anthony, Bas Arens, Deon van Eeden, Juan Gallego-Fernández, Tim Kana, Teresa Konlechner, Roy Lubke, Marisa Martínez, Luana Portz, Ken Pye, and Enzo Pranzini.

1

The Need for Restoration

1.1 The Problem

The shorelines of the world continue to be converted to artifacts through human actions such as eliminating dunes to facilitate construction of buildings and support infrastructure; grading beaches and dunes flat to facilitate access and create space for recreation; and mechanically cleaning beaches to make them more attractive to beach users. Beach erosion, combined with human attempts to retain shorefront buildings and infrastructure in fixed positions, can result in truncation or complete loss of beach, dune, and active bluff environments. The lost sediment may be replaced in beach nourishment operations, but nourishment is usually conducted to protect shorefront buildings or provide a recreational platform rather than to restore natural values (Figure 1.1). Sometimes nourished beaches are capped by a linear dune of uniform height alongshore designed to function as a dike against wave attack and flooding. Many of these dune dikes are built by earth-moving machinery rather than by aeolian processes, resulting in an outward form and an internal structure that differ from natural landforms. Dunes on private properties landward of public beaches are often graded and kept free of vegetation or graded and vegetated with a cover that may bear little similarity to a natural cover.

Transformation of coastal environments is intensified by the human tendency to move close to the coast (Smith 1992; Wong 1993; Lubke et al. 1995; Roberts and Hawkins 1999; Brown and McLachlan 2002; Brown et al. 2008; Romano and Zullo 2014). The effects of global warming and sea level rise and increased impact of coastal storms are added to human pressures (FitzGerald et al. 2008; Boon 2012; Cazenave and Le Cozannet 2013; Miller et al. 2013; Stocker et al. 2013; Vousdoukas et al. 2018; Barnard et al. 2019; Fang et al. 2020; Mathew et al. 2020), further amplifying the transformation. The most significant threat to coastal species is loss of habitat, especially if sea level rise is accompanied by increased storminess (Brown et al. 2008). Coastal development has already eliminated much natural beach and dune habitat worldwide (Defeo et al. 2009).

1

Figure 1.1 La Victoria Beach in Cádiz, Spain, showing lack of topography and vegetation on a nourished beach maintained and cleaned for recreation.

Sandy beach ecosystems could adapt to storms and sea level rise by retreating landward and maintaining structure and function over various spatial and temporal scales (Orford and Pethick 2006; Cooper and McKenna 2008; Berry et al. 2013). Retreat from the coast (managed realignment) would resolve problems of erosion and provide space for new landforms and biota to become reestablished. The advantages of adapting by retreat are acknowledged, but actual responses by removing human structures from developed ocean coasts are limited and often resisted by the public (Ledoux et al. 2005; Abel et al. 2011; Luisetti et al. 2011; Morris 2012; Niven and Bardsley 2013; Cooper and Pile 2014; National Research Council 2014; Costas et al. 2015; Harman et al. 2015). Retreat appears unlikely except on sparsely developed shores (Kriesel et al. 2004). Most local governments and property owners would probably advocate management options that approach the status quo (Leafe et al. 1998), even under accelerated sea level rise (Titus 1990). The great value of land and real estate on coastlines and the level of investment on developed shores are too great to consider anything short of holding the line (Nordstrom and Mauriello 2001; Niven and Bardsley 2013).

Post-storm evaluations reveal ample evidence of the vulnerability of shorefront houses and infrastructure to storm damage (Saffir 1991; Sparks 1991; Platt et al. 2002; Kennedy et al. 2011; Hatzikyriakou et al. 2016; Hu et al. 2016; O'Neil and Van Abs 2016). Despite this destruction, strategies for reducing the number of people and buildings at risk and redistributing the risks, benefits, and costs among stakeholders are rarely implemented (Abel et al. 2011; Rabenold 2013; National Research Council 2014). More often, buildings and support infrastructure destroyed by long-term erosion or major storms are quickly rebuilt in reconstruction efforts, and landscapes after the storm often bear a greater human imprint than landscapes before the storm (Fischer 1989; Meyer-Arendt 1990; FitzGerald et al. 1994; Nordstrom and Jackson 1995; Platt et al. 2002). Market and policy incentives for development and redevelopment in coastal communities can overwhelm attempts of planners to discourage development (Andrews 2016; Holcomb 2016).

Problems associated with conversion of coasts to accommodate human uses include loss of topographical variability (Nordstrom 2000); loss of natural landforms and their habitats (Beatley 1991; Garcia-Lozano et al. 2018; Pérez-Hernández et al. 2020); reduction in space for remaining habitat (Dugan et al. 2008) or elimination of biota from those surfaces (Kelly 2014); fragmentation of landscapes (Berlanga-Robles and Ruiz-Luna 2002; Masucci and Reimer 2019); threats to endangered species (Melvin et al. 1991; Maslo et al. 2019); reduction in seed sources and decreased resilience of plant communities following loss by storms (Cunniff 1985; Gallego-Fernández et al. 2020); loss of intrinsic value (Nordstrom 1990); loss of original aesthetic and recreational values (Cruz 1996; Demirayak and Ulas 1996); and loss of the natural heritage or image of the coast, which affects appreciation of environmental benefits or ability of stakeholders to make informed decisions on environmental issues (Télez-Duarte 1993; Golfi 1996; Nordstrom et al. 2000; Gesing 2019; Lapointe et al. 2020). Natural physical and ecological processes cannot be relied on to reestablish natural characteristics in developed areas without people allowing these processes to occur. Establishment of coastal preserves, such as state/provincial or national parks, helps maintain environmental inventories, but inaccessibility to these locations or restrictions on intensity of use do not provide the opportunity for many visitors to experience nature within them (Nordstrom 2003). Natural enclaves near regions that are intensively developed may be too small to evolve naturally (De Ruyck et al. 2001). Natural enclaves near regions with shore protection structures are subject to sediment starvation and accelerated erosion that alter the character and function of habitat from former conditions (Roman and Nordstrom 1988). Natural processes may be constrained within areas managed for nature protection because of the need to modify those environments to provide predictable levels of flood protection for adjacent developed areas (Nordstrom et al. 2007c). Designating protected areas to preserve endangered species may have a limited effect in reestablishing natural coastal environments unless entire habitats or landscapes are included in preservation efforts (Waks 1996; Watson et al. 1997). Establishment of coastal preserves may also have the negative effect of providing an excuse for ignoring the need for nature protection or enhancement in areas occupied by humans (Nordstrom 2003).

Alternatives that enhance the capacity of coastal systems to respond to perturbations by maintaining diverse landforms and habitats should supplement or replace alternatives that resist the effects of erosion and flooding associated with climate change (Nicholls and Hoozemans 1996; Klein et al. 1998; Nicholls and Branson 1998; Orford and Pethick 2006; Cooper and McKenna 2008). Restoring lost beach and dune habitat can compensate for environmental losses elsewhere, protect endangered species, retain seed sources, strengthen the drawing power of the shore for tourism, and reestablish an appreciation for naturally functioning landscape components (Breton and Esteban 1995; Breton et al. 1996, 2000; Nordstrom et al. 2000). Ecologists, geomorphologists, and environmental philosophers point to the need to help safeguard nature on developed coasts by promoting a new nature that has an optimal diversity of landforms, species, and ecosystems that remain as dynamic and natural as possible in appearance and function while being compatible with human uses (van der Maarel 1979; Westhoff 1985, 1989; Doody 1989; Roberts 1989; Light and Higgs 1996; Pethick 1996; Nordstrom et al. 2000). There is also a growing interest in trying to develop a new symbiotic, sustainable relationship between human society and nature (with its diversity and dynamism) and to value the natural world for the sake of relations between humans and nature (Jackson et al. 1995; Cox 1997; Naveh 1998; Higgs 2003; Gesing 2019).

1.2 Human Modifications

Coastal changes and economically driven human actions are now recognized as mutually linked, often in iterative cycles, such as erosion and mitigation (Lazarus et al. 2011, 2016; Murray et al. 2013; Gopalakrishnan et al. 2016). The increasing pace of human alterations and the increasing potential for people to reconstruct nature for human use require reexamination of human activities (Table 1.1) in terms of the many ways they can be made more compatible with nature. Some environmental losses are associated with every modification, even the most benign ones, but the losses may be small and temporary. Human-modified landforms are often smaller than their natural equivalents, with fewer distinctive sub-environments, a lower degree of connectivity between sub-environments, and

Table 1.1. *Ways that landforms and habitat are altered by human actions*

Eliminating for alternative uses
Constructing buildings, transportation routes, promenades
Constructing alternative surfaces
Mining

Altering through use
Trampling
Off-road vehicle use
Fishing and harvesting
Grazing
Extracting oil, gas, water
Laying pipelines
Extracting and recharging water
Military activities

Reshaping (grading)
Piling up sand to increase flood protection levels
Removing sand that inundates facilities
Breaching barriers to control flooding
Dredging channels to create or maintain inlets
Widening beaches to accommodate visitors
Eliminating topographic obstacles to facilitate access or construction
Removing dunes to provide views of the sea
Building more naturalistic landscapes

Altering landform mobility
Constructing shore protection and navigation structures
Constructing marinas and harbors
Placing structures between sediment sources and sinks
Introducing more- or less-resistant sediments into beach or dune
Clearing the beach of litter
Stabilizing landforms using sand fences, vegetation plantings, or resistant materials
Remobilizing landforms by burning or removing vegetation

Altering external conditions
Damming or mining streams
Diverting or channelizing runoff
Introducing pollutants
Saltwater intrusion

Creating or changing habitat
Nourishing beaches and dunes
Restoring sediment budgets (bypass, backpass)
Burying unwanted or unused structures
Creating environments to attract wildlife
Controlling vegetation by mowing, grazing, fires

Table 1.1. (*cont.*)

Removing or cleaning polluted substrate
Adding species to increase diversity
Introducing or removing exotic vegetation
Introducing pets or feral animals

(*Sources*: Ranwell and Boar 1986; Nordstrom 2000; Doody 2001; Brown and McLachlan 2002; Brown et al. 2008).

often a progressive restriction in the ability of the coast to adjust to future environmental losses (Pethick 2001). The challenge for restoration initiatives is to enhance the natural value of these landforms and increase their resilience through human actions.

Human actions are not always negative, especially if applied in moderation and with an awareness of environmental impacts. The effects of agriculture can range from "disastrous" to adding real ecological values (Heslenfeld et al. 2004). Intensive grazing can destroy vegetation cover and mobilize entire dune fields, but controlled grazing can restore or maintain species diversity (Grootjans et al. 2004; Kooijman 2004). Many sequences of vegetation succession, now assumed to be natural, appear to have been initiated by human activity when examined more closely (Jackson et al. 1995; Doody 2001). Many changes, such as conversion of the Dutch dunes to recharge areas for drinking water, introduced a new landscape, with different species utilization and different uses for nature appreciation (Baeyens and Martínez 2004). The protection provided by beaches and dunes altered by humans to provide flood protection has allowed more stable natural environments to form that have developed their own nature conservation interest and may even be protected by environmental regulations (Doody 1995; Orford and Jennings 1998). The messages are clear. Humans are responsible for nature, and human actions can be made more compatible with nature by modifying practices to retain as many natural functions as possible in modifying landscapes for human use. These two messages are increasingly applicable in natural coastal areas used as parks as well as developed areas.

1.3 Values, Goods, and Services of Beaches and Dunes

Beaches and dunes have their own intrinsic value, but they also provide many goods and services of direct and indirect benefit to humans (Table 1.2). In coastal environments, a good can range from nesting and incubation sites for commercial marine species to potable water supplied by dune fields. Services can include filtering of pollutants or natural beach accretion that provides recreational

Table 1.2. *Values, goods, and services provided by coastal landforms, habitats, or species*

Protection for human structures (providing sediment, physical barrier, or resistant vegetation)
Subsistence for local human populations (food, fuel, medicinal material)
Market for real estate and resort economies
Raw materials for construction and industrial use
Sites for active recreation
Aesthetic, psychological, therapeutic opportunities
Filtering pollutants
Source of groundwater (in dunes)
Denitrification
Ecological niche for plants adapted to extreme conditions
Habitable substrate for invertebrates
Refuge areas (e.g., invertebrates in wrack; rabbits in dunes)
Nest or incubation sites (e.g., turtles, horseshoe crabs, grunion, surf smelt in beaches)
Food for primary consumers (e.g., invertebrates in wrack)
Food for higher trophic levels (scavengers and predators)
Sequestering carbon
Reducing concentrations of greenhouse gases
Providing synergistic benefits of multiple habitat types (e.g., corridors)
Education and research
Intrinsic value

(*Sources*: Lubke and Avis 1998; Arens et al. 2001; Peterson and Lipcius 2003; Everard et al. 2010; Barbier et al. 2011).

opportunities. The term "value" used in this book indicates human value (that assigned to the beach/dune system as a result of the goods and services produced) and the natural value inherent to the beach/dune system independent of human assignation (intrinsic value). The term "natural capital" is often applied as an alternative term for goods and services. Debates over the natural capital concept may occur because it appears to reduce nature to its monetary exchange or instrumental value and cannot be justified in arguments about preservation or intrinsic value (Daly 2020; Des Roches 2020). For example, a wide beach and dune that provide storm protection to oceanfront property and recreation for tourists can be placed in economic terms (Lazarus et al. 2011; Taylor et al. 2015), whereas the significance for a noncommercial species or the therapeutic use of a landscape is more challenging. Optimum designs for beaches and dunes that enhance these alternative uses can be quite different, as will be seen in subsequent chapters. The natural capital concept has many applications useful to economic or social arguments and philosophical debate but appears less convincing as a rationale for managing natural resources than the terms "values," "goods," or "services" without disciplinary contexts and subtexts. Accordingly, we use these terms, favoring the word "value" for general application.

Ecosystem services often exhibit spatial and temporal variation within a given landscape, and they can have weak association with other services (Biel et al. 2017). Not all goods and services can be provided within a given shoreline segment, even in natural systems, but many may be available regionally, given sufficient space and alongshore variation in exposure to coastal processes and sediment types. It may not be possible to take advantage of some of the goods and services, even where the potential exists. One or a few goods and services can be overexploited, resulting in the loss of others (Pérez-Maqueo et al. 2014). Mining and several forms of active recreation may be incompatible with use of beaches for nesting. Alternatively, beaches and dunes can have multiple uses, such as protecting property from coastal hazards while providing nesting sites, habitable substrate, and refuge areas for wildlife, if human uses are controlled using compatible regulations (Nordstrom et al. 2011).

Not all uses that take advantage of the goods and services provided by beaches and dunes should be targets of practical restoration efforts. It makes little sense to restore minerals in a landform only to mine them later. Provision of new sources of fuel to a subsistence economy may be accommodated in more efficient ways than attempting to favor driftwood accumulation on a restored beach or planting trees on a dune. In these cases, restoration efforts may not be required to sustain a human use, but they may be required to reinstate the microhabitats and associated processes that were lost through previous exploitation. Components of ecosystems should not be seen as exchangeable goals that can be used and created again to suit human needs (Higgs 2003; Throop and Purdom 2006).

1.4 Approaches for Restoring and Maintaining Natural Landforms and Habitats

Coastal evolution may take different routes. The development process need not result in reshaping of beaches into flat, featureless recreation platforms (Figure 1.1) or totally eliminating dune environments in favor of structures. The trend toward becoming a cultural artifact can be reversed, even on intensively developed coasts, if management actions are taken to restore natural features (Nordstrom et al. 2011). Creating and maintaining natural assemblages of landforms and biota in developed areas (Figure 1.2) can help familiarize people with nature, instill the importance of restoring or preserving it, enhance the image of a developed coast, influence landscaping actions taken by neighbors, and enhance the likelihood that natural features will be a positive factor in the resale of coastal property (Norton 2000; Savard et al. 2000; Conway and Nordstrom 2003). Tourism that is based on

Figure 1.2 Two views of Folly Beach, South Carolina, USA, showing the effect of beach nourishment in reestablishing dunes.

environmental values can extend the duration of the tourist season beyond summer (Turkenli 2005). Restoration of natural habitats thus has great human-use value in addition to natural value.

The reasons for restoration may be classified in different ways (Hobbs and Norton 1996; Peterson and Lipcius 2003) such as (1) improving habitat degraded by pollutants, physical disturbance, or exotic species; (2) replenishing resources depleted by overuse; (3) replacing landforms and habitats lost through erosion; (4) allowing human-altered land (e.g., farms) to revert to nature; (5) compensating for loss of natural areas resulting from construction of new human facilities; and (6) establishing a new landscape image or recapturing lost environmental heritage. Restoration can be conducted at virtually any scale and used in a variety of ways.

The terms conservation and restoration have been distinct as policy directives, but programs for either will have elements of the other within them. Maintenance of the abundance and diversity of natural landforms and habitats depends on the adoption of several actions including (1) identifying remaining natural and semi-natural habitat; (2) establishing new nature reserves; (3) protecting existing reserves from peripheral damage; (4) making human uses in other areas compatible with the need for sustainability; and (5) rehabilitating degraded areas, including reinstating natural dynamic processes (Doody 1995). The first four actions may be defined as conservation, whereas the last action may be defined as restoration. This book addresses this last need in its many forms and scales, with special emphasis on regional and local actions. The goals for conservation and restoration become complementary in a changing world in which species are shifting distributions, communities are disassembling and reassembling in new configurations, extreme events are pushing systems beyond past thresholds, and outcomes of human actions are increasingly uncertain (Wiens and Hobbs 2015). Restoration of a lost habitat makes that habitat a target of future conservation; maintenance of habitat in a conservation area subject to degradation by human activities may require periodic restoration efforts; and successful evolution of a restored area may depend on the proximity of seed banks in a nearby conservation area. Many of the management principles appropriate to conservation or restoration will apply to the other and can be tailored to the specific context. Restoration goals can be compatible with conservation goals in improving degraded habitat. Alternatively, restoration goals can counteract conservation goals; for example, using restoration as compensation or mitigation for development of natural areas – a practice severely criticized because of the difficulty of replicating lost environments (Zedler 1991).

Many coastal environments are protected by international, national, and regional policies governing use of natural resources and establishment of nature reserves by governmental and nongovernmental organizations (Doody 2001; Rhind and Jones 2009). The EU Habitats Directive and its designation of the Natura 2000 network of protected areas covering Europe's most valuable and threatened species and habitats is key to conservation and restoration in much of Europe. Natural habitat types and animal and plant species whose conservation requires the designation of special areas of conservation are identified as are criteria for selecting eligible sites for inclusion. Natura 2000 sites are protected through a series of policy instruments put in place by the directives and subsequently translated into legislation by EU countries. For example, Article 6 of the Habitats Directive requires member states to take measures within Natura 2000 to maintain and restore the habitats and species in a favorable conservation status and avoid activities that could significantly disturb these species or result in damage or deterioration of their habitats or habitat types. Other EU initiatives provide impetus, rationale, and suggestions for maintaining and enhancing key environmental resources. The EUrosion study (European Commission 2004) promotes the need to work with natural processes; for coasts, this means working with the sediments and processes as the foundation for the habitats and the species they support (Rees et al. 2015). Every country has numerous policies and regulations affecting management of coastal resources (e.g., the Wildlife and Countryside Act, the Convention on Biological Diversity, and Biodiversity Action Plan in the United Kingdom). About 83 percent of the dune area in the United Kingdom has been protected under legislation because of biological interest (Williams and Davies 2001). The number of national regulations can be large. Ariza (2011) lists twenty-nine legal texts for beach management in Spain alone. The number of regulations is not necessarily the answer to nature protection in some locations. Local actions can disregard the principles and guidelines developed at higher levels of government (Cristiano et al. 2018; Jayappa and Deepika 2018).

Evaluation of the significance of local legal provisions is beyond the scope of this book, but examples of international policy guidelines for conservation and

Table 1.3. *Policy guidelines for coastal management applicable to beach and dune systems*

Protect human lives and settlements
Increase the width or volume of beaches and dunes

Protect the coastal strip
Establish a protected zone landward and seaward of the water line
Restrict activities that permanently change the landscape

Preserve natural coastal dynamics
Establish no-development zones for nature protection and for buffers against sea level rise
Restrict or modify protection structures outside settlements to favor natural environments
Restrict defense measures where active bluffs supply sediments
Use natural materials such as stone, sand, soil, or wood in coastal defense structures
Consider mutual relationships between physiographic, ecological, and economic
    parameters
Prevent habitat fragmentation
Create and maintain corridors and patches acting as stepping stones for migration of wild
    species

Provide sustainable and environmentally friendly tourism and development
Adjust to the carrying capacity of the environment
Orient and manage tourism according to conservation goals
Establish new tourism facilities on sites where tourism facilities already exist
Increase environmental awareness of tourists
Have zero net loss of coastal habitat
Apply the "user pays" principle for environmental management, monitoring, and shore
    protection
Treat the coastline as public domain

Protect endangered or threatened species, biotopes, landscapes, and seascapes
Prohibit deliberate capture or killing of threatened species
Prohibit deliberate disturbance of these species during breeding, rearing, hibernation, and
    migration
Add provisions for biotope protection, giving preference to endangered or threatened areas
Prohibit activities that damage biotopes or require mitigation or compensation for damages
Conduct restoration projects for biotopes
Prevent the introduction of alien species

(*Sources*: Helsinki Commission URL www.helcom.fi/Recomendations.html; European
Code of Conduct for Coastal Zones [Council of Europe 1999]; Habitats Directive;
Nordstrom et al. (2007c).

restoration are presented here for perspective (Table 1.3). These guidelines can be applied at nearly any spatial and governmental level with minor changes in wording. The need to protect human lives and settlements is likely to be an overriding concern, but many activities taken to protect people can be made compatible with guidelines to allow natural dynamics, provide for sustainable development, and protect biotopes and landscapes (Table 1.3), and can be accommodated within traditional shore protection designs. Examples include (1) allowing dikes or protective dunes to erode to expose more land to episodic inundation by the sea while increasing the structural integrity of new dikes farther landward; (2) restoring sediment transfers from bluffs to adjacent shore segments; (3) increasing sediment transport rates through groin fields; and (4) replacing exotic vegetation with native species. New strategies are emerging for innovative nature-based coastal protection and management as an alternative or complement to hard solutions. New terms such as "ecoengineering," "building with nature," "green infrastructure," and "living shorelines" reflect these softer approaches (Pontee and Tarrant 2017; Martínez et al. 2019; Morris et al. 2019). These strategies include recreating natural habitats, enhancing existing habitats, using more organic materials for new hard structures, and enhancing existing hard infrastructure by favoring colonization by marine organisms or combined hard structures with natural features to form hybrid solutions (Firth et al. 2014; Pontee and Tarrant 2017). Awareness of a landscape's aesthetic appeal can lead to designs that are considered attractive as well as functional (Gobster et al. 2007; Williams et al. 2016), with natural features providing much of the aesthetic appeal.

One of the impediments to restoration efforts is the cost of projects. Application of the "user pays" principle to the requirement of zero net loss of coastal habitat (Table 1.3) can make restoration feasible by defraying costs using funds required for compensation or mitigation of environmentally damaging actions elsewhere, such as construction of marinas (Nordstrom et al. 2007c).

1.5 Definitions

Different approaches to restoration practice can occur depending on the balance sought between humans and nature or the disciplinary specialties of the parties conducting the restoration (Swart et al. 2001; Nordstrom 2003). The word restoration implies ecological restoration to many managers and planners, but coastal restoration also includes reestablishment of sediment budgets and space for landforms to undergo cycles of erosion, deposition, and migration (Orford and Pethick 2006; Cooper and McKenna 2008) as well as of previous human values, including cultural, historical, traditional, artistic, social, economic, and experiential (Nuryanti 1996; Higgs 1997; Wortley et al. 2013). Other terms may be used as synonyms or surrogates for restoration, including "nature development," "rehabilitation," "reclamation," "enhancement," "sustainable conservation" (that implies ongoing human input) among others introduced and defined in Corlett (2016). Terms used in a cultural restoration context may be similar to some used for environmental restoration or include "renovation" and "reconstruction." Like the related term "ecological engineering," many synonyms can be used for restoration, and many subdisciplines can claim to practice it, with different emphases (Mitsch 1998).

An ongoing discussion can be traced in the literature about the appropriate term to use for what practitioners of restoration do, and the meaning of terms may become vague, shift in their meaning, or fade from use (Halle 2007). There seems to be little value in substituting an alternative general term for restoration, when that new term may itself be subject to reinterpretation, misinterpretation, or misuse. The vagaries associated with the term restoration can be eliminated for beach and dune restoration initiatives by identifying the goals and actions taken. These include (1) restoring the sediment budget to provide substrate for new environments or a seaward buffer against wave or wind erosion to allow landward environments to evolve to more mature states; (2) restoring a vegetation cover on a backshore to cause dunes to evolve to provide habitat, protect against flooding, or create a more representative image of a natural coast; or (3) restoring the process of erosion and deposition by removing shore protection structures or allowing them to deteriorate. Allowing an aging shore protection structure to deteriorate, rather than rebuilding it, is an example of use of the concept of restoration as a positive way to promote what is essentially a low-cost, do-nothing alternative.

Ecological restoration has been defined as an intentional activity that initiates or accelerates the recovery of an ecosystem with respect to its health, integrity, and sustainability (Society for Ecological Restoration 2002). Elements of this definition would be appropriate for restoration to establish non-living components of a degraded environment, but strict adherence to ecological principles would not be required. The goal of ecological restoration is understood to emulate the structure, function, diversity, dynamics, and sustainability of the specified ecosystem, but the requirement that it represent a self-sustaining ecosystem that is not reliant on human inputs would have to be relaxed for systems that remain intensively used by humans (Aronson et al. 1993). Ecological fidelity should be at the core of restoration, but successive layers of human context must be added to produce an expanded conception of good ecological restoration for locations where humans remain part of the system (Higgs 1997). Present underlying principles for ecological restoration (Table 1.4) recognize the importance of these socioeconomic inputs.

1.6 Categories of Restoration

Management options that can be implemented to achieve a more naturally functioning beach and dune system must be based on practices that are doable as

Table 1.4. *Underpinning principles for ecological restoration*

Engage stakeholders
Satisfy cultural, social, economic, and ecological values
Foster a closer and reciprocal engagement with nature
Lead to improved social and ecological resilience and human well-being

Draw on many types of knowledge
Traditional practitioner ecological knowledge
Local ecological knowledge
Scientific discovery through experiments and practitioner–researcher collaborations

Be informed by native reference ecosystems while considering environmental change
Use reference models, targets, and goals based on specific real-world ecosystems
Account for temporal dynamics, including models of future change
Anticipate multiple or sequential references to identify potential successional paths
Acknowledge social/ecological histories and traditions in some areas

Support ecosystem recovery processes
Use remnant species to kick-start regeneration
Consider additional treatments or research to overcome limitations to recovery

Assess using clear goals and measurable indicators
Use natural and social attributes
Monitor progress over time (including former, restored, and subsequent conditions)
Apply adaptive management to systematically improve restoration success

Seek the highest level of recovery attainable
Start at highest level in the partial recovery/full recovery continuum
Adopt a policy of continuous improvement informed by sound monitoring

Apply at large scales to gain cumulative value
Include actions at landscape, reach/drift cell, and regional scales
Address degradation from outside restored site (pollutants, sediment blockages, exotics)
Enhance connectivity (e.g., wildlife corridors)
Acknowledge broader stakeholder challenges or support

Make part of a continuum of restorative activities
Acknowledge the physical and social context
Recognize the role of the project within a broad sustainability paradigm

---

(*Source*: Gann et al. 2019).

documented by evaluations of topography, biota, and perceptions of stakeholders. General restoration goals for human altered areas include the following: (1) using active management to retain a desired characteristic; (2) getting a site off a human trajectory; or (3) restoring functions rather than restoring original conditions with all of the former species present (Palmer et al. 1997; White and Walker 1997). The former goal (using a target species or landscaping approach) is designed to create a certain look quickly, whereas the second and third goals are designed to lead to a system that is self-sustaining, requiring no inputs of energy or materials from managers or, at the least, requiring minimal intervention (Jackson et al. 1995). The first goal (using active management) is especially appropriate for restoring and maintaining stable backdune environments on private properties, where intensive actions at a local scale are feasible. The other two goals are more appropriate in areas managed by a government entity, where actions such as suspension of beach raking or use of symbolic fences can be undertaken over larger distances at low unit costs (Nordstrom 2003). Total restoration may be untenable where too many side effects would conflict with the interests of many stakeholders, but a program of phased restoration may be advantageous (Pethick 2001).

Restoration approaches are often placed in three or more basic categories based on the degree of human input. Aronson et al. (1993) use the terms restoration, rehabilitation, and reallocation to distinguish the increasing human role. Lithgow et al. (2014) use the terms conservation, restoration, and rehabilitation. Swart et al. (2001) use three categories they call wilderness, arcadian, and functional. Within each of these categories, stakeholders may have alternative perspectives on the way landscapes may be appreciated and valued, such as ethical, aesthetic, or scientific (van der Windt et al. 2007). The key features in the wilderness approach are biological and physical processes such as erosion, sedimentation, decomposition, migration, and predation that operate in a self-regulating system with little or no human influence. This approach requires relatively large areas to allow the physical and biotic free interplay (Swart et al. 2001). Although the overall goal may be to emulate aspects such as structure, function, and diversity, these components may not be part of the initial restoration effort, and direct management should be restricted as much as possible (Turnhout et al. 2004). On sandy coasts, the wilderness approach could be used to convert land covered in pine plantations back to dynamic nature preserves (van der Meulen and Salman 1996). The wilderness approach will be successful if functional interests do not compete and no opposition occurs, and it would need strong advocates or top-down steering methods requiring political ownership of land, finances, or authority if it is implemented in populated areas (Swart et al. 2001).

The arcadian approach involves maintaining the landscape as semi-natural and extensively used, with human influence considered a positive element that can enhance biodiversity and create a harmonious landscape (Swart et al. 2001). This approach would be appropriate in converting a graded and raked beach that lacks a dune into a more naturally functioning beach/dune system, with value for nature-based tourism rather than more consumptive forms of tourism. Aesthetic appreciation is important in this approach.

The functional approach involves adapting nature to the current utilization of the landscape, including urban functions, where nature is often characterized by species that follow human settlement (Swart et al. 2001). The functional approach may be necessary where dunes occur on private properties, and successful restoration may require compatibility with resident conceptions of landscape beauty that may be based on patterns, lines, and colors rather than on the apparent disorder of nature (Mitteager et al. 2006). Horticulture, rather than restoration, may have to be accepted as the means to retain a desired image in spatially restricted environments (Simpson 2005). Economic considerations are critical in this approach.

The degree of rigor used to restore landforms and habitats and the metrics used to assess success will differ based on the approach taken. Ideally, criteria by which success of ecological restoration projects can be judged include (1) sustainability without continuous management; (2) productivity or increase in abundance or biomass; (3) recruitment of new species; (4) soil and nutrient buildup and retention; (5) biotic interaction; and (6) maturity relative to natural systems (Ewel 1990). Some of these criteria (especially 1 and 6) may be unachievable in systems that have undergone considerable alteration by humans. One challenge in assessing the value of restored coastal resources is to determine the metrics that characterize the health and services provided by the ecosystem and the degree of departure from and rate of natural recovery to conditions that would have prevailed in the absence of the human modification (Peterson and Lipcius 2003). A major problem is that restored environments cannot be expected to evolve as they would have in a natural system that occurred in the past because conditions outside their borders have changed from that past state in response to continued human inputs.

1.7 The Elusiveness of a Time-Dependent Target State

Much attention is now focused on resilience. In a geomorphic sense, resilience can be defined as the ability of a geomorphic system to recover from disturbance and the degrees of freedom to absorb or adjust to disturbance (Stallins and Corenblit 2018). Resilience is used in many scientific fields, often ambiguously, causing confusion over terminology and leading to different interpretations, even within the same scientific discipline (Kombiadou et al. 2019). Resilience of a dune can be defined as recovery to a pre-disturbance height and volume (Houser et al. 2015), which is compatible with traditional engineering attitudes about stability (Stallins and Corenblit 2018). This definition is likely to be readily acceptable in developed communities, where stability is often the management goal (Jackson and Nordstrom 2020). Resilience can also be defined as the potential to recover around a different set of interactions that may not result in a return to the original state (Stallins and Corenblit 2018). These two definitions represent alternative ways to manage developed systems, and the meaning of resilience must be better defined in geomorphic and ecologic terms before it can be operationalized in specific management strategies. It can be argued that recovery after perturbation should not be restricted to regaining pre-disturbance morphological dimensions but should be viewed in terms of reorganization and adaptation in order to regain or maintain functions (Kombiadou et al. 2019). Unfortunately, arguments indicating that allowing natural processes to prevail makes coasts more resilient are relatively easy to document but may be of little value in guiding management efforts on developed coasts, where landform mobility places buildings and infrastructure at risk (Jackson and Nordstrom 2020). Coastal resilience must refer to the capacity of the combined socioeconomic and natural systems to cope with disturbances, including sea level rise, extreme events, and human impacts, with enhancement of socioeconomic resilience often coming at the expense of natural resilience, and vice versa (Masselink and Lazarus 2019).

The landforms on low-lying coasts, such as barrier islands, are created and shaped by the very processes that threaten infrastructure on them, restricting options to provide protection to human structures using natural solutions (Jackson and Nordstrom 2020). Framing small-scale soft coastal protection strategies as "working with nature" is also counterintuitive where the new landforms and habitats are temporary and cannot maintain themselves over time without ongoing human assistance (Gesing 2019). The implications are that restoration efforts in locations where natural dynamism is desired, such as dune reserves, may have to differ substantially from efforts in developed areas, where dynamism must be limited. We agree with Alexander (2013) and Masselink and Lazarus (2019) that resilience is a multifaceted concept that is adaptable to alternative uses and contexts in different ways but not one that is readily used as a model or paradigm or well defined in management contexts.

The many changes that occur to the coastal landscape through time make the selection of a target state for restoration difficult (Corlett 2016) and leave the appropriate restoration approach up for considerable debate (e.g., Delgado-Fernandez et al. 2019; Arens et al. 2020; Creer et al. 2020; Pye and Blott 2020). Reconstruction of beach and dune conditions in the past reveals a series of states (Piotrowska 1989; Provoost et al. 2011), all of which have characteristics that would be favored by a percentage of stakeholders. Human influence can be documented well into the past in Europe, and a target state for a natural system could be based on conditions many millennia ago or at a more recent time, such as just prior to the industrial revolution. In the Marche region of central Italy, many of the sandy barriers that are eroding and require shore protection efforts are not representative of the gravelly and sandy pocket beaches that alternated with active cliffs until about 2000 BC (Coltorti 1997). The sandy barriers resulted from delivery of sediment following deforestation caused by human use of the hinterlands. A landscape prior to Roman influence would be different from either a landscape prior to the industrial revolution, when the sediment-rich beaches and foredunes were largely unaffected by human construction at the coast, or a modern landscape, when beaches and dunes are spatially restricted and eroding as a result of dam construction.

Studies of the Americas point out the problem of using the pre-Columbian landscape as the target state for a natural environment, given new evidence for the many alterations made by Native Americans (Denevan 1992; Higgs 2006). Specification that restoration must resemble an original, pre-disturbance state can become an exercise in history or social science. Even if an initial state could be specified for a region, it is rarely possible to determine what the landscape or ecosystem looked like, how it functioned, or what the full species list looked like (Aronson et al. 1993). Historical data provides useful background information, but past states may not provide appropriate restoration goals for a contemporary dynamic environment subject to a new suite of processes (Falk 1990; Hobbs and Norton 1996; Wiens and Hobbs 2015). The reference model for restoration should describe the approximate condition the site would be in if degradation had not occurred; this is not the same as the historic state, which does not account for the system to change during intervening conditions (Provoost et al. 2011: Gann et al. 2019). Analytical approaches must acknowledge past conditions but also future scenarios when key process drivers may change (Stein et al. 2020).

Populations, communities, or ecosystems identified for restoration may now be in alternative states, and restoration targets are biased by shifting historical baselines (Peterson and Lipcius 2003). Adherence to specific standards based on native reference ecosystems may be an unattainable target for an increasingly large number of restoration sites (Higgs et al. 2018), and restoration goals may need to include novel regimes because of changing environmental and socioeconomic conditions (Jentsch 2007). Beaches and dunes vary greatly in size, location, and surface cover and they are inherently dynamic, undergoing cycles of complete destruction and rebuilding over periods of decades. This dynamism implies that adherence to a specific pre-disturbance state may be less important than the opportunity for these landscapes to achieve spatially and temporally diverse transitory states that need not be large.

The impacts of global climate change, sea level rise, and increases in number and intensity of coastal storms introduce a high degree of uncertainty in the way coastal zones will change, and suggestions for achieving desirable restoration states will have to address this uncertainty (Stocker et al. 2013; Wiens and Hobbs 2015; Corlett 2016; Kopp et al. 2019). The causes of potential coastal changes are reviewed elsewhere (FitzGerald et al. 2008; Cazenave and Le Cozannet 2013; Miller et al. 2013; Stocker et al. 2013). Our concern is with finding ways to mitigate or adapt to these changes by restoring and managing coastal landscapes in ways that are compatible with ongoing changes introduced by natural processes and human actions. Sea level rise can be assessed through static inundation studies that flood existing topography with projected sea levels or dynamic response studies caused by anthropogenic, ecologic, or morphologic processes that drive coastal landscape evolution (Lentz et al. 2016). Beaches and dunes are dynamic, and effects of inundation and erosion will be partly compensated by adjustments within the natural system and human attempts to favor or restrict these adjustments. Our premise is that uncertainty should not deter efforts to accept more dynamism in restoration solutions that allow natural processes some freedom to adjust to change.

Uncertainties may make it difficult to attain definitive goals, so initial goals may need to change as a restored landscape evolves. Adaptive management is a systematic "learning by doing" approach for improving restoration practice and is suggested as the standard approach for any restoration project (Gann et al. 2019). Adaptive management provides a formalized framework for moving from goals and objectives through the planning and implementation of management actions, followed by monitoring and analysis to see what happens, and then using what has been learned to adjust the management actions (Wiens and Hobbs 2015).

1.8 Types of Restoration Project

Landforms and habitats can be restored in many ways and at many scales in projects designed explicitly for the targeted environments or in projects that have other objectives, where the restored environments are by-products (Table 1.5). Restoration projects can be conducted where a natural environment has lost so much value that its restoration is considered cost effective or where the existing human environment has so little remaining value that efforts to enhance or protect existing uses are discontinued (Nordstrom et al. 2007c). Many restoration efforts are attempts to overcome the more damaging actions identified in Table 1.1.

Direct creation of landforms and habitats (Table 1.5) usually involves reconstructing a landform to a specific design shape, followed by planting target vegetation to achieve an end state quickly or by planting pioneer vegetation where time and space are available to allow more mature habitats to evolve naturally at a slower rate. Modifying surfaces of existing landforms to accommodate species requires planting or removing vegetation. These actions can affect large areas and have great impact on the surrounding environment at a relatively low cost.

Table 1.5. *Ways that landforms and habitats can be restored on beaches and dunes*

Directly creating or recreating landforms and habitats
Restoring barrier island systems (Penland et al. 2005; O'Connell et al. 2005; FitzGerald
    et al. 2016)
Restoring pocket beaches in urban environments (Shipman 2001)
Increasing landform and habitat size to provide environmental corridors, vegetation
    mosaics, and zones of growth and decay.
Rehabilitating landforms after substrate removal for commercial use (Lubke et al. 1996;
    Lubke and Avis 1998; Gómez-Pina et al. 2002)
Rehabilitating drilling areas and pipeline corridors (Ritchie and Gimingham 1989; Taylor
    and Frobel 1990; Soulsby et al. 1997)
Creating habitat to mitigate for construction projects (Cheney et al. 1994)

Modifying sediment surfaces or morphology to accommodate species
Direct reshaping for species enhancement (Schupp et al. 2013)
Replanting surfaces damaged by human actions (Baye 1990; Gribbin 1990)
Removing exotic species from dunes (Choi and Pavlovic 1998; Hesp and Hilton 2013).
Removing trees from afforested areas (Lemauviel and Roze 2000; Lemauviel et al. 2003)
Mowing or cutting vegetation to increase diversity (Grootjans et al. 2004; Kooijman 2004)
Reinstating natural dynamics in overstablized dunes (van Boxel et al. 1997; Arens et al.
    2013a)

Modifying hard structures to reestablish landforms or accommodate species
Adding sediment or burying structures (Bocamazo et al. 2011; Nordstrom 2019)
Decreasing size or barrier effect (Pranzini et al. 2018b)
Changing slopes or surface complexity (Moschella et al. 2005; Chapman and Underwood
    2011)
Complete or partial removal (Zelo et al. 2000; Toft et al. 2010; de la Vega-Leinert et al.
    2018)
Allowing structures to deteriorate (Nordstrom et al. 2016)

Changing land use or human activities to allow nature to evolve
Removing or condemning threatened buildings (Rogers 1993)
Removing shorefront roads (Gómez-Pina et al. 2002; Portz et al. 2018)
Restricting/preventing raking or driving on the beach (Priskin 2003; Dugan and Hubbard
    2010)
Restricting mechanical dune grading (Nordstrom et al. 2011; Kelly 2014)
Controlling trampling of dunes (Eastwood and Carter 1981)
Favoring native species for landscaping on private lots (Mitteager et al. 2006)

Restoring sediment budgets
Nourishing beaches and dunes (Valverde et al. 1999; Hamm et al. 2002; Hanson et al.
    2002)
Reinstating sediment contributions from bluffs (Nordstrom et al. 2007c; Moore and Davis
    2015)
Increasing transport past groins and breakwaters (Donohue et al. 2004; Pranzini et al.
    2018b)
Installing sand bypass and backpass systems (Dean 2002; Nordstrom et al. 2002)

Table 1.5. (*cont.*)

Addressing effects emanating from outside the location to be restored
Reestablishing river hydraulic/sediment regimes (Willis and Griggs 2003; Warrick et al. 2019)
Mitigating pollutants (Willams and Tudor 2001; Kooijman and Smit 2001)
Mitigating sea level rise and erosion by beach fill (Houston 2017)
Mitigating sea level rise and erosion by retreat (Rhind and Jones 2009; Nordstrom et al. 2016)

Restoring human values with side benefits for natural landforms and habitats
Nourishing beaches for shore protection or recreation (Nordstrom et al. 2011)
Building dunes for shore protection (Freestone and Nordstrom 2001; Taylor et al. 2015)

Only a few references are provided here. Additional references are in appropriate chapters where themes are elaborated.

Changing land use to provide space for nature to evolve (Table 1.5) can result in fully functioning natural environments, especially where human structures that restrict natural processes are removed. This option may be difficult to implement because compensation for loss of previous uses may be costly, and previous users may be reluctant to relinquish control, regardless of compensation. Restoring sediment budgets may be a more palatable alternative to changing land use because it allows natural environments to form seaward of existing structures (Figure 1.2). The new environments will be spatially restricted unless the previous rate of sediment input is exceeded, allowing accretion to occur. Restoring sediment budgets can be the most environmentally compatible method of retaining environments on eroding shores where retreat is not an option, depending on the rate of introduction of the sediment and the degree to which the sediment resembles native materials. Maintaining natural rates of sediment transport can be costly because of the expense of maintaining permanent transfer systems and the difficulty of finding suitable borrow materials that are not in environmentally sensitive areas.

Restoration actions outside the coast (Table 1.5) can have windfall side benefits on beaches and dunes. Reestablishing hydraulic regimes in rivers by removing dams would increase the delivery of sediment to coasts (Warrick et al. 2019). Reducing pollutant levels in runoff or in the atmosphere would favor return of vegetation to former growing conditions. Actions taken to restore human use values can have significant side benefits. Nourishing beaches for shore protection or recreation provides space for natural features to form, and building dunes for shore protection provides substrate that natural vegetation can colonize. The barrier against salt spray, sand inundation, and flooding provided by a foredune built to protect human infrastructure allows for survival of a species-rich backdune, even on spatially restricted dunes on intensively developed coasts.

Windfall benefits may accrue regardless of the rationale for the many projects that become de facto restorations. Making the goals of each project broader by identifying these benefits may attract the interest and support of a greater number of stakeholders. A case can be made for projects that restore ecosystem functions rather than a specific species or set of species or a cosmetic landscape surface (Choi 2007), but in developed systems any change toward a more naturally functioning coastal system that is compatible with human use is desirable.

1.9 Scope of Book

The focus of this book is on developed coasts, where restoration to a previously undisturbed state is not feasible, and arcadian and functional approaches (Swart et al. 2001) are the most applicable. The differences between a natural beach (Figure 1.3a) and developed alternatives (Figure 1.3b–d) are obvious. The challenge is to maximize natural features to the extent allowed by natural processes

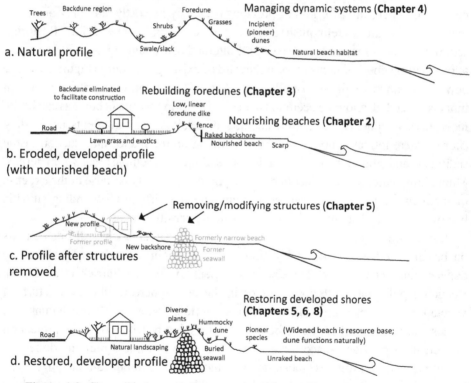

Figure 1.3 Generalized cross-shore topographic profiles, revealing alternative scenarios for developed coasts and chapters where detailed discussions of alternatives are found.

and human demands. The working premise is that any attempt to return a natural system or system component to a more natural state or cause stakeholders to develop a natural landscape ethic is a good thing, unless the attempt is accompanied by more damaging side effects. Tradeoffs may be required in reinstating natural environments in the face of economic constraints, spatial restrictions, or the desires of conflicting interest groups. There may be a need to accept less-than-ideal states for restored environments or to conduct ongoing human efforts to maintain them where existing natural landforms are being altered to accommodate human uses.

Restoration activities can be conducted by participants in the public (government) and private sectors and at any level, from the highest government authority to an individual property owner. Within the private sector, coastal restoration often occurs as a result of mitigation requirements for construction projects that remove or alter coastal habitat. National, state/provincial, and local governments in countries set policies, regulations, and land use controls to guide actions that influence the morphology and ecology of coastal reaches. Funding for projects can be supplied by any participant, but is usually done by higher levels of government for large projects, such as regional beach nourishment operations. Nonprofit organizations have played an increasingly active role in the development of smaller projects to restore beaches and dunes that were lost to extreme events and restore ecological function. These groups can work together in coordinating and providing advice for restoration actions to meet multiple stakeholder interests (Kar et al. 2020), especially in lands that are only partially developed and have conservation interest. Private residents can initiate restoration activities on their properties, often guided by public and private sector expertise and self-help documents (Mitteager et al. 2006).

Coastal resources can be restored or maintained despite an ever-increasing human presence, if beach nourishment plays an increasing role (de Ruig 1996; Nordstrom et al. 2011). Beach nourishment may be a contentious issue because of its cost, finite residence time, and potential detrimental effect on biota (Pilkey 1992; Nelson 1993; Lindeman and Snyder 1999; Greene 2002), but nourishment projects continue to increase in numbers and scope, requiring increased attention to their resource potential (Nordstrom 2005). The nourished beaches depicted in Figures 1.1 and 1.3b are not ideal, but much can be done to enhance their restoration potential. The types of nourishment projects as currently practiced are identified in Chapter 2, along with evaluations of their advantages and disadvantages, the mitigating and compensating measures used to overcome objections to them, and alternative practices that can make nourishment more compatible with broader restoration goals.

Dune building practices for locations where dunes have been eliminated or reduced in size are identified in Chapter 3. Case studies are provided, including examples of locations where beaches and dunes modified to provide limited human uses have evolved into habitats with greater resource potential than envisioned in original designs. The resulting habitats provide evidence that the principles discussed in subsequent chapters can have achievable outcomes, especially on intensively developed coasts where beaches are narrow and dunes have great significance for shore protection.

Much of the problem of managing beaches and dunes is that a healthy natural coastal system tends to be dynamic (Cooper and McKenna 2008; Hanley et al. 2014), but humans want a system that is stable to make it safe, maintain property rights, or simplify management (Nordstrom 2003; Bossard and Nicolae Lerma 2020). Managers are beginning to reevaluate the desirability of stabilizing coasts and are examining ways to make stabilized coastal landforms more mobile to enhance sand transfers from source areas to nearby eroding areas, reinitiate biological succession to increase species diversity, or return developed land to a more natural condition (Nordstrom et al. 2007c). Chapters 4 and 5 discuss the tradeoffs involved in restricting or accommodating dynamism in coastal landforms and identify human actions that can be taken to restore natural dynamism, expanding on some of the actions identified in Table 1.5. Chapter 4 evaluates options in locations where restoration efforts are not constrained by preexisting or planned shore protection structures, and many of the issues are in dune fields, such as depicted in Figure 1.3a. Issues in these locations include maintaining or restoring sand supply, encouraging localized destabilization, reducing the impact of nutrient enrichment, optimizing water supply, reinstating appropriate dune woodland, removing nonnative vegetation, recreating or reinvigorating slacks, controlling scrubs, and establishing appropriate levels of grazing (Rhind and Jones 2009). Human alteration of the landscape in many of these areas should be at a minimum and limited to locations that experience significant human impact (Pye et al. 2014).

Chapter 5 evaluates options for accommodating natural processes where they have been severely restricted by human structures or where structural solutions would normally be considered appropriate. Accommodation includes altering the effects of shore protection structures to allow for shoreline features to migrate landward (as depicted in Figure 1.3c) or designing strategies that combine natural and structural solutions (as depicted in Figure 1.3d). Rethinking structural approaches is gaining increased attention given the predicted increases in sea level rise and impact of coastal storms and acknowledgment of the need to make coasts more resilient.

The dimensions of natural cross-shore gradients of topography and vegetation are becoming increasingly difficult to maintain as shorelines continue to retreat and people attempt to retain fixed facilities landward. The kinds of cross-shore environmental gradients achievable under spatially restricted conditions (e.g., Figure 1.3d) are identified in Chapter 6, along with a discussion of the tradeoffs involved in truncating, compressing, expanding, or fragmenting these gradients. Chapter 6 underscores the need for municipal governments and private property owners to develop strategies to maximize opportunities where human structures are a dominant feature and retreat from the coast is not presently considered an option.

Restored environments evolve by natural physical processes and by initial and ongoing human actions (Burke and Mitchell 2007), and their use by humans will change because of changing human perceptions and needs. Assessments of restored landscapes must consider natural and human processes and landforms as integrated, coevolving systems (Gann et al. 2019) requiring integration of the needs of stakeholders and the physical constraints of coastal processes. Chapter 7 identifies differences in stakeholder preferences and some of the compromises required in resolving these differences, focusing on approaches by municipal managers, developers, property owners, environmental interest groups and agencies, engineers and scientists.

A program for beach and dune restoration should include a target state; environmental indicators and reference conditions to evaluate success; realistic guidelines for construction and maintenance; and methods for obtaining acceptance and participation by stakeholders. Chapter 8 discusses how these components can be implemented, with examples of actions that can be taken by municipal managers and private landowners.

A vast array of methods is available to address restoration needs at many different levels of management, from property owner to national government. The number of alternatives and the suitability of each are dependent on local constraints imposed by availability of sediment, space, technical expertise, money, and social and political will. Time is an important consideration because restored coastal environments are expected to change. Restoration plans should be living documents that allow for changing landscape trajectories. Site- and time-specific constraints make it impossible to prescribe a specific solution for individual sites, so we have concentrated on providing the rationale for restoration in generic terms. Local managers must employ concepts and strategies to suit stakeholder needs and capabilities. Despite the large literature on beach and dune restoration, many issues remain to be resolved in future studies. Chapter 9 identifies ongoing research needs.

2

Beach Nourishment and Impacts

2.1 The Potential for Restoration

Many natural values and human–nature relationships that are lost in the process of converting undeveloped coasts to developed coasts can be regained by nourishing beaches with compatible sediment and allowing natural processes to reestablish landforms and biota (Figure 2.1). Beach nourishment (also termed recharge, replenishment or fill) is used extensively in the United States and Europe (Davidson et al. 1992; Hanson et al. 2002) and is also used in many other countries around the world including Mexico (Mireille et al. 2020), Australia (Cooke et al. 2012), India (Chittora et al. 2017), China (Liu et al. 2019), Sri Lanka (Ratnayake et al. 2019), Japan (Shibutani et al. 2016), and South Korea (Choi et al. 2020).

Beach nourishment has both positive and negative impacts (Table 2.1). There is a difference between nourishing a site for erosion control and nourishing a site for ecological enhancement. Benefits to natural systems are not automatic byproducts of nourishment operations because nourished beaches may be built too high and wide to allow wave and swash processes to rework the backshore or the beaches may be graded flat or raked to enhance recreation, thereby eliminating the opportunity for naturally functioning landforms to evolve (Nordstrom 2000; Jackson et al. 2010) (Figure 1.1). Sediment imported from outside the fill area (exotic sediment) can change the morphology, grain size, chemical composition, and evolutionary trend of the beach. If dunes are included in designs, they are often built as linear dikes by earth-moving equipment rather than by aeolian processes. Beach nourishment is often called beach restoration, but the functions of a natural beach system are not restored simply by placing a volume of sediment on a coast (Nordstrom 2000). Many other actions may be necessary to optimize the value of the new resource.

The considerable shore protection and recreational values of nourished beaches are well documented (Dean 2002; Gómez-Pina et al. 2004; Reid et al. 2005;

Figure 2.1 The nourished beach at Pellestrina Island, Veneto, Italy, in 2006. No beach existed seaward of the seawall prior to the placement of fill ending in 2001. The *Tamarix gallica* trees at far left were planted to form a windbreak, but the backshore is evolving naturally.

Houston 2013, 2017). This book focuses on the way existing management practices can be evaluated and modified to achieve restoration goals. Beach nourishment can have many impacts beyond those originally intended. No attempt is made to discuss social or economic advantages and disadvantages or synthesize the myriad studies that evaluate traditional fill practices, determine fill longevity, evaluate beach profile evolution, or develop models that predict changes to beach shape or fill volumes. Many of those studies provide little insight into the significance of beach nourishment as a goal for restoring natural environments. The primary focus here is on evaluating the potential for beach nourishment to provide space for landforms and habitats to form in locations where human development has eliminated them (Box 1). We acknowledge that more attention should be given to debates about the long-term viability of nourishment (e.g., Houston 2017; Parkinson and Ogurcak 2018), especially considering long-term sea level rise. More consideration should also be given to the feasibility of alternatives to nourishment, such as land use controls, construction setbacks, and retreat

Table 2.1. *Potential positive and negative impacts of a traditional beach nourishment design*

Positive impacts
Creates beach and dune where lost because of erosion or development
Reestablishes conditions for return of natural aeolian transport
Protects stable dune habitat from wave erosion
Provides habitat for rare or endangered species
Provides wider space for full environmental gradients to form and evolve
Protects human facilities
Buries less suitable beach substrate or incompatible human structures
Enhances reputation of beach resorts
Reinvigorates local economies
Counteracts effects of sea level rise

Negative impacts
Increases turbidity and sedimentation
Changes morphology and surface sediment characteristics of borrow areas
Changes grain size characteristics and morphodynamic state of beaches
Increases salinity levels in sediments or in aerosols associated with placement by spraying
Removes feeding and spawning areas
Buries habitat
Alters abundance, biomass, richness, average size, and composition of species
Alters character of migratory corridors
Displaces mobile species
Disrupts foraging, breeding, and nesting species
Enhances undesirable species that find new habitat more accommodating
Changes evolutionary trajectories of beach, dune, or species

policies to provide space for landforms and habitats to form and prevent the losses caused by removing sediment from borrow areas and depositing it in fill areas (Greene 2002). We do not assume that nourishment programs will be the most successful alternative for addressing shoreline erosion. Our emphasis on beach nourishment in a restoration context acknowledges the great likelihood of nourishment being used for shore protection and recreation enhancement and the difficulty of implementing alternatives to nourishment to achieve restoration goals in developed areas, especially because of lack of subsequent management for nature.

Reviews of characteristics and designs of nourishment projects for shore protection are found in Houston (1996); Trembanis and Pilkey (1998), Capobianco et al. (2002), Dean (2002), Finkl (2002), Hamm et al. (2002) Hanson et al. (2002), Campbell and Benedet (2007), Moses and Williams (2008); Cooke et al. (2012),

Box 1
Improving resiliency through beach nourishment at Myrtle Beach, USA

Myrtle Beach, South Carolina, USA, provides a lesson in coastal resiliency applicable to many developed beaches. This intensively developed beach in the 1970s faced chronic erosion and laissez-faire construction of seawalls or revetments to protect encroaching development. Usable beach area was diminishing along with the aesthetic character of the shorefront. City leaders made the difficult decision to prohibit new seawalls and committed to sand nourishment – a soft engineering solution that was not embraced by many property owners in the 1980s. Almost forty years later, and after five nourishment events, all seawalls are buried under a continuous dune, and native grasses provide an appealing transition between the wide dry sand beach and upland landscaping. The cost of beach restoration and advancement for the 14 km oceanfront (in 2020 dollars) averaged ~$200 per meter of coast per year (1985–2020). Figures 2.2 and 2.3 reveal that beach nourishment enhanced the appearance of the shorefront, provided a storm buffer, and reduced property damage by storms. The renourishment interval is about ten years between projects using offshore deposits placed via hopper dredges. Hurricane Matthew (in 2016) reduced storm berm width, destroyed the foredune and its vegetation, and partially damaged the secondary dune. Four years later, storm berm widths and dune vegetation recovered to levels comparable to pre-storm conditions. The primary dune is growing with the help of sand fencing and vegetation plantings, while the secondary dune appears similar to its pre-storm condition. Importantly, beach maintenance is now widely embraced by property owners and the millions of tourists who visit each year. See Kana and Kaczkowski (2019) for more information.

Contributor: Tim Kana, Coastal Science and Engineering, Inc.

Figure 2.2 Myrtle Beach, SC, March 1985.

Figure 2.3 Myrtle Beach, SC, September 2001.

Luo et al. (2016), and Pinto et al. (2020). Assessments of socioeconomic effects and long-term feasibility include Thieler et al. (2000), Klein et al. (2004), Ariza et al. (2008), Burgess et al. (2016), Parkinson and Ogurcak (2018), and Catma (2020). Greater amounts of information are available on the negative environmental effects of nourishment than on the positive environmental effects. Accordingly, much of this book is devoted to finding ways to enhance positive effects. Dune building operations associated with beach nourishment and ways nourished beaches can be managed to have greater environmental values are examined in subsequent chapters.

2.2 General Design Considerations

Many nourishment projects follow similar general design characteristics, although Roberts and Wang (2012) and Bitan and Zviely (2020) note the importance of tailoring projects more closely to site-specific constraints. Nourished beaches are designed to have a width, profile shape, and fill volume calculated to provide the necessary protection to landward infrastructure, and an advanced fill volume to account for profile adjustment and beach loss between renourishment intervals. The beach loss may be related to the background erosion rate – an accelerated rate caused by using sediment that is more readily mobilized and transported than native sediment – or spreading losses as the shoreline bulge in the fill area is smoothed by longshore transport. The most common means of sediment delivery is by hydraulic means through pipelines.

Large-scale operations most frequently use sediment from offshore sources. Dean (2002) estimates that 95% of all sand volumes placed in beach nourishment projects are through dredging from offshore because large quantities of suitable sand are often available 1–20 km offshore and unit costs for delivery are relatively low. In contrast, sediment used for nourishment in Australia is transported by land, and projects often rely on using sand from the same coastal compartment (Cooke et al. 2012); sediment used for nourishment operations in Portugal is mostly dredged from inlets and channels associated with ports (Pinto et al. 2020).

Sediment from offshore is usually obtained by either (1) pipeline dredges using a suction head, with an open pipe and either an agitation jet or cutterhead with rotating steel blades to mobilize sediments; or (2) hopper dredge, usually a moving ship that conveys the sand to a location near the fill area and discharges it through the bottom of the hull, through a moored pipeline, or through the air as a sand and water slurry (Dean 2002). Trucks are often used for small-scale operations, e.g., $<200,000$ m^3 (Muñoz-Perez et al. 2001), although these operations can be quite large if inland source areas are more suitable, e.g., the 652,400 m^3 project at Myrtle Beach, South Carolina (Kana and Kaczkowski 2019). Trucking operations can avoid the large mobilization costs of using dredges, but they interfere with traffic and damage roadways, and the resulting sediment may be less compacted when placed on the beach (Dean 2002). Borrow areas for trucking operations include nearby beaches, dunes, river sediments and quarries, including natural sedimentary deposits and crushed rock. Sources of fill may change through time due to depletion, changing environmental concerns, changes in technology, or availability of opportunistic sources, such as dredge spoil.

The deposited sediment is usually reworked by bulldozers to the design profile, which is often a trapezoidal shape. Great differences in the shape, location, and grain size characteristics of the fill can occur because of the different methods used. The traditional approach has been to create a flat subaerial berm because it is easy to construct with earth-moving equipment; it facilitates determination of volumes for paying contractors; and it creates a wide recreation platform that makes the economic investment appear worthwhile. The beach may be built higher than a natural beach to provide enhanced protection against wave runup and flooding and provide an extra volume of sediment for a given cross-shore distance. A higher beach may also be considered aesthetically pleasing to tourists (Blott and Pye 2004). Constriction of swash uprush at the upper limit of wave reworking at high tide often results in a conspicuous vertical scarp that may persist unless eliminated by mechanical grading or storm wave attack.

Subaqueous nourishment, including shoreface nourishment (van Duin et al. 2004; Van Leeuwen et al. 2007), is increasing in use and is the principal method of nourishing beaches in the Netherlands (Stive et al. 2013). These operations can have a lower unit cost than subaerial nourishment, but more sediment would be required to eventually achieve the same volume on the upper beach (Mulder et al. 1994). Subaqueous nourishment provides little opportunity for local managers to make direct use of the sedimentary resource, but it does slow the rate of beach retreat and may be important in the long-term maintenance of restored beach and dune habitat. Subaqueous nourishment is less intrusive on beach use and can be more acceptable to the public (Stive et al. 2013). The development of dunes would be less likely than if a wide backshore is produced by subaerial nourishment (van Puijenbroek et al. 2017), which could be considered a negative effect, but the nuisance of excessive aeolian transport off the backshore and into coastal infrastructure would be reduced.

Beach nourishment changes the volume of sediment on the beach but not the basic transport processes. Longshore currents will continue to transport sediment downdrift and out of the project area. Bypassing or backpassing projects can be conducted to complement nourishment operations. Bypassing is a way of transferring sediment past obstructions to longshore transport, such as at inlets where maintenance dredging or jetties prevent sediment from moving through the system and can be a key component in regional sediment management (Clausner et al. 1991; McLouth 1994; Dean 2002; Castelle et al. 2009; Beck and Wang 2019). Bypass operations may be conducted using sediment from updrift or using sediment from an alternative source (Lin et al. 1996). Alternative sources can include navigation channels or subaerial sites (O'Brien et al. 1999; Marinho et al. 2018; Spodar et al. 2018), and the sediment characteristics can depart dramatically from sediment in the longshore transport system. Use of sediment already in the system ensures compatibility and avoids the problem of dredging more stable borrow areas.

Backpassing (also called recycling) is a way of transferring sediment from accreting downdrift areas back to eroding areas updrift, usually in small amounts and with trucks and earth-moving equipment available at the local level (Mauriello 1991; Nordstrom et al. 2002; Phillips 2009). Backpassing can also occur where sediment moved by natural processes to offshore sediment sinks is returned to the beach (Bray and Hooke 1998), or where sediment deposited at inlets or at the depositional end of spits is returned to updrift beaches (Cialone and Stauble 1998; Doody 2001; Alexandrakis et al. 2013). The number of backpassing operations may increase in the future as suitable local borrow areas are depleted (Arthurton 1998; Nordstrom 2000).

2.3 Mega Nourishments

Mega nourishments involve placement of a massive quantity of sediment and can have a dramatic effect on changing the coastal landscape, but their great implementation costs have restricted them to few locations. These nourishments can be placed at a single location, such as at the Sand Motor, the Netherlands (Stive et al. 2013), or be extended alongshore, as in the proposed project for the Chandeleur Barrier Arc, USA (FitzGerald et al. 2016). In a sense, the magnitude and number of separate projects along some shorelines such as in the Netherlands and the entire ocean shore of the state of New Jersey, USA, can be considered mega nourishments. The commitment to continually nourish more spatially restricted areas could be included, such as along the island of Sylt, Germany, where 51 million m^3 has been deposited (Hofstede 2019), or on the coast of Lincolnshire, UK, where 15 million m^3 was deposited over a length of 26 km, with a commitment to add 500,000 m^3 annually (Burgess et al. 2016). Nevertheless, the term has become associated with the Sand Motor (Figure 2.4), a 21.5 million m^3 project conducted south of The Hague on the coast of the Netherlands. The advantages of a localized mega nourishment is that sediment can be placed at relatively low cost per m^3; new fill would only be required every decade or two; the initial perturbation and associated ecological stresses would be localized; and the slow diffusion of sediment downdrift would favor morphologic

Figure 2.4 The Sand Motor south of The Hague, in the Netherlands, 2019, showing the widened beach and south end of the lagoon. Artificial dune cuts, like the one in the foreground, are discussed in Chapter 4.

change and habitat development at relatively natural rates of longshore transport (Stive et al. 2013; Hoonhout and de Vries 2019).

Mega nourishments can be a beach extension type designed to remain in place by subsequent nourishment to preserve safety levels and improve resource value or a feeder type designed to erode to nourish beaches downdrift (Tonnen et al. 2018). The proposed Chandeleur Barrier project is an interesting plan to allow for coastal retreat but retain barrier island integrity (FitzGerald et al. 2016). The Sand Motor design included a 2.4 km long beach up to ~1 km wide, shaped like a cuspate foreland with a hook-shaped peninsula protecting a shallow lagoon intended to provide aqueous habitat (Stive et al. 2013). The peninsula itself had a sediment volume of 17 million m^3 (de Schipper et al. 2016). The construction height of the deposit is well above the height of traditional projects, which is another unusual aspect (Nolet and Riksen (2019) and an issue that has great importance in assessing the alternative values of the project as either a feeder beach with a high backshore or a source for local dune building with a low backshore (Hoonhout and de Vries 2017). One of the issues is how the coastal system adapts after creating a large unnatural shape (de Schipper et al. 2016). The site performed well in delivering sediment alongshore (de Schipper et al. 2016), but it did not resemble a natural coastal landscape about eight years after placement (Figure 2.4), and it has not been as effective as a source of sand for aeolian transport and dune building as it could, considering the large potential source width (Hoonhout and de Vries 2019). Part of the lower-than-expected aeolian transport rates may be attributed to coarse surface lag deposits on higher parts of the beach (Hoonhout and de Vries 2017, 2019), which is an aspect addressed in Chapter 3.

An example of a beach extension–type mega nourishment is the section of coast between Petten and Camperduin in the Netherlands. There, a new beach and dune ridge were created seaward of the dike (the Hondsbossche Zeewering, Figure 2.5) that was previously the main line of defense. The added sediment was placed in a configuration to achieve greater topographic and ecological diversity and consists of a widened beach, a new dune ridge, with the low ground between the new dune ridge and dike functioning as a slack. The project provided new beach and dune habitat fronting the dike and an ecological link alongshore between the dunes existing to the north and south of the dike. The new beach and dune are allowed to evolve naturally, with the old dike forming backup protection.

Figure 2.5 New beach and dune ridge created seaward of the dike, Hondsbossche Zeewering, The Netherlands.

A mega nourishment was modeled for implementation on the coast of the English Channel (Brown et al. 2016) and revealed that the concept has potential long-term value, but a mega nourishment at that location would be less useful than smaller projects for addressing erosion at multiple sites alongshore. There is a clear need for evaluations of the value of the Sand Motor concept in other locations, particularly where the potential for delivery of sediment to downdrift areas can be clearly demonstrated or where the accretion bulge can be designed to match natural analogs. A mega nourishment would not be effective as a feeder beach at the updrift end of a coast with increasing rate of longshore transport downdrift, despite the intuitive appeal of this strategy. Caldwell (1966) demonstrated how this would only serve to widen the beach in the updrift sector without benefiting each sector downdrift. Restoration along a shore with an increasing rate of transport would involve placing fills at intervals along the whole eroding coast.

Accreting forelands and deltas at the mouths of streams are somewhat analogous to the Sand Motor in terms of overall shape. Large deltas, including the Ebro, Po, Arno, and Ombrone, formed at stream mouths in the Mediterranean Sea in the past, accentuated by delivery of sediment caused by conversion of land use to agricultural in tributary basins. This accretion trend subsequently reversed because of reduced sediment flux caused by catchment reforestation, retention within reservoirs, fluvial regulation, and dredging, leading to loss of beach, dune, and barrier/lagoon environments (Pranzini 2001; Anthony et al. 2014). Placement of a large cuspate-shaped deposit at these locations would provide the sediment and space for new landforms and habitats. The slow accretion that created the multiple beach and dune ridges and swales in the past could not be replicated by natural processes operating across a greatly widened beach placed over a period of weeks to months. Artful placement of the fill or sand fences or reshaping by bulldozers would be required to recreate the preexisting topography. Examination of the characteristics of the Sand Motor eight years after placement (Figure 2.4) reveals little evidence of the kind of morphology or habitat that would occur if a feature like this evolved by slow accretion through natural processes alone.

2.4 Sediment Characteristics

The cost of obtaining, transporting, and placing fill sediment on beaches often leads to use of materials that are readily available nearby but differ from native sediment in grain size, sorting, shear resistance, moisture retention, grain shape, compaction, or mineral composition. Sediment on the active foreshore is quickly reworked and will become better sorted by size, shape, and density and mixed with in situ sediment to appear more like native material. Sediment within the inactive fill area on the backshore will retain many of the characteristics it had when emplaced. Sediment on the backshore that is more poorly sorted will usually develop a conspicuous lag deposit of shell or gravel as the finer sediment is blown away. Fill sediment that is somewhat coarser than native material can enhance the longevity of the nourished beach (Bitan and Zviely 2020), although considerations for biota would argue for approximating the characteristics of the native material. Fill operations appear to have a less adverse effect where the grain size characteristics of nourishment sediments are a good match (Nelson 1993; National Research Council 1995; Cahoon et al. 2012).

Gravel beaches may be nourished with sand because it is readily mined and transported by pipe. Where gravel (also called shingle or coarse clastic sediment) is available and cheaper than sand, it may be used to nourish sandy beaches. Interest in nourishing beaches with gravel as an alternative to sand occurs because the coarser particles are more stable (Johnson and Bauer 1987; Komar et al. 2003), but beach users normally prefer sand (Morgan 1999). Eroding gravel beaches should be nourished with gravel, where available, to retain the natural habitat and heritage. Gravel nourishment projects are often conducted at small scale, in keeping with the generally small size of gravel beaches in nature, e.g., 60,000 m^3 at Torre del Porto in Italy (Altomare and Gentile 2013). The gravel nourishment projects at Nice, France, since 1976 were relatively large cumulatively, totaling about 600,000 m^3 (Anthony and Sabatier 2013; Box 2). Nourishment projects using mixed sand and gravel in England were as great as 300,000 m^3 at Hurst Spit (Dornbusch 2017), 500,000 m^3 at Hayling Island (Zarkogiannis et al. 2018), and the large-scale Lincolnshire project noted in Section 2.3 (Blott and Pye 2004). Gravel borrow areas can be more limited in extent than sand borrow areas, placing doubt about their long-term availability in some regions (Arthurton 1998). Gravel beaches are addressed in greater detail later in this chapter.

Box 2
Artificial nourishment of the beach at Nice, France

The resort city of Nice, capital of the French Riviera, prides itself on its famous seafront promenade (The English People's Promenade) constructed in 1864 and its 4.5 km–long gravel beach. Protection of the promenade and urban front by a seawall and massive urbanization of the river channels that supplied gravel to the beach resulted in chronic beach erosion. Since 1976, the beach has been artificially nourished with about 600,000 m^3 of gravel (Figure 2.6), rendering this operation one of the most important for gravel beaches in the world. Nourishment ranged from nil in certain years to a peak of over 97,000 m^3 in 2000. Despite the scale of nourishment, beach width has remained stable. There is no possibility for alongshore gravel loss on Nice beach, nor backshore retreat of the beach over time. Fill material is lost offshore through small

Box 2 (cont.)

submarine canyons impinging on the beach and cut during the Messinian salinity crisis when isolation of the Mediterranean from the Atlantic Ocean between 5.6 and 5.3 million years ago caused a large fall in Mediterranean water level. Loss of gravel is favored by artificial beach flattening and widening in spring (Figure 2.7) in order to enhance the summer "carrying" capacity of this highly touristic beach. See Anthony et al. (2011) for more information.

Contributor: Edward Anthony, Aix-Marseille University, Aix en Provence, France

Figure 2.6 Deposition of gravel fill at Nice.

Figure 2.7 Grading to enhance recreation at Nice.

2.5 Potential Negative Impacts of Nourishment Operations

Many studies in peer-reviewed journals examine relationships between biota and sediments and the environmental requirements of organisms that use or live on the beach (Burger et al. 1977; Diaz 1980; Connors et al. 1981; McArdle and McLachlan 1992; Defeo and McLachlan 2005; Jackson et al. 2008). These studies are often used to assess potential effects of nourishment projects, although they may not themselves be written in a nourishment context. Some of these studies may have been conducted to assess impacts of beach nourishment projects but may be published as basic research contributions. Articles that are detached from their applied contexts are not as valuable in determining negative effects as articles that directly evaluate nourishment projects.

Studies of environmental impacts of the beach and borrow areas are frequent and conducted on an ongoing basis (Drucker et al. 2004). Articles in the peer-reviewed literature that assess the effects of dredging and nourishment on species include, among others, Gibson and Looney (1994) and Gibson et al. (1997) on vegetation colonization; Crain et al. (1995), Steinitz et al. (1998), Rumbold et al.

(2001), and Brock et al. (2009) on sea turtle nesting; Peterson et al. (2000, 2014), Leewis et al. (2012), and Manning et al. (2014) on invertebrates; Manning et al. (2013) on surf fish; Stull et al. (2016) on zooplankton abundance; Hill-Spaknik et al. (2019) on microalgae biomass; and Kenny and Rees (1994) on macro benthos in borrow areas. Studies that assess responses to actual nourishment in the field are supplemented by controlled lab and field experiments. Reviews are found in Rakocinski et al. (1996), Peterson et al. (2000), Jones and Magnun (2001), Posey and Alphin (2002), and Speybroek et al. (2006). Detailed protocols for physical and biological monitoring are available for offshore dredging operations in Minerals Management Service (2001). Despite the growing literature and increased interest in addressing adverse effects of nourishment, some projects are still conducted without conducting studies to anticipate environmental impacts or answer questions raised in preliminary studies (Pezzuto et al. 2006; Gore 2007; Cooke et al. 2012).

Methods of assessing potential impacts in borrow areas are evaluated in a thematic issue of the *Journal of Coastal Research* (Drucker et al. 2004). Examples of sampling protocols are identified in Minerals Management Service (2001) and Nairn et al. (2006). Sampling and analytical methods for borrow and fill areas are summarized in Nelson (1993), and guidelines for both physical and biological aspects of beach nourishment are presented in Stauble and Nelson (1985). The potential impacts of dredge and fill projects vary with many factors, including the timing of the activities, the methods used, the design template of the beach, sediment compatibility, and the mitigating measures employed (US Fish and Wildlife Service 2002).

2.5.1 Loss of Habitat and Displacement of Mobile Species in Offshore Borrow Areas

Dredging borrow areas may have deleterious effects on the resident infaunal assemblages that may, in turn, adversely affect commercially and ecologically important finfish that utilize these areas for migrating or foraging (US Army Corps of Engineers 2001; Drucker et al. 2004). Dredging can decrease abundance, biomass, taxa richness, average size of the dominant species, and composition of species and biomass (US Army Corps of Engineers 2001). Many infaunal species are opportunistic and adapted to recovery from natural disturbance, so benthic soft sediment communities may recover from dredging relatively quickly if there are no major changes to the underlying substrate (Posey and Alphin 2002). Abundance of some benthic species may recover within a year; recovery of biomass and taxa richness may occur within a year or take somewhat longer; and changes in biomass composition may take longer (US Army Corps of Engineers 2001). Some species may depend on early colonizers to prepare an environment before showing full recovery (Kenny and Rees 1996).

Dredging borrow areas creates bathymetric depressions (all termed pits). Characteristics of borrow pits, and methods of determining their location and mining them are presented in Dean (2002). Typical plan areas for borrow sites reported by Dean (2002) are 1–10 km^2 with typical excavation depths 2–10 m. Mielck et al. (2019) report depths of approximately 12 m, with deepest post-dredging depths up to 23 m. Future borrow areas may be sought in deeper water as environmental concerns for shallower habitat increase. Dredging in depths of up to 40 m and even at depths of 80 m has been suggested to avoid disturbing *Posidonia* sea grass (van der Salm and Unal 2003). Dredging in deeper water also reduces the geomorphic effect of pits on waves and thus on adjacent beaches (Benedet et al. 2013).

Borrow pits that are relatively shallow and fill rapidly appear to allow rapid recovery of diversity and abundance of infauna, but recovery of the functional structure of the assemblages may take much longer (US Army Corps of Engineers 2001). Crowe et al. (2016) found that the original sandy habitat characteristic of pre-dredge conditions had increases in silt and clay, decreases in $CaCO_3$ content, and changes to finer sand sizes in pits up to 3.5 m deep in 8–12 m water depths. Benthic community changes were largely loss of species associated with coarser sands and shell and recolonization by species associated with finer sands, silt, and clay. The pits showed little evidence of recovering 6–8 years after dredging. Their study suggests that selection of borrow areas near estuarine sediment plumes may increase the likelihood of fine grain sedimentation in the pits. Deeper pits have steeper slopes that contribute to slope failure and erosion, while sluggish bottom currents within the pits prevent sand transport from the margins and allow mud to settle (Mielck et al. 2019). Gonçalves et al. (2014) found that a 5 m deep pit in 16–18 of water depth filled about 17% in the first 4 years, but they predicted full recovery in 38 years, given the significance of storm events in filling the pits.

Long-lasting and deep borrow pits that are depositional areas for fine sediments attract a different biological community than found on the pre-dredged substrate or surrounding unaffected area, and restriction of water movement in these deep pits results in lower water quality (hypoxic or anoxic conditions) and a depauperate community (US Army Corps of Engineers 2001). Recovery of infaunal assemblages is expected to be rapid where borrow areas are on shoals that are bathymetric peaks on the sea bottom and are in areas of strong currents and sand movement, causing pits to fill rapidly (US Army Corps of Engineers 2001). Use of these shoals as borrow areas for nourishment operations is likely to be favored for other reasons, because dynamic shoals are likely to yield high-quality sand suitable for beaches. On the negative side, ridge and shoal features are unique habitats and often high-quality fishing areas that require more study before they are extensively mined on a long-term basis (Drucker et al. 2004)

2.5.2 *Disturbance by Burial and Turbidity*

The effects of burial and turbidity are not confined solely to the immediate vicinity of borrow and fill areas but may occur kilometers away (Newell et al. 2004; Pezzuto et al. 2006). Temporary loss of infaunal beach communities through sand burial during nourishment is expected and considered largely unavoidable; the more important issue is often the recovery rate after project completion (National Research Council 1995). Sandy organisms show a high degree of plasticity (Brown 1996; Soares et al. 1999; Jaramillo et al. 2002). Intertidal beach communities are adapted to large-scale sediment disturbances that occur during major storms, but that presupposes reworking of native materials and may be aided by events that occur outside times of peak recruitment of invertebrates. Massive losses of beach infauna can occur as a result of storms that can be especially critical in developed areas where their populations are restricted (Harris et al. 2011), so the concept of "adapted" does not mean that impact will be insignificant (Manning et al. 2013). Recolonization or spawning on nourished beaches can be impeded if the texture of fill material is not well matched to local beach sand; if the timing of the fill operation is not well matched to recruitment periods of fauna; and if the fill is repeated frequently (Manning et al. 2014; National Research Council 2014; Peterson et al. 2014; Viola et al. 2014; Martin and Adams 2020). Impacts of nourishment on invertebrates, in turn, propagate up the food web to reduce beach use by predatory shorebirds for foraging (Peterson et al. 2014).

Some studies reveal little evidence that beach nourishment has significant negative impact on benthic fauna of the nearshore of exposed beaches (Gorzelany and Nelson 1987), although macro benthic assemblages occurring deeper than 3 m occupy a relatively stable environment that may take longer to recover if buried. Beach nourishment activities appear to have no significant impact on zooplankton abundances in the surf zone, suggesting that surf zooplankton are already adapted to physical disturbances (Stull et al. 2016). Nourishment may degrade the intertidal and shallow subtidal foraging habitat for demersal surf fish by inducing mortality of macro benthic infaunal prey through rapid burial, by introducing excessive amounts of coarse shelly sediments incompatible with some key benthic invertebrates, or by incorporating excessive amounts of fines, inhibiting detection of prey (Manning et al. 2013).

Use of beach fill on low-energy estuarine beaches will bury a portion of the shallow low-tide terrace that fronts them, covering more stable benthic habitat and eliminating shallow water areas that some aquatic plants require for solar radiation. The problem of nearshore burial is likely to be more acute here than on exposed (high-energy) beaches, where the bottom is subject to frequent disturbance and where biota are similar to biota on the lower foreshore. Low-energy environments may also have a higher proportion of slow-growing species that may take longer to recover (Newell et al. 1999, 2004). Burial of bay-bottom habitat is frequently mentioned as a problem in reviews of potential fill operations by agencies charged with maintaining estuarine ecosystems, and the perceived losses have been cited as a reason for lack of acceptability of nourishment projects in estuaries (US Army Corps of Engineers 1980; US Army Corps of Engineers 1986).

A problem with burial also occurs on nearshore hard-bottom habitats on exposed shores (Castelle et al. 2009). Hard bottom is found in a wide range of environments, from coral reefs in the south Pacific to cohesive shores in the Great Lakes (Larson and Kraus 2000). In southeast Florida, hard bottoms occur at depths of 0–4 m, and they are often buried or indirectly affected by sedimentation from nearby fill areas. Hard-bottom habitat is rarer than sandy bottom, making it difficult for species to find. Many hard substrate organisms are sessile and cannot burrow up through the sediment overburden (Nelson 1989). Burial creates an entirely different kind of substrate, so confining operations to times when fish populations are low does not reduce the impact to populations that would have returned to the hard substrate. Abundance of organisms on hard bottoms can greatly exceed nearby sandy bottoms, and these areas can support thriving recreational fisheries. Since virtually nothing is known about the tolerance of members of the hard-bottom community to stresses associated with beach nourishment, a more cautious approach is suggested than on sandy-bottom communities (Nelson 1989).

2.5.3 Change in Beach Morphology and Dynamics

The morphology of natural beaches can range from high-energy dissipative beaches with gently sloping foreshores and offshore bars to low-energy reflective beaches with steep foreshores and planar offshores (Wright and Short 1984; Masselink and Short 1993). Nourished beaches are commonly designed with higher berms and wider backshores for storm protection and tourism. These designs when placed on a gently sloping beach can increase erosion (Benavente et al. 2006). Changes in grain size can result in a change in beach morphology (Benedet 2004; Anfuso and Gracia 2005; Benavente et al. 2006). Variability in the abundance and life histories of beach macrofauna is controlled by effects of mean grain size and beach slope (McLachlan 1996; Caetano et al. 2006; Fanini et al. 2007), and beach width, slope, and elevation are important in nest site selection by sea turtles (Wood and Bjorndal 2000). The morphology of a nourished beach can have a pronounced effect on landward vegetation. Colonization can be relatively slow where the post-nourishment profile is low and flat, contributing to continuous salt spray, desiccating effects of the wind, and inundation from storm overwash that can uproot germinating seedlings or kill species intolerant of salt water (Looney and Gibson 1993; Gibson and Looney 1994). Alternatively, this inundation can remove coarse lag deposits and provide a mechanism for natural reworking of the fill, and species involved in primary succession may result from strandings caused by storm wave uprush and inundation (Looney and Gibson 1993). These considerations point to making provisions for a dune in project designs to provide a buffer between pioneer species on the beach and more stable communities landward.

Dredge spoil placed at elevations higher than existing natural dunes can reduce the salt spray that is important in germination and zonation and is the primary source of nutrients for foredune species (Looney and Gibson 1993). The new conditions can contribute to growth of species not found in this niche under natural conditions or create an imbalance in species normally found there. Natural growth of vegetation on beach fill placed at a greater height than the berm at Sandy Hook, New Jersey, USA, for example, created a monospecific stand of seaside goldenrod (*Solidago sempervirens*). This species is native to the area, and it is an expected colonizer of coastal dunes, but the sole presence of *Solidago* on the backshore rather than the dune resulted in an unnatural appearance.

2.5.4 Introduction of Noncompatible Sediment

Noncompatible sediment may be used intentionally to take advantage of a nearby source, dispose of unwanted sediment in a more productive way, extend the longevity of the fill (e.g., gravel added to a sand beach), or change the aesthetics of the beach for recreation. The scarcity of beach-compatible borrow material may increase the potential for using dredge spoil as fill material (Trembanis and Pilkey 1998; O'Brien et al. 1999; Frihy et al. 2016). Disposal of appropriate spoil on the beach is considered a beneficial use of the sediment (Silveira et al. 2013), but the term "beneficial" can be a misnomer as a generalized concept. Dredge spoil is rarely the same as native beach sediment and may contain unacceptable amounts of shells, fine material, seeds of plants not found on the beach, or pollutants. A better use for some of the unpolluted material may be found, such as placement of fine material on inundated portions of wetlands (Manning et al. 2014).

Silts and clays are often introduced as a subfraction of sandy sediments and are of little value in a beach environment because they are removed when reworked by waves. The fines may also be more associated with pollutants (Tondeur et al. 2012).

Pollutants may occur through beach disposal of unwanted debris under the guise of beach nourishment (Cooper et al. 2017; Li et al. 2020). Shell hash, vegetal debris, and dead macrofauna can be conspicuous components where estuarine sources are used (Pezzuto et al. 2006). Much of the gravel used in nourishment projects is from upland sources, including quarries (Pacini et al. 1997; Shipman 2001), and can significantly alter beach sediment texture and lithology (Kilibarda et al. 2014; Box 3). In some cases, the gravel represents a by-product derived from road and tunnel construction, building sites, or mining sites (Bourman 1990; Dzhaoshvili and Papashvili 1993; Anthony and Cohen 1995; Nordstrom et al. 2004; Cooper et al. 2017). Beach disposal of colliery waste along the coast of northeast England provides one of the more impressive examples. There, tens of millions of tonnes of waste were deposited on foreshores; much of the material was shale with coal and associated trace metals; the resulting material isolated the former bluffs from wave erosion, but the granular state of the material became altered in situ to a more consolidated clayey condition more resistant to erosion than the constituent grains, creating and maintaining beach scarps and reducing the number of natural beach species (Cooper et al. 2017).

Gravel from external sources may be processed to make its average grain size and sorting values more compatible (e.g., by sieving) to native sediment. The resulting sediment may still be less rounded. The abrasion rate of beach sediment depends on hardness as well as time exposed to transport (Dornbusch et al. 2002; Cammelli et al. 2006). Sediment that remains on the active foreshore surface gradually becomes rounded, but sediment that is quickly stranded at locations high on gravel ridges will remain angular. Grading the beaches to place sediment back in the active swash zone can reinitiate the rounding process, but there should be little need to do this unless there is great recreational demand for the beach.

The degree to which sediment intended as fill (rather than disposed for convenience) can depart from native material is seen in the marble beaches created in Japan (Wiegel 1993; Deguchi et al. 1998) and Italy (Cammelli et al. 2006; Nordstrom et al. 2008). Marble beaches are not likely to form and survive under natural conditions. Marble is relatively rare, and carbonate bedrock is subject to karst formation and subsurface drainage, leaving little surface runoff to deliver sediment to streams and then to the coast. Carbonate sediment is a compatible source where beaches typically have a portion of beach material derived from shell fragments, but carbonates are less durable than silica and eventually wear down to fine sizes that are washed away from beaches (Houston 2017). If a marble beach did form from direct wave erosion of a marble formation, it would be subject to rapid loss from abrasion. Marble gravel is used in Tuscany (Figure 2.10) to protect shorefront roads, protect sand beaches managed and groomed for intensive recreation, protect a nature reserve, and recreate beaches seaward of seawalls in resorts (Nordstrom et al. 2008). The marble beaches do not provide habitat similar

Box 3
Artificial gravel beach at Elba Island, Italy

Cavo (Elba Island, Italy) lost its beach in the 1970s because of reduced sediment input from abandoned agricultural and mining activity, combined with marina construction. The coastal road was protected by a revetment and a detached breakwater, which greatly restricted bathing (Figure 2.8). In 1999, the offshore breakwater was removed; two sand-trapping groins were built and the beach was nourished along ~500 m of coast, widening the beach about 11 m. Borrow sediments were from waste remaining from former iron mining. The poorly sorted fill became an impermeable hard surface from wave reworking, and the sea turned red from iron-oxidized fine sediment released during and after fill. The submerged *Posidonia oceanica* offshore was also threatened by burial. Removal of the fill sediment would have resulted in increased pollution, so a project to cover it with gravel was carried out in 2006–2007. Groins were raised to accommodate a higher beach and extended offshore as submerged segments to reduce longshore transport but with limited landscape impact. The beach was nourished with 30,000 m^3 of carbonate quarry gravel crushed to 15–20 mm diameter, placed in a layer 4 m thick, with a 15 m wider berm. After thirteen years, most fill remained in place (Figure 2.9) and the gravel was rounded by wave reworking. The beach is now highly used by beachgoers, and interviews demonstrate their satisfaction with the new resource. The increased beach stability allowed additional facilities to be built (toilets, kiosks). The coastal road was widened, and a promenade was added, greatly increasing the tourist appeal.

Contributor: Enzo Pranzini, University of Florence, Italy

Figure 2.8 Cavo without beach, June 1995.

Figure 2.9 Cavo after gravel fill, April 2017.

Figure 2.10 Marina di Pisa, Italy, where marble gravel was placed where the beach seaward of the seawall had eroded.

to the native beaches (Fanini et al. 2007), and they do not provide long-term protection for human facilities on high-energy coasts because the sediments are too easily abraded. They are acceptable to beach users because the rapid abrasion of marble makes particles round and smooth and comfortable for walking and sitting; their whiteness makes both the exposed surface of the beach and the water appear clean; the surface is cool in the summer due to its high reflectivity; and the beaches have cachet, resulting from the visual appeal of marble and its use in sculpture and expensive floors and wall panels. Visitors to the beach often collect marble for use at their homes. Observations of the recreational use of marble gravel beaches that occur near sandy beaches indicate that the marble beaches are not selected only as a fallback option when sandy beaches are full (Nordstrom et al. 2008). A case can be made for using marble to nourish a resort beach such as Marina di Pisa (Figure 2.10), where recreational demand is great. Use of less attractive fill materials on a resort beach, such as the tailings from iron mining first used at Cavo on Elba Island (Box 3), can be considered undesirable on recreational as well as environmental grounds. The significance and challenges of addressing stakeholder interests in restoration projects are discussed further in Chapter 7.

2.5.4.1 Effect of Mineralogy

Light-colored sediment may be introduced to improve the aesthetics of a beach and heavy mineral sands can be used to extend the longevity of beach fills (Eitner 1996), but changes in the color of sediment generally result from use of an inexpensive, readily available source. The differences in color caused by nourishing beaches with sediment of different chemical composition can be conspicuous, but changes in grain size, shape, roundness, sorting, abrasion, and fracturing may be as important.

Alterations to the compaction, density, shear resistance, color, moisture content, and gas exchange of beach sands can influence nesting and hatching success of turtles and alter nest chamber geometry and nest concealment, while formation of scarps due to shear resistance can increase the number of false crawls (Crain et al. 1995; National Research Council 1995). Some effects (compaction and scarps) can be reduced or eliminated by grading operations, and some effects (lowered nesting frequency and false crawls) may be reduced in subsequent seasons as the beach equilibrates to natural processes (Crain et al. 1995; National Research Council 1995; Rumbold et al. 2001). Other adverse effects can be superficially inconspicuous but significant. For example, hatching and emergence successes of sea turtles in native (silicates) and imported (aragonite) fill materials indicate that the imported sand provides suitable substrate, but the small differences in temperature regimes can change hatchling sex ratios (Milton et al. 1997).

Shell fragments can cause lithification of sediment as they dissolve and form new crystals in the pore spaces. Shells also appear to discourage colonization of the beach by some species (Speybroeck et al. 2006). Sediment delivered to the dune from a fill site can change the mineralogical composition of the dune (e.g., making it more lime rich) and thus change the potential for vegetation growth (van der Wal 2000).

2.5.4.2 Coarse Sediment

Increasing sand size on a sandy beach can convert a dissipative beach to a more reflective beach (Anfuso and Gracia 2005) and cause a decrease in species richness and abundance (McLachlan 1996). Nourishing sand beaches with gravel can result in profound changes initially, although these changes may be short term as gravel is incorporated into the sand beach by wave reworking. Gravel has been placed on shores that have lost their beaches, on sandy beaches, on mixed sand and gravel beaches, and on gravel beaches where it differs from native gravels (Shipman 2001; Colantoni et al. 2004; Nordstrom et al. 2004; Cammelli et al. 2006; Nordstrom et al. 2008; Anthony et al. 2011; Bertoni and Sarti 2011; Kochnower et al. 2015; Grottoli et al. 2017; Shu et al. 2019). The fill may be pure gravel or mixed sand and gravel

(Horn and Walton 2007). Early reviews of gravel and mixed sand and gravel beaches are presented in Kirk (1980), Carter and Orford (1984), Mason and Coates (2001), and Jennings and Shulmeister (2002). More recent studies of mixed sand and gravel beaches have been added (e.g., Bergillos et al. 2016; Hemmingsen et al. 2019; Matsumoto et al. 2020), but field studies are limited, and the great variety of sediment characteristics of these mixed beaches and differences in processes in their site-specific settings complicate easy categorization.

Differences in the permeability and transport characteristics of sand and gravel create differences in the shape/sorting of sediments, the form of the beach profile, and its response to changes in wave energy (Carter and Orford 1984; Pontee et al. 2004). These differences will be more pronounced in the end members of these size ranges. Gravel beaches are more stable than sand beaches because of larger particles that are less easily entrained and rough surfaces that dissipate wave energies (Carter and Orford 1984). The greater space between particles increases percolation of water, leading to greater transport capacity on the wave uprush than backwash, enhancing deposition on the upper beach (Everts et al. 2002; Austin and Masselink 2006). This deposition creates higher, steeper foreshores on gravel beaches and more conspicuous microtopography, including storm berms and cusps. Swash runup is relatively shallow, and small change in water volume due to infiltration greatly decreases the energy available for sediment transport; as a result, gravel beaches can dissipate over 90% of all wave energy (Ibrahim et al. 2006). Uprush (particularly during storms) increases the elevation of the berm crest (Lorang 2002), and the steep slope of the foreshore above mean sea level provides protection against runup from future storms. Coarse gravel (cobbles, boulders) is less likely to move than sand or pebbles, increasing stability of shoreline position and retention of fill materials (Everts et al. 2002; Komar et al. 2003; Shu et al. 2019). Although stable in the cross-shore direction because of percolation, longshore transport of gravel fill is common during storms and can result in shoreline realignment, creating narrower beaches on the updrift side of a gravel fill area and wider beaches downdrift (Cammelli et al. 2006; Shu et al. 2019). Longshore transport rates are likely to increase in the future, with sea level rise and greater storm inputs (Dornbusch 2017). This problem can be partially offset by enhancing the effectiveness of groins within the length of the fill area, backpassing the fill updrift, or renourishing frequently.

Placing gravel on a sandy bottom, even if no subaerial beach exists, can cause sand to migrate onshore, creating a mixed sand and gravel beach (Cammelli et al. 2006). The vibratory mechanism of swash and backwash results in kinetic sieving where finer sediment infiltrates the pore spaces in the gravel, creating a gravel surface layer with mixed sand and gravel layers below (Carter and Orford 1984).

The poorly sorted subsurface layer on mixed beaches can contribute to formation and persistence of scarps (Zarkogiannis et al. 2018). As sand is mixed with gravel (which occurs relatively soon after gravel placement) or as the gravel abrades (a longer term process), the beach will behave more like a sand beach hydrodynamically and may develop a lower gradient (McLean and Kirk 1969; Carter and Orford 1984; Mason and Coates 2001; Blanco et al. 2003). Placing poorly sorted sediment on a gravel barrier beach also can be counterproductive if the new sediment prevents infiltration of backwash from storm waves on the barrier face and swash action erodes into the fill causing headward erosion of the barrier crest (Orford and Jennings 1998). Finer grain sizes can enhance the likelihood of vegetation development on a gravel beach (Walmsley and Davey 1997a; Gardner and Burningham 2013).

Some beaches nourished with gravel have become primarily sand (Caputo et al. 1993). The amount of gravel placed on a sand beach and the elapsed time are critical in the form, function, and aesthetic appeal of the beach. The beach will retain its value for shore protection if the upper layer is still gravel and remains highly permeable and a high barrier to overtopping. Conversion to a primarily sandy beach may not occur where there is limited sand in the sediment budget, and new gravel is introduced at the site or at a feeder beach updrift.

Adding small amounts of gravel to an existing subaerial sand beach to improve its value for shore protection may be counterproductive because isolated gravel particles may be readily entrained from, and moved across, a finer-sized bed (Carter and Orford 1984; Aminti et al. 2003; Nordstrom et al. 2008), and gravel will be rapidly displaced alongshore or washed onto the backshore. Sand that moves into the gravel interstices may not increase the overall size of the fill, so any reduction in volume of sand offshore of the foreshore can be considered a loss in the coastal sediment budget (Cammelli et al. 2006).

2.5.4.3 Fine Sediment

Silts and clays may occur in source sediment that is primarily sand. Some sand-size aggregate sediments may disintegrate into finer particles after fill is placed, such as the calcium and silicon aggregates described in Chiva et al. (2018). Fine sizes increase turbidity during placement or when the settled fines or aggregated particles are reworked from the fill on the backshore during storms. Turbidity can adversely affect species with light requirements or low silt tolerance, although the visual impact of turbidity can sometimes seem more of a problem than its actual impact on biota (Smith et al. 2002). Fine sediments alter the structure of habitats after they settle into the beach matrix by changing bulk density, shear resistance, compaction, moisture retention, and flow in the beach water table. They may be more resistant to burrowing organisms, and the hydraulic conductivity in the beach may decrease, causing lower rates of water table discharge (Jackson et al. 2002).

Fill deposits reworked by waves and swash may become similar to native beach materials, but the depth of reworking on low-energy beaches is limited, and fill deposits on these beaches that are not reworked may be closer to the surface than the depth reached by burrowing organisms. For example, the depth that horseshoe crabs (*Limulus polyphemus*) deposit their eggs may be greater than the wave-reworked active layer on an eroding nourished estuarine beach (Jackson et al. 2002). Pretreatment to remove fines will avoid this problem as well as the problem of offshore deposition of fines removed from the fill by wave action.

2.5.5 Effect of Equipment Use

Direct physical disturbance can occur to flora and fauna by deployment of equipment, including laying pipelines, driving on the beach and dune to reach disposal sites, or bulldozing fill deposits into desired shapes. Indirect visual and auditory disturbance can occur to fauna, especially foraging birds, by operating machinery or using lights at night. Some pollution from exhaust fumes or fuel leaks may occur (Speybroeck et al. 2006). Use of mechanized equipment should be minimized to avoid these disturbances. Reshaping of deposits directly into a final form could be restricted to emergency situations, allowing natural processes to create the internal structure and external form of new landforms. Subaqueous nourishment can avoid many of the disturbances associated with use of equipment on the backshore.

2.5.6 Aesthetic and Recreational Problems

Many beach nourishment operations are conducted primarily to improve recreational use (Andrade et al. 2006). "Improvement" is usually perceived as a wider backshore, although some managers or users may complain about overly wide beaches caused by nourishment (Pranzini et al. 2018a), and nourishment can eliminate surf breaks desired by surfers (Castelle et al. 2009). The environmental costs of creating a wider backshore and the missed opportunities in using it only as an erosional buffer and recreational platform are major themes in this book, but the visual aspects of nourishment are also important. Observations of aesthetic effects of nourishment operations are often peripheral to the main purposes of project evaluations, but stakeholder complaints about the impact of exotic fill on turbidity and discoloration of nearshore waters and the alien appearance and value of a beach for recreation can lead to calls for more aesthetically appealing sand (Chandramohan et al. 1998; Nordstrom et al. 2004; Klein et al. 2009; Pranzini

2009; Pranzini and Vitale 2011; Pranzini et al. 2016; Parkinson and Ogurcak 2018). Gravel and shells that develop a surface lag decrease the aesthetic characteristics of the beach and its use for recreation (Klein et al. 2009), revealing the value of using well-sorted fill sediments (Marcomini and López 2006). Stranded dead organisms representing losses during dredging or losses resulting from suffocation by fine sediment and organic matter in the fill can raise concerns among beach users and environmental agencies (Pezzuto et al. 2006).

Acceptability of fill sediment for recreational use is critical to decisions about future nourishment operations, including mitigating actions to overcome undesirable effects (Pranzini et al. 2010, 2018a). The most readily available and least costly sediment sources may not be perceived as aesthetically suitable (Malvárez García et al. 2002; Nordstrom et al. 2004; Parkinson and Ogurcak 2018). Color of beach materials is a critical factor in stakeholder satisfaction, and beach nourishment can significantly modify original beach color (Pranzini et al. 2010). Care must be taken to match the native materials if local stakeholders take pride in the color of their beaches (Dean 2002), even when the fill materials have considerable appeal of their own (Arba et al. 2002). Plans for some nourishment operations have required that the color of nourished sand be within specified levels (Dean 2002). Color can be quantified prior to fill using a Munsell Chart, but color changes can occur to fill after exposure to the sun and atmosphere, sediment mixing, and selective transport of sediments of different mineralogy (Dean 2002; Berkowitz et al. 2018). Simply rinsing sample borrow materials is likely to underestimate changes in color, and the time necessary for color equilibration to occur may be unknown prior to emplacement.

Some fill sediments have an aesthetic appeal and recreational value that may be more acceptable to users than the native material. Nourishment operations using these materials can be catalysts for reconfiguring stakeholder attitudes about nature. Using gravel fill on a beach formerly composed of sand will change the use of the beach for recreation. Beach users may dislike the coarse, angular sediments; steep foreshore slopes; and high berm heights that make access difficult. In contrast, gravel fill on sandy coasts adds topographic diversity and new user options, such as listening to the sound of gravel rolling in the swash, collecting interesting rocks, throwing rocks into the water or using them as building blocks to create an alternative to sand sculptures. Gravel beaches are good for wave watching because the steep, permeable beach quickly dissipates swash uprush, and beach users can sit close to the breaking waves without being wetted by the swash. Gravel does not hold water or facilitate capillary rise like sand, so it is a drier platform for sitting, and it does not adhere to skin and clothes. Sitting on granules and small pebbles can be preferable to sitting on sand, although sand may be preferred over larger gravel. Disadvantages of gravel beaches are the more difficult vertical access, the high berm heights that can obscure views of the sea from inland, and the coarse grains that make walking barefoot uncomfortable and deploying beach umbrellas difficult (Nordstrom et al. 2008).

The use of marble as beach fill provides a striking example of how nature can be reconfigured. The attractiveness of marble combined with its importance in the regional economy and its historic significance make the marble beaches in Tuscany an evocative symbol of urban nature and a unique recreational experience, but the natural environmental heritage is sacrificed for an artificial socioeconomic one (Nordstrom et al. 2008).

2.5.7 Increasing Incentives for Constructing Buildings and Infrastructure

The intensity of development that results in a high benefit/cost ratio for protection projects makes beach nourishment an economically practical solution to coastal erosion (Kana and Kaczkowski 2019), but nourishment itself may increase demand for new development or incentive to build larger structures for human use (Klein et al. 2009; McNamara et al. 2015; Armstrong et al. 2016). The great boom in construction along the coast of Spain is perhaps the most dramatic impact of the value of beach nourishment as an economic incentive (Benavente et al. 2006). Applications for permits to build new structures farther seaward and make temporary structures permanent can be expected. Increase in levels of investment can be considered an advantage of nourishment (Wright and Butler 1984), but this gain should not come without a matching gain in natural beach and dune resources (Nordstrom et al. 2011). Beach nourishment projects can also be accompanied by shore protection structures designed for alternative protection or to aid retention of fill (Anthony and Sabatier 2013; Sancho-García et al. 2013; Luo et al. 2016; Utizi et al. 2016). Nourishment projects are also conducted near preexisting protection structures. Cooke et al. (2012) found that 48% of the beaches that received nourishment in Australia also had hard engineering structures. Shore protection structures can interfere with natural sediment transfers and landform migration, and they fragment natural components of the landscape.

2.5.8 Unknown Long-Term Implications for Biota

In the past, detrimental environmental effects of nourishment were often considered temporary (US Army Corps of Engineers 2001), with site-specific assessments indicating that impacts are minimal because (1) fishery value of affected species is unimportant; (2) a sufficient amount of undisturbed bottom exists nearby; (3) problems can be mitigated in follow-up actions; and (4) projects can be conducted when populations of biota are at their lowest. These assumptions were rarely supported by hard data (Lindeman and Snyder 1999), and little is known about many subtle or complex impacts (Gibson and Looney 1994; Rakocinski et al. 1996; Gibson et al. 1997). Alternatively, it is relatively easy to document differences between borrow areas and nourished sites in comparison to previously undisturbed sites or nearby control areas, but it is difficult to determine the significance of these differences. Studies that point out differences without providing future guidance do not contribute much to the conversation.

It is too early to make informed statements about the long-term, cumulative implications of large-scale projects or maintenance nourishment programs that result in repeated elimination or burial of species and change the persistence and resilience of communities (Lindeman and Snyder 1999; Posey and Alphin 2002; Speybroeck et al. 2006; Manning et al. 2014; Peterson et al. 2014). Most biological studies quantify the elimination and early recovery of fauna, but thresholds for species and environments should also be specified, as well as the significance of the relative proportions of the different returning species. A further complication is that different species of macrofauna have different preferences for grain sizes (Van Tomme et al. 2013).

The longest recovery times for biota appear to be when there is a poor match between the grain size characteristics of the fill materials and original substrate (Reilly and Bellis 1983; Rakocinski et al. 1996; US Army Corps of Engineers 2001). Accordingly, longer study periods with more frequent sampling and more cautious estimates of the magnitude, variability, and duration of impacts of beach replenishment appear warranted (Wooldridge et al. 2016). The great cost and long time frame required to conduct high-quality studies of ecological impacts and the difficulty of applying findings from one location to another will result in many unanticipated and unwanted environmental impacts. These kinds of impacts do not deter nourishment projects in most locations where protection and recreation projects are economically justified.

2.6 Alternative Practices to Minimize Environmental Losses and Enhance Values

2.6.1 Addressing Adverse Effects

The first step in reducing the impacts of beach nourishment projects on biota is to anticipate the effect of borrow and fill activities and eventual beach characteristics as well as potential mitigation techniques that can be used for each type of fauna and at each stage in use of the borrow and fill areas. Table 2.2 provides examples of the many factors that could be considered.

Table 2.2. *Checklist of factors to consider in assessing impacts of dredging and filling on fauna*

Characteristics of biota
Species type, abundance, average size
Biomass, community structure
Lifespan
Preferred substrate type
Feeding mode
Mobility/type of movement
Number of spawns per year and time of spawns
Life stage (egg, larva, hatchling, adult)

Potential effects on biota
Direct mortality
Disturbance
Disorientation
Nest site selection
Nesting success
Alterations to embryonic development
Time to recovery

Project design
Type of dredge
Time of dredging and placement (seasonal, day/night, nesting times)
Need for, and placement of, pipelines and equipment
Method of fill disposal
Actions to contain runoff/spillage/turbidity
Relationship of design profile to final landforms reworked by waves and winds
Need for reshaping with bulldozers
Method of constructing dunes
Need for planting or seeding vegetation
Provisions for monitoring and adaptive management

Potential alterations to borrow/fill areas
Creation of depressions, furrows
Alteration of wave climate, refraction patterns, current patterns and velocities
Change in mobility of substrate and bedforms
Change in beach morphology, erosion rate
Increased turbidity

Sediment characteristics
Alternative sediment sources
Grain size, sorting, shape
Silt/clay content
Chemical composition of sediments
Pollutants, including seeds of exotic vegetation
Expected effect on gas diffusion, moisture content, groundwater flow
Hardness and roundness of particles and compaction of fill after placement
Potential for scarp formation
Color (initial and change after exposure)

Table 2.2. (*cont.*)

Mitigation/management techniques
Beach tilling
Scarp leveling
Moving nests
Controlling uses of new resource (including opportunistic endangered species)
Adaptive management plans and funding mechanisms

(*Sources*: Diaz et al. 2004; Nairn et al. 2004; Dickerson et al. 2007).

Many potential adverse impacts of beach nourishment to biota can be avoided or minimized by employing existing technologies and practices but often at greater costs (US Fish and Wildlife Service 2002). The timing of the operations can be critical (Lawrenz-Miller 1991). Conducting operations around the activities of fauna is possible but not without difficulty. The need to avoid the nesting and hatching season of loggerhead turtles (*Carretta carretta*) at Folly Beach, South Carolina, required operations to be conducted in the most extreme weather conditions in winter (Edge et al. 1994).

Increasing the likelihood of recolonization in borrow areas may be possible if small, unmined "refuge patches" are left undisturbed within the larger mined area (Hobbs 2002). Other suggestions for borrow areas include dredging fewer but larger areas or multiple small areas; dredging in patches or strips; or dredging shoals in longitudinal rather than transverse strips (Cutter et al. 2000; Minerals Management Service 2001). Recolonization of fill areas may be enhanced if smaller quantities are deposited more frequently (Bishop et al. 2006). Frequent deposition of large amounts would be counterproductive because ecosystems could be completely eliminated and not have time to recover (US Fish and Wildlife Service 2002). Fill sand could be placed in the shallow surf zone rather than the intertidal beach, allowing slow accretion on the upper beach (Schlacher et al. 2012).

Some modifications are relatively inexpensive, such as using bulldozers to construct shore parallel berms to control turbidity by preventing discharged water from returning unimpeded to the sea (Dean 2002). Tilling can make sediment less compact to make it more suitable for turtle nesting, although the effect can be temporary and may not affect nest site selection (Davis et al. 1999).

2.6.2 Improving Habitat

Beach nourishment, like other soft engineering projects for environmental management, offers the opportunity for creating or enhancing biogeomorphological systems (Naylor et al. 2002). Changes in beach fill practices can improve habitat as a direct goal of construction or a side benefit as the nourished area evolves (Nordstrom et al. 2011). The rationale for alternative designs is identified in this chapter. Actions that can be used by local managers to increase the potential for natural habitats to evolve after the fill is placed are presented in subsequent chapters. Many nourishment projects can be modified to provide better conditions for biota, but opportunities will be lost until local stakeholders develop a renewed appreciation for the intrinsic value of nature and the values, goods, and services provided for human use (Table 1.2). True restoration may not occur in the absence of a detailed restoration plan that includes control of subsequent human activities (Nordstrom and Mauriello 2001; Nordstrom et al. 2011).

New wildlife habitat can be created when beach fill is placed where armoring has eliminated the upper beach (Starkes 2001) or where undesirable hard clay substratum, marsh peat, or human obstructions have been uncovered by erosion of natural sediment (Smith et al. 2020). Alternatively, smaller volumes of sediment with special characteristics may be used to enhance habitat for specific macrofauna, such as sea turtles (*Caretta caretta, Chelonia mydas, Dermochelys coracea*), piping plovers (*Charadrius melodus*), horseshoe crabs (*Limulus polyphemus*), capelin (*Mallotus villosus*), grunion (*Leuresthes tenuis*), surf smelt (*Hypomesus pretiosus*), and Pacific sandlance (*Ammodytes hexapterus*)(Jackson et al. 2020).

Another option would be to add criteria for enhancing biota to protection or recreation projects to convert them into multipurpose projects. The greatest opportunity for multiple-use projects may be large-scale nourishments, where adequate sediment and space are made available. One example is the Sand Motor, where co-biodiversity was enhanced by creating the hook-shaped deposit and sheltered lagoon (Figure 2.4), resulting in large-scale heterogeneity of sandy-beach morphology (van Egmond et al. 2018). One caveat is that large-scale projects that create novel habitat such as the Sand Motor and Cape May Meadows (Figure 2.11) may provide options for enhancing certain aspects of the environment, but these special cases may not be accurate representations of natural environments and may not be feasible in standard nourishment operations.

Some positive by-products of beach nourishment have been unanticipated, such as recolonization of new beaches by rare or threatened species that were not specific targets of the nourishment, including seabeach amaranth (*Amaranthus pumilis*) and piping plovers (*Charadrius melodus*) (Nordstrom et al. 2000). Dune development is enhanced on nourished beaches (van Puijenbroek et al. 2017), including desirable increases in the level of dune mobility (Rhind and Jones 2009), but the size, location, and function can differ greatly, depending on attitudes and actions of local managers (Nordstrom and Mauriello 2001). Many reports of windfall benefits are in the form of simple observations, with little data or explanation of the specific reasons these nourishment operations were successful in

Figure 2.11 The nourished beach and bulldozed dune at Cape May Meadows, New Jersey, USA, designed to provide nesting area for piping plovers and protect freshwater habitat landward. The beach and dune are far larger than would occur naturally on an estuarine beach.

this new context. The subsequent sections identify some of these benefits, along with suggestions for how they can be incorporated directly into project designs. General suggestions for conducting and evaluating nourishment projects to enhance ecological values are presented in Table 2.3. These suggestions must be expanded and placed in specific contexts for each nourishment operation.

Each beach nourishment operation is likely to result in alterations to the landscape that can be considered positive (e.g., nesting sites for threatened species) or negative (e.g., invasion of exotic species or predation by feral cats or foxes). There is still much that can be done to evaluate and improve the way beach fill projects are managed post fill to better achieve restoration goals (Nordstrom et al. 2011). Many of the unintended beneficial outcomes of nourishment reveal great potential as intended outcomes of future projects, while unintended negative effects provide caveats that should be addressed in those plans.

Restoration actions in Puget Sound in Washington, USA, provide perspective on the degree to which shorefronts that have been armored by protection structures can be converted to functioning beach environments using beach fill and modifying structures. An interesting aspect of these projects (four of which are described in Chapter 5) is that they involve placing sediment on top of existing

Table 2.3. *Suggestions for conducting and evaluating beach nourishment projects to enhance ecological values*

Design considerations
Add nature appreciation and intrinsic value of wildlife as justification for projects
Develop guidelines for ecological considerations in construction and post-construction
  practices
Identify complex interactions, thresholds, long-term effects in impact studies for borrow
  areas
Design a beach elevation that allows for wave runup across the backshore
Place sediment as close to natural elevation and shape as possible
Use a dune, not an unusually high beach, to provide flood protection
Deposit smaller quantities of fill more frequently
Use containment berms to control turbidity
Use earth-moving equipment to facilitate, not replicate, natural processes

Sediment characteristics
Evaluate sediment for future biological compatibility as well as stability and longevity
Use sediment similar to native sediment in size and chemical composition unless change in
  habitat is desired
Consider pretreatment to make suboptimal fill more compatible

Addressing losses of biota
Address impacts within and outside dredge and fill areas
Keep borrow pits shallow to allow for rapid refilling and faunal recovery
Leave local refuge and recruitment sites within borrow and fill areas
Avoid placing fill on bay-bottom or hard-bottom habitat
Adjust timing of nourishment operations to activities of fauna, including predators

Enhancing natural values
Identify ways to attract target species using sediment, topography, and vegetation as cues
Identify ways to change sediment composition to enhance viability of organisms in beach
Expand target species programs to encompass other species
Find ways to expand cross-shore environmental gradients
Allow nature to evolve by avoiding raking, grooming, or excessive trampling
Renourish before restored habitats have been lost

Follow-up actions
Monitor and apply adaptive management (using funding set aside in original project
  design)
Develop programs for education and outreach
Do not allow nourishment to provide incentive for new coastal construction

---

intertidal habitat, resulting in an offshore shift of tide lines. These actions have occurred in a regulatory environment that normally discourages covering bay bottom in order to reduce impacts on biological resources (Shipman et al. 2000).

Enhancing or adding habitat as a result of beach nourishment has far greater potential than has occurred in the past. Widening beaches that become degraded through time takes on added significance in staging areas for migratory birds that appear to be faithful to specific regions (Doody 2001). Wider beaches allow shorebirds to place nests farther from the high tide line, where they are less subject to flooding and disturbance by people using the beach. The greater space decreases competition for nesting resources and reduces aggressive behaviors by some species toward other species, reducing the likelihood of colony abandonment (National Park Service 2005).

Existing beaches could be selectively replenished with size components that are conspicuously lacking. Size, shape, and sorting characteristics of gravel affect mobility, and there is a difference in kinds of habitat and value of habitat associated with the different types of gravel used in nourishment operations (Williams and Thom 2001). The most productive and diverse portions of the estuarine beaches in Puget Sound appear to be in mixed sand, gravel, mussel bed, and boulder substrate rather than sand (Armstrong et al. 1976). Coarse sand and small gravel added to beaches can increase the likelihood of spawning by fish such as surf smelt and sandlance. The design for Samish Beach in Puget Sound included a gravel berm using several thousand m^3 of sediment 2–76 mm in diameter with a surface layer of 1140 m^3 of 2 mm diameter to enhance spawning for surf smelt and 200 m^3 of sand to represent the type of surface that occurs on the backshore (Zelo et al. 2000).

The state of Delaware, USA, took a proactive approach to improving estuarine beaches for horseshoe crabs as part of its shore protection program. One of their projects used beach fill to increase the amount of gravel on a beach where sediment was considered too fine to attract spawning crabs (Jackson et al. 2007). Gravel and coarser sand were added to make the beach resemble other, more gravel-rich, local beaches that had attracted more crabs in previous years. Coarser grain sizes appeared to be important for site selection, but finer sizes are more important for egg viability because of moisture retention. Small amounts of gravel were quickly incorporated into the wave-reworked sand in the beach, causing no change in hydraulic conductivity and egg viability, while gravel clasts remained conspicuous on the surface (Jackson et al. 2007). The conclusion was that the source of sediment for nourishing an estuarine beach should be compatible with the need to increase both spawning rates and egg development, with a gravel subfraction to mimic natural estuarine beaches and a dominant mode of medium sand to favor egg development.

Gravel or shell could be added to the surface of sandy backshores of ocean beaches to encourage use by nesting shorebirds that prefer gravelly surfaces to make their eggs less conspicuous (Maslo et al. 2011). Changing a few of the surface characteristics with judicious use of alternative materials can be effective in small-scale projects at relatively low cost. Planting certain species of vegetation could also be used to provide feeding and refuge areas or cues about predator–prey relationships.

Determining the optimum conditions for enhancing habitat is complicated by the need to accommodate different species. Different species of sea turtles, for example, show different preferences for nest site selection related to open sand and vegetated areas as well as berm elevations (Wood and Bjorndal 2000). Habitat enhancement shows great promise, but the way the beach is modified is often species-dependent, and the effects of changing beach characteristics on factors such as nest site selection and other uses are still speculative.

The lag in recovery of species after nourishment can be advantageous in some cases. Creation of new islands for bird nesting provides sites initially free of mammalian predators (Erwin et al. 2001). Beaches that have been recently nourished will have fewer infauna that feed on eggs laid in the beach, so eggs of species that spawn on beaches will undergo less predation. Methods of sand delivery that mimic inputs by natural processes, such as slow addition of sediment or deposition in active surf and swash zones, would likely not reduce the influence of predators.

Most nourishment operations are designed to minimize longshore transport of fill because of high project costs, but sediment does move alongshore to adjacent areas (Beachler and Mann 1996; Houston 1996; Ludka et al. 2018; Spodar et al. 2018), and longshore transport can be an intended component of some nourishment projects (Brutsché et al. 2015; de Schipper et al. 2016). An increase in the sediment budget downdrift should enhance the likelihood for landforms to survive or grow in those locations and increase topographic diversity and viability of habitat in a way that is less obtrusive than by direct nourishment (Nordstrom 2005). Documentation of effects of nourishment outside project areas can change the conception of nourishment from a local operation designed to protect a few shorefront homes to a comprehensive regional strategy for environmental restoration.

Alternatively, nourishment can result in erosion of adjacent beaches, depending on the complexity of the new shoreline (Slott et al. 2010), or sediment inputs to downdrift areas can be considered negative, if rates of transport are too high and beaches are too wide, compromising beach use (Castelle et al. 2009). Policy-makers should make potential impacts explicit so they can be included as positive or negative effects in evaluations of project success (van Koningsveld and Lescinski 2007).

2.7 Alternative Designs for Beach Fills

2.7.1 Changing Shapes

The vertical erosional scarp that forms on nourished beaches that have been built too high (Figure 1.3b) can restrict cross-shore movement of biota including nesting turtles (Crain et al. 1995), foraging shorebirds (Maslo et al. 2011), and spawning fish (Martin and Adams 2020). An alternative design, more compatible with natural habitat requirements, would include a beach with the height of a natural berm or even lower to allow natural processes to form the final berm (Dean 2002). A dune could be created landward of the lower berm to provide flood protection. This design can be retrofitted on beaches where fill was placed too high by mechanically grading the beach to lower the backshore, eliminate the scarp, and create a gentler overall slope. A gentler beach slope will allow wave swash to create a wider foreshore and allow the backshore to be reworked during storms. The size of the dune can then be increased naturally by aeolian transport across the wider foreshore and reworked backshore that provides a better source of finer sand or by scraping some of the beach sand into the dune as an interim form of protection. The effects of eliminating the scarp without reducing the elevation of the entire backshore would be temporary, and this maintenance action would be less desirable than constructing the initial nourished beach at a lower elevation and building a new dune to provide the flood protection function.

Modifications to the foreshore and backshore using beach fill are relatively easy to accomplish using earth-moving equipment, and their effects are readily seen. Changes to offshore topography are less commonly reported. Potential actions include creating or modifying projects to enhance surfing (Dally and Osiecki 2018; Vieira da Silva et al. 2020), filling gaps in bars, and creating artificial bars to form wave shadow areas landward and supply sediment for onshore transport (Kuang et al. 2019). These adjustments hold promise for future implementation, but studies on successes and failures of nearshore changes caused by nourishment activities are generally lacking.

2.7.2 Changing Project Size or Spatial Aspects

Large nourishment projects reduce the unit cost of beach fill and reduce the proportion of fill caused by end losses (Dean 1997), but large projects may have greater ecological impacts than projects that are small or introduce sediment at a slower rate (Bilodeau and Bourgeois 2004). The rationale for reducing amounts or rates of fill is not simply to avoid burial to excessive depths. Fewer turtles nest on a nourished beach in the first year, when beaches are wide, than when the beaches are narrowed to more natural widths by erosion. Invertebrate species without pelagic larval stages depend on gradual recolonization of fill from the edges, so shorter nourished beaches should have faster recovery (US Fish and Wildlife Service 2002).

Creative ways can be found to overcome some of the problems of larger projects. One suggestion is to divide the fill area into noncontiguous smaller areas, allowing nourished segments to function as sediment feeder sites for unnourished segments and unnourished segments to function as feeder sites for biota; nourished and unnourished segments can then be alternated in subsequent maintenance operations (US Fish and Wildlife Service 2002). This is analogous to leaving strips of vegetation within portions of dunes mined for heavy metals so a species pool of potential colonizers is available for regenerating habitats (van Aarde et al. 2004).

Small, yearly replenishment projects that match annual losses can be more economically viable than larger, less frequent projects and provide more suitable sediment if trucks can be used to haul sediment from nearby beaches and dunes (Muñoz-Perez et al. 2001). Small projects are also appropriate where predictions of beach change are questionable and thus initial projects are experimental (Cooper and Pilkey 2004).

2.8 Restoring Sediment Characteristics

Uncontrolled deposition of inexpensive or readily available sediment, such as quarry waste, can dramatically alter the characteristics of a beach, diminishing its natural and recreational value (Buchanan 1995), but mining river channels and flood plains (where legal) can yield sediment with grain-size distributions similar to those naturally carried to the river mouth and thus available to the beach (Cipriani et al. 1992). These source areas would have to be just inland from a beach to have similar characteristics. Transporting sand from an accreting portion of beach to an eroding one in backpassing operations could be made within a single littoral cell so the sediment budget and grain-size characteristics of a cell are not changed (Anfuso and Gracia 2005).

Identifying how exotic beach sediment can be effectively utilized is becoming increasingly important as more beaches are altered from their natural state. This requires better understanding of how the sediment becomes more natural in appearance and accepted and used by stakeholders as a result of abrasion and sorting by waves or human actions (Nordstrom et al. 2004). Sediment derived from opportunistic sources that is not initially suitable for placement in beaches and dunes can be processed prior to nourishment by sieving through screens (Nordstrom et al. 2008) or washing to remove fines (Pacini et al. 1997). The sediment can be made more compatible after nourishment by bulldozing it into active surf and swash zones to allow for natural reworking (Blott and Pye 2004; Nordstrom et al. 2008). Post-nourishment modification of beach fill to remove unwanted coarse sediment can be accomplished manually using rakes or mechanically using earth-moving equipment.

Well-sorted sand is preferable to poorly sorted sediment for recreational use of the beach but not necessarily for use as habitat. Sand is often used as fill on beaches backed by eroding bluffs composed of till (mix of silts, clays, sand, and gravel). Sand used as fill would not replicate the boulder-strewn surface with intermittent exposure of clay substrate in the nearshore. Beach fill for these locations would have greater value in restoring the habitats and natural heritage value if it were composed of more poorly sorted materials, with a gravel subfraction that would form a narrow, high beach with little likelihood of offshore transport that would bury preexisting substrate.

Restoring the boulder fields and surface pavements that once characterized segments of the nearshore can be of concern. Boulder fields on the Baltic coast of Germany are rarer than in the past because many boulders delivered by cliff erosion were removed for construction of buildings (Reinicke 2001). The practice of removing boulders from the beach was legally stopped in the nineteenth century, but by that time, the hard-bottom habitat that represented the main substrata for macroalgae development and contributed to the biodiversity of the coast was eliminated in many locations. The use of sand as beach fill can cover the remaining hard substrate, contributing to further loss of habitat. Restoration of hard-bottom habitat is considered a future option in Germany, but no plans have been finalized. Boulders are not easily mined and transported as fill materials. Restricting use of beach fill in adjacent areas or using groins to block fill sediment entering the region may have to be implemented to avoid excessive sedimentation.

The axiom that beaches should be nourished with sediment similar to native materials applies to gravel beaches as well as sand beaches. Sand added to a gravel beach in sufficient quantity (>25%) can change morphodynamics from that of a gravel beach to that of a sand beach (Mason and Coates 2001), resulting in instability of an originally stable gravel beach (Pagán et al. 2018). Changing the elevation and breach potential of gravel barriers (as would occur with sand added) can change the susceptibility of the backbarrier zone to marine influences and determine whether that zone will have freshwater or saltwater wetlands (Orford and Jennings 1998). Gravel beaches can have distinctly different suites of species from sandy beaches because they are dry, nutrient-poor and heterogeneous in physical characteristics (Walmsley and Davey 1997b; Gauci et al. 2005). Sand and humus provide the medium for roots to take hold and the moisture for growth on sandy gravel sites (Scott 1963), but few purely gravel beaches are vegetated, except high up on gravel ridges or in inactive landward ridges where there is sufficient stability and protection from wave action (Fuller 1987; Doody 2001).

Even if gravel beaches are devoid of vegetation, they can have a rich invertebrate fauna (Doody 2001). Gravel beaches can retain more wrack than sand beaches and can differ in the species composition of fresh wrack input (Orr et al. 2005). Some taxa, such as the amphipod *Pectenogammarus planicrurus*, are found only on coarse gravel (Bell and Fish 1996). Generalist plants, such as *Eryngium maritimum*, can be indifferent to grain size characteristics of substrate, but typical gravel species, such as *Rumex crispus* ssp., *Crambe maritima*, and *Glaucium flavum* can perform better on gravel substrate, presumably due to better drainage (Walmsley and Davey 1997b).

The ecology of gravel beaches is discussed in Scott (1963) and Packham et al. (2001). Restoring vegetation on gravel beaches is discussed in Walmsley and Davey (1997 a,b). Coarse-textured gravel provides substrate for established plants and helps maintain the characteristically sparse nature of vegetation; patches of up to about 20% sand appear to aid regeneration from seed and maintain the diversity of the dune species that are commonly associated with gravel beaches (Walmsley and Davey 1997b). Where sand is unavailable or inappropriate for nourishment, finer gravel sizes can help increase the water retention capacity of the substrate, stop seeds from sinking too deep for establishment, increase germination rates, and help plants survive drought (Fuller 1987).

2.9 Monitoring and Adaptive Management

The original design of any beach nourishment program should be adapted through time to reflect information and understanding gained from ongoing studies to ensure that project goals are achieved; compensating and mitigating measures are addressed; and unnecessary monitoring costs are avoided (Minerals Management Service 2001). The adaptive management component of a beach nourishment program is a place for introducing new restoration initiatives not envisioned in the original plan. Monitoring should be extended to the evolving landscape outside immediate project limits, including dunes that may be by-products of the fill. Monitoring of sites downdrift is most appropriate for mega nourishments, where part of project design is to increase the alongshore sediment budget. Documentation of beneficial collateral effects of nourishment projects that are not specifically designed as feeder beaches is also important to help overcome the criticism that sediment leakages are losses.

Scientific review boards and advisory boards may be required to evaluate projects that encompass large areas or are part of ongoing programs of state/provincial or national governments. Factors important in these projects include (1) altering the duration and frequency of sampling; (2) adding or eliminating monitoring elements; (3) monitoring new locations; (4) synthesizing and interpreting data generated over many years at many project sites; (5) detecting regional differences in impact or recovery; and (6) responding to needs before they become critical (Minerals Management Service 2001). The Minerals Management Service (2001) recommends a benthic ecologist, physical oceanographer or process sedimentologist, a biostatistician, and a marine fisheries specialist on their scientific review/advisory board for evaluating impacts in borrow areas. They also suggest having a national-level science review committee to (1) deal with multisite issues; (2) develop nationwide guidelines; (3) determine conformance of monitoring programs with the guidelines; (4) approve changes to monitoring programs recommended by the regional groups; and (5) be independent of work done at specific sites.

2.10 Concluding Statement

Beach nourishment can help restore lost landforms and habitats or change the entire character of a coast and its natural and human use values. Nourishment with compatible sediment provides the sedimentary resources and space for natural features to evolve with minimal human input, providing subsequent human actions are compatible, as discussed in subsequent chapters. The initial design should allow for as much wave and wind reworking as is acceptable given the need for the landforms to protect against coastal hazards. A relatively large deposit of subaerial fill (but not a mega nourishment) may be appropriate to create a beach that is wide enough to provide space for a dune, but subsequent nourishment operations may be smaller and more frequent to mimic natural sediment transfers. Use of gravel fill on a sandy shore or sand fill on a gravel shore or use of sediment of a different mineral type can create a beach with different usefulness for shore protection and recreation, although departures in habitat characteristics from the former characteristics or those of adjacent habitats may be questionable.

Damage to biota occurs in borrow and fill areas, with unknown long-term effects and often without adequate compensation. Nourished beaches should be considered as more than consumptive resources built for shore protection and recreation. Greater attention should be paid to the many different implications and meanings of beach nourishment, and new research should be initiated to demonstrate the scientific, technologic, and economic feasibility of methods more compatible with maintaining or restoring natural landforms and habitats. Maximizing the value of beach nourishment projects may require expanding the spatial and temporal scales of investigation, evaluating the significance of alternative beach and sediment types, identifying new uses for fill, evaluating alternative technologies and practices, and implementing adaptive strategies for post-fill management.

3

Dune Building Practices and Impacts

3.1 Characteristics of Human-Altered Dunes

Beaches and dunes are connected by sediment exchanges that are cyclic and related to storm events. Erosion of the dune by storm waves is followed by delivery of sand from offshore to the beach after storms and then to the dune during onshore and obliquely onshore winds. Delivery of sand by aeolian processes restores dune width and volume, allowing dunes to provide natural barriers against overwash, flooding, wind stress, sediment transport, and salt spray during small storms, which helps maintain the overall integrity of habitats that remain landward of them. Variations in topography within dunes result in local differences in exposure to these processes, creating a variety of microhabitats. The values, goods, and services that dunes can provide (Table 1.2) include protection of human structures, niches for plants, habitable substrate and refuge areas for fauna, and nest sites.

Dunes in human-altered areas differ from naturally evolving forms, depending on the way they are managed for alternative human uses. Dunes may be directly eliminated or reduced in size to facilitate construction of buildings, provide views of the sea, or provide access to beaches. They may be remobilized through destruction of the vegetation cover by grazing or trampling, or they may be built up to increase levels of protection against coastal storms. If built up to provide protection, they are often designed as linear features with a uniform crest height (Castelle et al. 2019; Bossard and Nicolae Lerma 2020). Where human infrastructure has encroached on the beach, the rebuilt dune is often placed farther seaward than a natural dune and is lower, narrower, and less variable in topography, vegetation, and mobility (Nordstrom 2000). Restoration projects for dunes can involve rebuilding them where they have been eliminated, increasing their size consistent with the amount of space available, or allowing natural processes to rework them into topographically diverse landforms with a variety of habitats.

The primary emphasis of this chapter is on building dunes where they have been eliminated, with a focus on foredunes. Subsequent chapters identify strategies to manage existing dunes of different sizes and with different relationships to human structures and uses. Chapter 4 addresses issues that are more relevant to wide dunes and dune fields, where space and time are not major constraints and dune mobility is acceptable. Chapter 6 addresses issues associated with foredunes in locations where human infrastructure restricts opportunities for maintaining dunes as dynamic systems.

Dunes may be built up on the seaward side, on the landward side, or on top of existing foredunes to create a larger barrier to erosion, flooding, and overwash (van Bohemen and Meesters 1992). They may also be rebuilt after existing foredunes are modified by mining (van Aarde et al. 2004), excavation for pipelines (Ritchie and Gimingham 1989), or removal of pollutants (Nordstrom 2000). Building up the top of the dune occurs where the level of protection from flooding and overwash must be increased (Matias et al. 2005). Extending the seaward side of the dune is more common and occurs where (1) human land use restricts extending the dune landward; (2) a higher dune is undesirable because views of the sea would be restricted; or (3) the surface cover of the existing foredune is considered valuable. Nourishment of the landward side of the foredune may be appropriate where (1) beaches are too narrow to supply sediment or provide adequate protection against wave uprush; (2) the existing dune is vegetated and considered more resistant to wave erosion than a newly formed dune; or (3) the existing dune is already small and too close to the sea to last throughout a storm.

Dunes can be constructed or enhanced by allowing them to form by natural aeolian processes (usually on nourished beaches where sediment source widths are sufficient) or by constructing them using bulldozers, sand fences, or vegetation plantings. Artificial dunes are often built to unnatural shapes, resembling linear earthen dikes (Feagin 2005; Nordstrom et al. 2007c), but actions can be taken to make these dunes appear and function more naturally (Chapter 4).

3.2 Dunes Built by Aeolian Transport from Nourished Beaches

Adequate dry beach width is needed for incipient dunes to develop into a foredune by natural processes (McLean and Shen 2006). The most natural way of restoring dunes is by providing a beach with sediment sizes that can be reworked by waves and winds (Nordstrom 2005), and conspicuous dune growth can occur following nourishment (Bocamazo et al. 2011; Silveira et al. 2013; Kaczkowski et al. 2018). A new dune that results from transport off a nourished beach (Figure 3.1) can then have the grain size characteristics, internal stratification, topographic variability, surface cover, and root mass of a natural dune.

Figure 3.1 Evolving foredune on the nourished beach at Ocean City, New Jersey, USA, in March 2007, fourteen years after construction of the protective dune at the far left began. The height and volume of the foredune ridge reflect the absence of severe storms in previous years.

Nourishment of the backshore alters aeolian transport as a result of changes in (1) beach width; (2) grain size characteristics; (3) moisture conditions (because of a higher surface or change in permeability); and (4) shape of the beach or dune profile. The extra width of a nourished beach provides a wider source of sand for dune building and greater protection from wave erosion for the dunes that form (van der Wal 1998, 2004). The widened beach increases the potential for a wider cross-shore gradient of physical processes to occur, with a greater number of distinctive habitats, from pioneer species seaward to woody shrubs and potentially trees landward (Freestone and Nordstrom 2001).

Aeolian transport can be excessive and considered a nuisance when beach fill materials are well sorted and fine, e.g., using dune sediment as an opportunistic source (Marqués et al. 2001). If poorly sorted sediment is placed on the beach, removal of the fines can leave a coarser shell or gravel lag surface that resists aeolian transport (Davis 1991; Psuty and Moreira 1992; van der Wal 1998; Marcomini and López 2006). After initial surface deflation, a nourished beach may play a limited role in transport rates unless the surface is disturbed to exhume finer sand and eliminate the scarp that forms a trap for blowing sand. Field observations indicate that a shell lag surface can form within weeks and can remain in areas that are too high to be periodically flooded (van der Wal 1998; Hoonhout and de Vries 2017). Storm reworking of the backshore or disturbance by beach cleaning and grading of scarps can initiate new cycles of increased aeolian transport.

Raking the beach to remove litter will break up the surface lag, expose finer sand to deflation, and reinitiate aeolian transport, but there may be little need to rake an overly high beach where waves cannot deposit litter. Raking to keep beaches clean cannot be counted on to create a good source of surface sediment because few beaches are raked in the winter, when the strongest sand-transporting winds occur in many locations.

If a beach is built at low elevation and allowed to evolve naturally, wind-blown sand will accumulate near the seaward part of the vegetated surface (Hesp 1989; Kuriyama et al. 2005), at the base of existing dunes (Kuriyama et al. 2005), or at the wrack lines that form on the nourished beach, especially the upper storm wrack line (Nordstrom et al. 2007b). The accumulation of wrack plays an important role in dune formation and is discussed in greater detail in Chapter 4. Large-scale nourishment projects can result in formation of embryonic dunes seaward of former foredunes or sand transport onto and over foredunes resulting in a dynamic dune environment with development of blowouts (Arens et al. 2013a). The elevation of a nourished beach and its effect on development of a lag surface may play a role in the kind of dune that develops. Large-scale nourishment projects also contribute to greater potential for dune building in unnourished segments downdrift because of longshore sediment transport (Nolet and Riksen 2019).

Embryonic (incipient) dunes will continue to form at seaward locations (Figure 3.2a) until the rate of wave erosion balances aeolian deposition, at which time the most seaward dune may grow into an established foredune (Figure 3.1). Restoration of the morphology and vegetation assemblages of natural foredunes can take up to ten years (Woodhouse et al. 1977; Maun 2004; Houser et al. 2015). Wind-blown sand may accumulate on landward human infrastructure if there are no obstructions, such as bulkheads or seawalls. The hazard of blowing sand and the threat of flooding argue for building a stabilized dune near the infrastructure to provide an interim protective barrier as incipient dunes evolve naturally seaward of it. This stabilized dune can be initially built by machines or by using fences and vegetation plantings.

3.3 Building Dunes by Deposit of Fill from External Sources

Dunes may be constructed by directly depositing fill sediment from nearby subaqueous source areas, such as offshore borrow areas, inlets, and navigation

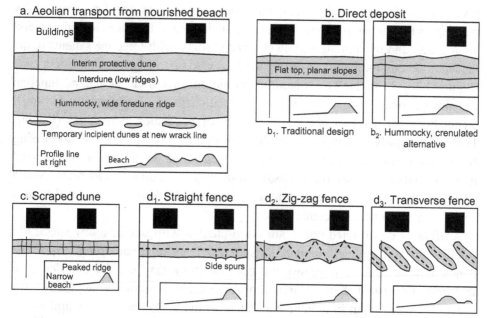

Figure 3.2 Characteristics of dunes built by alternative methods.

channels, or from inland sedimentary deposits and reshaping the fill using earth-moving equipment (Baye 1990; Matias et al. 2005; Bocamazo et al. 2011). These dunes are usually built to optimize a flood protection function and are designed for ease of construction or management. They are essentially dikes, often with a flat top and planar sides of consistent slope with little topographic diversity (Figure 3.2b$_1$). Dunes built as protection structures may retain their artificial form through time if they are rebuilt to the same template when they erode or if they are protected on the seaward side by groins and beach fill that restrict modification by wave processes. They may eventually evolve to resemble natural dunes in form as a result of aeolian transport and vegetation growth if they are not repeatedly nourished and reshaped.

The grain size characteristics, rates of change, and characteristic vegetation of dunes built by mechanical placement usually differ from landforms created by aeolian deposition (Baye 1990; van der Wal 1998; Nordstrom et al. 2002; Matias et al. 2005), but they can have well-sorted sands that resemble dune sediments if suitable borrow areas are used. Use of sand from the sea may be preferable to sand from terrestrial sources because it may not have harmful soil organisms (van der Putten 1990) and it is likely to be better sorted.

Dunes built for flood protection can provide habitat for faunal species that would occur in a natural dune environment (Latsoudis 1996), and actions can be taken to enhance this value. Plant diversity is strongly related to topographic diversity (Tukiainen et al. 2019). Patchiness of habitats can be increased by creating an undulating foredune crest, resulting in local differences in drainage and wind speed, and by creating a more crenulated shape with foredune salients in the landward subenvironment, converting the landward boundary from a line to a zone (Figure 3.2b$_2$). This type of contouring would enhance both the natural function and image of the dunes (Nordstrom et al. 2007c). Even relatively small crenulations can enhance the visual quality of vegetation boundaries (Parsons 1995). Frequently, managers take the opposite approach, using earth-moving equipment to smooth out residual clumps of vegetation or gaps that promote wind funneling (Ranwell and Boar 1986). Reshaping to enhance variability would recognize the value of the landform as habitat and aesthetic resource rather than solely as a protection structure.

Artificially constructed dunes are often planted with one or a few species to stabilize the surface, e.g., *Ammophila* spp., *Panicum* spp., and *Spinifex* spp. The dunes may be built as interim protection, but they may last for decades on a widened, nourished beach, and aeolian accretion seaward of them may eventually place them in a niche equivalent to the backdune on a natural shore. Frequent sand inundation is important in enhancing the vigor of *Ammophila* and other coastal plants adapted to burial because of increased soil volume, increased nutrients, enhanced activity of mycorrhizal fungi, and reactive plant growth (Maun, 1998). Dunes constructed using fill and planted with *Ammophila* can remain depauperate for years if sand inputs are precluded by dunes that form seaward of them. The 3 m high dune at Ca'Savio, Italy (Figure 3.3), was artificially constructed for flood protection using bulldozers and then planted with *Ammophila littoralis*. A sand fence on the seaward side of the bulldozed dune trapped the sand blowing off the widened beach, creating a foredune that further decreased the likelihood of sand blowing inland. The lack of fresh sand on the bulldozed dune caused the planted *A. littoralis* to die, leaving a bare surface with exposed shell fragments that enhance the unnatural look of the flat-faced dike. A similar problem occurred in Avalon, New Jersey, USA, where the height of the foredune crest fronting the artificially nourished dune was increased using multiple lifts of sand fences (Nordstrom et al. 2002). The lack of vigorous vegetation can make the artificial appearance and coarse-grained surface of artificially filled dunes readily apparent.

If no barriers to aeolian transport are created seaward of dunes built by direct deposit, deposition of fresh sand can create surface characteristics similar to natural dunes. Accumulation of wind-blown sediment on artificially filled dunes can be accelerated by using fill materials that are compatible with aeolian transport (well-sorted sand), employing sand-trapping fences on (not in front of) the nourished dune, and creating a gentle seaward slope in the initial fill deposit (Matias et al. 2005). This degree of mobility would likely not be acceptable where the bulldozed

Figure 3.3 Filled dune dike and fronting fence built at Ca' Savio, Veneto, Italy, showing the lack of success of *Ammophila littoralis* planted on the fill due to prevention of aeolian deposition by use of sand fences seaward.

dune is created as primary protection on a narrow beach, but it would be achievable on a nourished beach.

3.4 Building Dunes by Beach Scraping

Building dunes from in situ beach sand using earth-moving equipment (variously termed bulldozing, grading, scraping, reshaping) is a common option in some locations (Conaway and Wells 2005; Cooke et al. 2012). Scraping can move sand from its natural source area to the dune landward of it at a rapid pace without causing a dramatic alteration of sediment budgets. Scraping is capable of quickly recreating dunes on spatially restricted eroding beaches, where new dunes could not form or survive long enough to provide protection against storms. As such, it provides a way to protect infrastructure until more comprehensive shore protection programs are implemented (Kana and Kaczkowski 2019). Scraping is also a way of quickly burying structures to provide a more natural surface (Nordstrom 2019). Small scraping operations conducted at the municipal level are common because many municipalities have ready access to earth-moving equipment, and other options for shore protection may be too costly or may take too long (Nordstrom 2000). Scraping may also be conducted at each property (Conaway and Wells 2005; Harley and Ciavola 2013; Nordstrom 2019) where it is affordable to property owners and local regulations allow it (Box 4).

Box 4
Recovering dunes for hurricane protection on the Mexican Caribbean coast

Every year, the coasts of Mexico, especially the Caribbean, are affected by hurricanes that cause severe damage to built infrastructure, with high economic costs. This is a problem for hotels and the tourist industry. In the last decade, hotel owners learned the benefits of coastal dunes and their vegetation for protection from flooding and erosion. Thus, the hotels and facilities are now farther inland. The hotel site in Puerto Morelos (south of Cancun) that is pictured in Figures 3.4 and 3.5 had an artificial protective dune built seaward for protection by bulldozing sand from local sources. Native plants from nearby natural dunes – *Tournefortia gnaphalodes*, *Canavalia rosea*, and *Coccoloba barbadensis* – are planted on the artificial dune. These species accumulate sand and are natural dune builders that are tolerant of salinity and wave impact. Other plants native to the area are allowed to grow by natural dispersal and colonization, resulting in a semi-natural foredune. The dunes provide nesting space for sea turtles, which is also beneficial as a tourist attraction. Access of hotel guests to the beach is concentrated on a wooden walkway to avoid trampling on the dune. Coconut palm trees are planted on the beach as a tourist amenity, but they do not affect sand movement into the dune. These trees are not native to Mexico, but palms are extensively used on tourist beaches.

Contributor: M. Luisa Martínez, Instituto de Ecología A.C., Mexico

Figure 3.4 Planted palm trees at Puerto Morelos, Mexico.

Figure 3.5 Bulldozed and planted dune at Puerto Morelos, Mexico.

Scraped sediment is usually placed as a peaked ridge that may be slightly hummocky, resulting from incremental deposition by bulldozer blades. The shapes could be worked to resemble engineered dikes or more natural shapes, but there is little reason to do this to emergency interim landforms. Scraping initially results in an unvegetated surface and a loosely packed substrate. The surface is susceptible to high rates of aeolian transport, leaving the surface deposits coarser and more poorly sorted than sediment in nearby dunes (Conaway and Wells 2005). Greater aeolian transport landward of the dune crest may be expected initially, but these losses from the dune will be minor when compared to wave-induced losses (Conaway and Wells 2005).

The advisability of scraping operations as a form of sediment management is ambiguous (Tye 1983; McNinch and Wells 1992; Ellis and Román-Rivera 2019). Scraping from the beach to the dune changes the morphology of the beach/dune system and the sorting and internal characteristics of the deposit rather than the cross-shore volume of sediment. The shapes of the resulting ridges are often created ad hoc or using guidelines that do not always consider subsequent sand transport or the shapes and profiles of natural dunes (Smyth and Hesp 2015). The loosely compacted bulldozed dune placed seaward of the previous dune line can contribute to increased erosion of the new deposit with increased sediment transport downdrift (Kratzmann and Hapke 2012).

Scraping can be done to build temporary ridges on the backshore to restrict storm wave runup (Harley and Ciavola 2013; Hanley et al. 2014; Gallien et al. 2015). These features are alternatively called dunes or flood control berms – a term that better defines their purpose. Flood control berms have a single use function as a seasonal shore protection structure; they are not in equilibrium with wave-created beach berms; and they bear little resemblance to a dune that would form and survive farther landward. Flood control berms have no value as natural habitat or a proper image of nature and appear to offer no advantage to using the sediment to enhance the characteristics of existing dunes or create a new dune in an appropriate position on the landward edge of the backshore. These features make sense as a one-time emergency measure, but they should not be institutionalized as a standard annual practice, when more effective methods can be implemented.

Restrictions may be placed on the location and depth at which sediment can be removed from the beach because the implications of changes in beach/dune configuration are not well known. The state of North Carolina, USA limits the depth of scraping on the beach to 0.3 m and prohibits scraping seaward of low tide for most permit holders (Conaway and Wells 2005). Scraping depth is limited to 0.3 m at locations above high water level in New Jersey, USA (Nordstrom 2019). Scraping design at Fire Island National Seashore, New York, USA, does not allow removal of sand seaward of the berm crest (Kratzmann and Hapke 2012).

Scraping can have an adverse effect on plants on the backshore. Kelly (2014) found that scraping reduced beach vegetation by 91% compared to protected sites where scraping and raking did not occur. Some colonization of beach plant species can occur where propagules are included in the sand deposited on the face or foot of the dune (Kelly 2014), although the resulting vegetation is minimal relative to the losses on the beach. Even where backshore vegetation is not directly removed, the vegetation can be destroyed by earth-moving equipment passing from the foreshore to the dune.

Planting vegetation on scraped dunes is recommended (Smyth and Hesp 2015; Ellis and Román-Rivera 2019), and dense covers can occur where the constructed dunes last (Box 4). Planting may be counterproductive if beach widths are narrow and the cycles of wave erosion and mechanical deposition are too frequent to allow planted vegetation to evolve into a dense cover. Fences may be used to limit deflation of the new temporary dune deposits (Conaway and Wells 2005; Kratzmann and Hapke 2012) and may be a more cost-effective option than planting in locations where dune erosion is frequent. Scraped dunes on narrow beaches have little restoration potential, but they can serve as reminders that dunes are important in providing flood protection, and they can help reduce political pressure to build seawalls (Gesing 2019; Kana and Kaczkowski 2019). Care must be taken to inform stakeholders that creating bulldozed dunes in narrow spaces provides only a temporary solution, and expectations and declarations of success should be adjusted accordingly (Gesing 2019).

3.5 Building Dunes Using Sand Fences

Sand-trapping fences were used as early as 1423 in the Netherlands and Germany (Cordshagen 1964; van der Laan et al. 1997). Fences can be used to (1) build a dune where no dune exists; (2) fill gaps in the crest of existing dunes; (3) create a higher or wider dune, making it a more effective barrier to wave runup, wind and wind-blown sand, and salt spray; (4) keep the dune or blowing sand from moving inland; or (5) build a new dune ridge seaward of an existing dune. In all of these cases, fences increase the likelihood that species less well adapted to the stresses of wind, blowing sand, and salt spray will survive closer to the water. Sand fences are one of the most important human adjustments affecting the morphology and vegetation on sandy coasts because they are one of the few structures permitted seaward of the dune crest in many jurisdictions; they are inexpensive and easy to emplace; they are usually deployed at the highly dynamic boundary between the beach and dune; and they are often deployed in nature conservation areas where other human structures are prohibited (Nordstrom 2000).

The method of emplacement of sand fences is often according to the whim of managers despite the existence of technical assessments and guidelines (e.g., Coastal Engineering Research Center 1984; Ranwell and Boar 1986; Hotta et al. 1987, 1991) and many websites. Recommendations for the most effective fence configurations vary among sites (Miller et al. 2001), and considerable differences can occur alongshore in the number and configurations of fences (Grafals-Soto and Nordstrom 2009). The degree to which fence placement should be standardized is not clear, especially since configurations that accommodate movement of fauna by including gaps alongshore may be incompatible with configurations designed to provide flood protection by creating a continuous barrier (Nordstrom et al. 2011).

Fencing materials include canes, brush, and tree branches that can be inserted into sand individually (Figure 3.6, Box 5), and wooden slats, plastic, and jute fabric that are attached to fence posts. Using sand fences with different characteristics and in different configurations (Figure $3.2d_{1-3}$) can result in considerable variety in foredune topography and vegetation along and across the shore. Straight fence alignments that are parallel to the shore are common and seem to provide the most economical method of building protective dunes (Coastal Engineering Research Center 1984; Miller et al. 2001), but they may create slopes that are too steep for planted vegetation to become established. Placing paired fences can create a foredune with a broad base and rounded crest that can look more natural and make planting vegetation easier (Schwendiman 1977). Shore parallel rows of fences can be placed to create a cross-shore dune topography of ridges and swales with local differences in exposure and sheltering that contribute to vegetation diversity (Grafals-Soto 2012).

Zigzag fence configurations (Figure $3.2d_2$) are frequently used. Paired zigzag fences can create wider, more gently sloping dunes with more undulating crestlines than straight alignments, resulting in a closer approximation to the shapes of natural dunes (Snyder and Pinet 1981). Zigzag fences can provide a more effective trap than straight shore parallel fences for winds blowing alongshore (Marcomini and López 2006). Side spurs perpendicular to straight alongshore alignments and separate shore perpendicular fences described as "air groins" have been suggested to increase trapping rates in locations of strong longshore winds (Anthony et al. 2007; Ruz and Anthony 2008).

Multiple lifts of fences can create a higher dune with much greater volume than single lifts (Coastal Engineering Research Center 1984; Mendelssohn et al. 1991; Miller et al. 2001), although once a foredune has been initiated, placing additional fences on the foreslope or at the dune toe may be unnecessary. The dune can become the obstacle that traps sand, so adding fences may have little shore protection value (Grafals-Soto and Nordstrom 2009), and the fences can interfere with movement of fauna.

Box 5
Restoration of the Laida Dune, northern Spain

Restoration of the Laida Dune began with placement of three parallel lines of brush sand fences spaced 7 m apart along 250 m shoreline length in October 2001. Two additional lines of fences were added in 2002 to increase dune height (Figure 3.6). 150,000 individual dune builder plants (*Ammophila arenaria* and *Elymus farctus*) were planted in November 2002. Additional plantings occurred in 2003–2004 where initial plantings were unsuccessful. A symbolic fence (foreground of Figure 3.6) was placed around the perimeter to prevent visitor trampling, and information signs were placed. Natural colonization of plant species in the restored dune was monitored over seven years. Throughout this period, the vegetation dynamics in the restored dune followed a process of primary succession, with a progressive increase in species number, cover, and heterogeneity (Figure 3.7). Forty-two plant species became established on the dune, of which eighteen were dune-exclusive species, representing 62.1% of the total number of species of the same type in the region. Five of these species are considered to be rare or threatened. Comparison with reference dune systems in the region revealed the success of the restoration of characteristic plant communities and the way the process took place over time at the regional coastal landscape scale (Gallego-Fernández et al. 2011).
   Contributor: Juan Gallego-Fernández, Universidad de Sevilla

Figure 3.6 Laida Dune in December 2002, showing brush fences and symbolic fence.

Figure 3.7 Laida Dune in July 2006, showing established vegetation cover.

Sand accumulation efficiency and morphologic changes depend on fence porosity, height, inclination, scale of openings, shape of openings, wind speed and direction, sand characteristics, number of fence rows, separation distance between fence rows, and placement relative to existing topography (Coastal Engineering Research Center 1984; Hotta et al. 1987, 1991; Li and Sherman 2015). Fencing with a porosity of about 50%, with space between open and closed areas of less than 50 mm, appears to work well and fill to capacity in about a year where appreciable sand is moving, with the dune achieving a height similar to fence height and a slope of 1 on 4 to 1 on 7 (Coastal Engineering Research Center 1984). A single row of fences is cost effective at low wind speeds, but a double row (spaced about 4× fence height) may trap sand faster at high speeds (Coastal Engineering Research Center 1984).

A fence placed at the seaward limit of natural vegetation or foredune line may be far enough landward to survive wave attack during small storms and have a wide source of wind-blown sand seaward of it (Coastal Engineering Research Center 1984). A nourished beach that is artificially widened lacks these diagnostic features. A conservative approach would be to place fences close to infrastructure to provide a protective dune quickly and allow natural processes to eventually create a more dynamic dune seaward. Studies suggest that dunes built with fences have a high likelihood of surviving storms when they are at least 100 m landward of the active foreshore (Dahl and Woodard 1977, Miller et al. 2001). The decision about where to place fences on a nourished beach or a beach widened by storm wave overwash could also be made on whether it is desirable to create a dune field with fewer ridges and greater space devoted to intervening low moist habitats (slacks) or a higher, drier, and more continuous dune field.

Use of fences in new dune building operations is often a secondary action taken after the primary protective dune is created using bulldozers (Mauriello 1989; Matias et al. 2005). Fences then help stabilize the bare sand surface and prevent sand inundation of property landward of the dune, while allowing the surface near the new fence to evolve by aeolian deposition. The locations of added fences are critical in determining whether the surface of the bulldozed landform can evolve to a more naturally functioning foredune or backdune. Minimizing fences or placing them close to the bulldozed dune will cause it to function as a foredune. Placing fence lines farther seaward will shelter the bulldozed dune and contribute to vegetation more common on a backdune. Vegetation plantings will still be important to stabilize the surface (Coastal Engineering Research Center 1984) and establish a natural trajectory.

The numbers, locations, and configurations of sand fences and the dunes they create change through time. Fences may deteriorate, be destroyed by wave uprush, buried by aeolian accretion, repaired, removed, or replaced. The number of fences increases as new ones are built to replace those that are weathered or end up far from the original zone of active sand transport (Grafals-Soto and Nordstrom 2009). Densities and numbers of fence rows can be great in places, and fences can remain as conspicuous elements in stable dunes (Figure 3.8). The main intended effect of sand fences to build protective dunes obscures their great significance in altering habitat by changing topography. Artificially maintaining a dune for protection of

Figure 3.8 Ocean City, New Jersey, USA, showing sand fences stranded in the dune following accretion on a beach widened by artificial nourishment.

human facilities against flooding, salt spray, and wind-blown sand results in greater species richness than would be possible in the restricted space available on developed coasts and allows more natural cross-shore gradients of processes and vegetation types to occur (addressed in Chapter 6). Use of different fence configurations, number of rows, and separation distances can diversify topography and encourage diversity of dune vegetation. Local relief in ridge and swale topography creates small-scale differences in sheltering and proximity to the water table that enhance the variety of habitat over short cross-shore distances, but the significance of fences to habitat is virtually unknown to local managers (Grafals-Soto 2012).

Alternatively, sand fences can have a negative effect on fauna. They can provide barriers to movement by fledgling birds and turtles that nest on the backshore, but they may be too low to exclude their predators, such as feral cats and foxes (Robley et al. 2007). The main value of fences appears to be in the early stages of dune building or in locations where beaches are narrow and cycles of dune destruction and rebuilding are frequent. Fences are not required in locations where human development is far landward and foredunes already exist. Field studies in locations where dunes are not spatially constrained indicate that mobility of dunes and crest irregularities need not be a concern where dunes are not required for shore protection

(Nordstrom et al. 2018). Emplacement of fences on the seaward side of foredunes can prevent natural dunes landward of them from growing vertically (Itzkin et al. 2020). The more dynamic approach of allowing dunes to evolve without fences appears to have a negative effect on the volume of the foredune right at the backshore in these locations but enhances the environmental quality of the dune by increasing the characteristic plant species associated with active dunes (Nordstrom et al. 2012; De Jong et al. 2014). Removing or altering fences to make them more compatible with natural values is addressed in Chapter 4.

3.6 Building Dunes Using Vegetation

Vegetation helps build and stabilize dunes by trapping wind-blown sand and holding it in place. Vegetation can also help attenuate wave and swash energy and resist erosion because of the above and below ground structure, although field evidence of the significance of vegetation in resisting erosion is lacking and effectiveness can vary depending on species type and storm intensity (Feagin et al. 2015, 2019; Silva et al. 2016; Charbonneau et al. 2017; Sigren et al. 2018). Vegetation was used to build and stabilize dunes in locations of mobile dunes for centuries (Sharp and Hawk 1977; Skarregaard 1989), with efforts in Great Britain as far back as the fourteenth and fifteenth centuries (Ranwell 1972; Clarke and Rendell 2015). Most of these early projects were likely done to stabilize mobile dunes, an issue treated in Chapter 4. Vegetating dunes to increase their value as protection structures became important in recent decades as coastal development made more locations subject to wave runup and overwash, and beach widths became more restricted.

Many experiments on stabilizing dunes using vegetation were conducted in the 1960s and 1970s, resulting in the production of many planting guidelines for building and stabilizing dunes (Woodhouse and Hanes 1967; Bilhorn et al. 1971; Graetz 1973; Adriani and Terwindt 1974; Woodhouse 1974; Davis 1975; Seltz 1976; Dahl and Woodard 1977; Schwendiman 1977; Woodhouse et al. 1977; Broome et al. 1982; Salmon et al. 1982; Ranwell and Boar 1986). Government agencies and departments developed their own guidelines, often based on these reports (Coastal Engineering Research Center 1984; Hamer et al. 1992; Skaradek et al. 2003). Subsequent field studies of planting projects for building and vegetating dunes and evaluations of success include van der Putten (1990), Mendelssohn et al. (1991), Schulze-Dieckhoff (1992), van der Laan (1997), Freestone and Nordstrom (2001), Miller et al. (2001), Feagin (2005), Hanley et al. (2014), Bessette et al. (2018). These studies are supplemented by studies of projects to revegetate existing dunes after removal of exotic species or degradation by human uses (addressed in Chapter 4). New studies continue to shed light on the results of earlier experiments.

Table 3.1. *Factors important in planting vegetation to build foredunes or stabilize mobile dunes*

Factor	Significance
Native or exotic	Compatibility with other species; heritage value
Perennial or annual	Effectiveness in the non-growing season
Adaptability to burial	Placement relative to beach and mobile surfaces
Nitrogen-fixing capability	Aid in establishing species in backdunes
Height and width of stalks	Sand-trapping ability, dune size, shelter for other species
Optimum time for planting	Plant survival
Speed of germination	Growth rate and ability to survive weather changes
Rate of growth	Time needed to achieve stability; morphology of dune
Ability to spread underground	Survive harsh surface conditions; morphology of dune
Ability to compete with weeds	Survival potential; need for adaptive management
Ability to attenuate wave energy	Effectiveness in resisting erosion
Density of growth	Effectiveness of surface stabilization and sand trapping
Susceptibility to disease	Survival potential
Susceptibility to fire	Survival/hazard potential
Periods of dormancy	Aeolian transport potential
Value as food or shelter	Use by target fauna
Availability (on site or in nurseries)	Cost effectiveness
Method of installing (culms, seeds)	Cost effectiveness
Length and depth of plantings	Survival potential; initial trapping ability; cost
Spacing between plants	Sand-trapping ability; cost effectiveness
Fertilizer requirements	Plant survival; speed of growth
Soil moisture needs	Method/depth of planting

(*Sources*: Schwendiman 1977; Hesp 1989; Feagin et al. 2015).

This section can only provide a broad overview of options for use of vegetation. Different planting schemes and treatments to enhance growth can be used to meet specific restoration goals (Long et al. 2013). Many guidelines for planting dunes are now conveniently placed on websites. Many of these guidelines are geared toward actions appropriate for local use, although some guidelines reflect general principles developed decades ago or in other locations.

Factors important in planting success are identified in Table 3.1. Some factors, such as susceptibility to fire hazard and nitrogen-fixing capability, are more relevant to uses in interior dunes (Chapter 4). The importance of each factor in determining the methods for each restoration project is site specific and is not discussed in detail here.

The biological and geomorphological processes forming new or incipient natural foredunes are reviewed in Hesp (1989; 1991). The plants that thrive in the dynamic beach/dune environment are adapted to stresses associated with salt spray, sand blasting, sand burial, swash inundation, periodic ponding, dryness, high light intensity, high temperatures, strong winds, sand salinity, and nutrient deficiency (Hesp 1991). Among the most useful species for building the initial foredune and maintaining it after inundation are *Ammophila arenaria* in northern Europe; *Ammophila littoralis* in southern Europe; *Spinifex* spp. in Australia and New Zealand; *Ammophila breviligulata* on the east coast of the USA; *Panicum amarum* and *Uniola paniculata* on the southeast and Gulf of Mexico coast of the USA; and *Panicum racemosum* in southern Brazil. Vines that occupy the backshore in tropical regions, such as *Ipomoea pes-caprae* in India, Mexico, Brazil, and the Pacific islands, are excellent stabilizers. Their low growth form does not make them as effective as grasses in building up the height of the dune, but planting them has been suggested as a way of controlling wave erosion (Ellison 2018; Odériz et al. 2020).

Basic dune stabilizing vegetation should be (1) easy to propagate, harvest, store, and transplant with a high survival rate; (2) commercially available at local nurseries at relatively low cost; and (3) able to grow in a variety of microhabitats on a spatially restricted foredune (Feagin 2005). Adult plants may be removed from healthy populations nearby and replanted in needed sites, but use of nursery plants avoids the potential fragmentation of donor beds on existing dunes (Balestri et al. 2019). Plantings should be placed far enough from the water to avoid erosion by waves before they become established.

Optimum planting times differ for different species. In southeastern USA, *A. breviligulata* is best planted in the winter; *U. paniculata* is best planted in the spring (Broome et al. 1982). In the more sheltered portions of the foredune in the Netherlands, cuttings of *Festuca rubra*, *Carex arenaria*, and *Elymus athericus* are more successful if planted in November than in April, although there may be no time-dependent effect using seedlings (van der Putten and Peters 1995).

Rates of accretion at vegetation plantings are slow relative to rates using sand fences, and dunes may show little accretion in the first year after planting (Mendelssohn et al. 1991; Miller et al. 2001). If rapid dune growth is required, sand fences may be necessary, but inevitably, beach nourishment is required to maintain a healthy, well-vegetated dune on an eroding shore (Mendelssohn et al. 1991). Sand fences tend to create steep dune faces that may be incompatible with use by fauna (Melvin et al. 1991), so if time is not critical, use of vegetation plantings alone is recommended to build a dune with internal and external characteristics representative of natural dunes.

Ammophila is especially well suited to sand burial and is the primary dune builder in regions with a temperate climate (van der Laan et al. 1997). Deposition of wind-blown sand supplies nutrients, eliminates competitors, and enables the plants to produce new roots in a substrate free of soil pathogens (nematodes and fungi); when sand deposition diminishes, the *Ammophila* degenerates (van der Putten and Peters 1995). Many cases of poor success in planting *Ammophila* occur in locations where there is little sand deposition (van der Putten 1990), and it should not be planted if barriers to sand transport are placed seaward (Figure 3.3). Some backdune species that would occur in a natural environmental gradient may appear, but the surface can lack the remnant beach grass that normally fills the area between these plants. The ecological advantages of having beach grass interspersed with later stages of vegetation are missing. Even dense swards of planted *Ammophila* can become degenerate after sand inputs are cut off from seaward, but this is not a problem where the loss is coincident with colonization by later successional species (Vestergaard and Hansen 1992). *Ammophila* can even be planted as an exotic species if it facilitates stability and growth of native species without becoming dominant (Box 6). Alternatively, successional species can be planted where sand deposition cannot be reinitiated (van der Putten and Peters 1995).

Ammophila reproduces naturally by rhizomes deposited on the beach after plants are eroded from the dune, and it may be planted as culms or by disk-harrowing rhizome fragments. It may also be sown, if the sand surface is temporarily stabilized, but the most traditional method is planting of bundles of culms (van der Putten 1990; van der Putten and Kloosterman 1991).

Ammophila occasionally reproduces from seed in moist slacks (van der Putten 1990), but other species may be more appropriate to plant in wet slack environments. In the USA, *Spartina patens* offers a good equivalent to *Ammophila* in dune swales that may be subject to periodic drought (Feagin 2005). Planting annuals or other ruderal species is not as helpful or necessary in initial planting efforts because they will eventually colonize the dune opportunistically (Feagin 2005).

Planting programs often use a single species. Stabilizing the sand surface with a dominant perennial species, such as *Ammophila* spp., can ameliorate environmental extremes and facilitate establishment of species that are less adapted to stressful environments (Mauriello 1989; Bertness and Callaway 1994; Callaway 1995; Martínez and García-Franco 2004; De Lillis et al. 2004; Feagin 2005, Schreck Reis et al. 2008; Emery and Rudgers 2010; Gallego-Fernández et al. 2011; Teixeira et al. 2016). Other species, including endangered ones, can colonize restored areas opportunistically if small populations are present nearby and if dispersal mechanisms in the form of wind, water, or animals are effective (Huxel and Hastings 1999; Snyder and Boss 2002; Grootjans et al. 2004; Redi et al. 2005). Colonization can occur even in locations where dunes are small and isolated by headlands if there are enough dune systems at the regional level (Gallego-Fernández et al. 2011).

Box 6
The Hout Bay Dune Rehabilitation Project, Cape Town, South Africa

The Hout Bay Dune Rehabilitation Project is an initiative aimed at managing the movement and migration of dunes, which had accumulated because of a dysfunctional headland bypass dune field. Extensive housing development precluded the restoration of the natural headland bypass system, as described by Tinley (1995). Topographic profiling, wind netting, and planting of dune-specific vegetation have assisted in restricting the movement of sand, which had been a problem for many years. The Cape Town City Coastal Management branch of the Environment Management Department contracted Vula Environmental Services to undertake the project. The process involved an integrated and combined approach of sand removal, wind net installation, *A. arenaria* planting, dune thicket vegetation planting and seeding, selective irrigation, and intensive maintenance. Early, emergency nets were erected in 2016 during dune reshaping, but these failed to stabilize the sand without vegetation and irrigation (Figure 3.9). In 2017, wind nets were placed at 5 m intervals and *A. arenaria* and indigenous dune pioneers and shrub thicket species were planted (Figure 3.10). Three winter seasons of planting and maintenance in this winter rainfall region were necessary to achieve the initial establishment and stabilization of the dune system, which covers approximately 55 000 m². By 2019 (Figure 3.11), a restored indigenous dune community, similar to the natural strandveld, was well established, and sand accumulation on nearby infrastructure and private properties was reduced significantly.

Contributors: Deon van Eeden, Vula Environmental Services, Cape Town, South Africa

Roy Lubke, Department of Botany, Rhodes University, Grahamstown, South Africa

Photos by Deon van Eeden

Figure 3.9 Bare sand and wind nets at Hout Bay in 2016.

Figure 3.10 Wind nets and plantings at Hout Bay in 2017.

Box 6 (cont.)

Figure 3.11 Dune pioneers and thicket at Hout Bay in 2019.

Each type of vegetation has unique features that can retard erosion during storms or make it valuable for other reasons (Morton 2002; Wootton et al. 2005; Charbonneau et al. 2016, 2017). Some species grow rapidly and stabilize surfaces quickly but are susceptible to pests, whereas other species are free of pests and provide greater, long-term stability but grow slowly (Woodhouse et al. 1977). Some species may create steeper or gentler slopes that affect movement of sediment and fauna.

Mixtures of two or more species can be planted to improve long-term viability of vegetation cover and dune stability (Woodhouse et al. 1977). The slow growth or susceptibility to disease of one early growth species can thereby be counteracted by neighboring species. Species characteristic of later growth stages may be added to diversify the vegetation and provide protection against wind, add alternative habitat, reduce the likelihood of invasive nonnative weeds, increase aesthetic value, or control visitors. Species other than initial dune builders may be susceptible to loss on active foredunes, especially in initial stages of dune building when physical stresses are great (Belcher 1977; Mauriello 1989; Long et al. 2013), so it may be best to delay planting of these species.

Indigenous species can be costly, time consuming to plant, slow to become effective (Avis 1995; Lubke 2004), and less effective in resisting erosion than exotic species (Charbonneau et al. 2017). These aspects led to use of exotic species in the past. Some exotic species, notably *A. arenaria* in South Africa and Australia, were recently used by government agencies to build dunes and stabilize surfaces but only where the species do not exclude recolonization of native species; *A. arenaria* declines in vigor with increasing sand stability and is replaced by indigenous plants, and planting or sowing seeds of indigenous species among new *A. arenaria* plantings helps ensure its rapid replacement (Hertling and Lubke 1999). Considering the problems associated with exotic species in many locations and the need for programs to eliminate them or make over-stabilized landscapes more dynamic (Chapter 4), there appears to be little value in using exotics for stabilization, except in rare cases.

Vegetation representative of later stages of succession may be planted in sheltered areas to accelerate succession and increase species richness, although species that are not planted or seeded will eventually invade these areas if adequate seed sources are available nearby (Avis 1995; Lubke 2013). The need for source areas for the rarest species is especially important to maintain species richness in locations where species turnover is high (Snyder and Boss 2002). From a nature conservation viewpoint, the unusual and specialist species that depend on the most natural aspects should rank higher than more common species that often comprise the bulk of new vegetation in altered habitats (Doody 2001). Doody (2001), citing others, identifies many rare and declining species in European dunes, including *Liparis loeselii* in southern Wales; *Dactylorhiza incarnata* and *Parnassia palustris* on the Wadden islands; and *Centaurium littorale*, *Eryngium maritimum*, *Vicia lathyroides*, and *Carex ligerica* in Lithuania. Many species typical of backdunes would not survive if placed close to the beach under conditions of a naturally evolving dune, but they may be able to survive close to the beach given ongoing human efforts. Opportunities for restoring backdune species close to the beach in human-altered environments are assessed in Chapter 6.

Newly emplaced vegetation may need additional human efforts, even in relatively undeveloped areas (Miller et al. 2014). A variety of techniques can be used in addition to sand fences to facilitate survival and growth. Vegetative mesh and containers can stabilize substrate, prevent excessive sand inundation or scour, and reduce excessive desiccation (Teixeira et al. 2016; Balestri et al. 2019). Wheat straw or sargassum can be spread on the surface to add nutrients or to act as a cover to reduce sand movement or retain moisture (Williams and Feagin 2010; Hooton et al. 2014). Inoculation with arbuscular mycorrhizal fungi has the potential to accelerate stabilization of dunes and vegetation (Gemma and Koske 1997; Koske et al. 2004). Using native soil microbial amendments to facilitate restoration outcomes is a current interest, although there are lingering questions that should be answered before their widespread adoption (Crawford et al. 2020). Some of these techniques may be more appropriate to restoring interior dune landscapes after removal of exotic species or after mining (Chapter 4) than in quickly building storm-damaged foredunes in developed areas, where prevention of inundation is more critical than restoring habitat.

3.7 Building Dunes Using Multiple Strategies

The habitat value of foredunes built primarily to protect human facilities can be greatly increased by allowing them to achieve greater topographic and biological diversity. Natural form and function should be goals for dune building where human structures are not in imminent danger. Results at Ocean City, New Jersey, USA (Figure 3.1), reveal the advantages of allowing natural aeolian deposition to occur following use of sand fences and plantings of *A. breviligulata*. The shore protection project there had no comprehensive restoration goals, but it illustrates how naturally functioning landforms and vegetation, with their inherent dynamism, can evolve after beaches and dunes are built to provide protection to human structures. A federal project using 6.6 million m^3 of fill dredged from a nearby inlet was used to create a beach 30 m wide. The municipality then placed two rows of sand-trapping fences 5 m apart on the landward part of the backshore and planted the space between with a single species (*A. breviligulata*). The nourished beach provided an excellent source of wind-blown sediment and protected the foredune from wave damage during small storms. The municipality sequentially placed rows of sand fences on the seaward side of the dune to encourage horizontal rather than vertical growth, so residents could retain views of the sea. Designation of nesting sites for piping plovers by the state endangered species program resulted in prohibition of raking, driving, or pedestrian use of the beach, leading to local colonization of the backshore by plants and growth of incipient dunes that survived several winter storm seasons and grew into new foredune ridges seaward of the dunes maintained by sand fences. The seaward portion of the dune was dynamic, but the landward crest built by the sand fences and initial *Ammophila* plantings retained its integrity as a protection structure.

A field study of vegetation on a fenced foredune near the one depicted in Figure 3.1 revealed 16 plant taxa 7 years after efforts to construct a protective dune began and 22 taxa the following year (Nordstrom et al. 2007a). The presence of natural seed sources nearby enabled succession to occur, even as some portions of the dune retained their dynamism. The species are typical of the dunes remaining in undeveloped shoreline segments in the region. Having nearby reservoirs of seed sources and plant material is important in maintaining habitats, even in nature reserves (Castley et al. 2001). Restoring new natural enclaves in developed areas places new reserves closer to sites targeted for restoration in the future.

Local erosional "hot spots" occurred at Ocean City where deteriorated fences or poorly vegetated gaps in artificially created foredunes contributed to increased sand inundation of adjacent vegetated areas, reversing the trend toward more stable habitat (Nordstrom et al. 2007a). Gaps like these can be allowed to evolve or sand fences can be used to build a seaward ridge to prevent sand inundation and increase the speed of succession (Nordstrom et al. 2007a). Avoiding use of sand fences to stabilize small dynamic zones will locally reproduce cyclic sequences of dune evolution and favor a mosaic of habitats that Castillo and Moreno-Casasola (1996) find valuable for species reproduction. Depending on how and when sand fences (or vegetation plantings) are employed to seal gaps, a foredune could reflect the differences in topography and cyclic reversals in vegetation succession found on a more dynamic landform or the more uniform sequences found on the stable landforms that people often prefer. Adding new fences is a more traditional and conservative approach but one that should be evaluated more carefully considering its lower restoration potential. A wait-and-see approach would be preferable to categorical stabilization of small erosional hot spots.

Time and space are critical to the evolution of vegetation gradients after planting of the initial stabilizer (Freestone and Nordstrom 2001; Lubke and Hertling 2001; Vestergaard 2013). The artificially widened beach and use of sand fences and vegetation plantings to help create and stabilize the initial dune crest at Ocean City allowed the dune to quickly function as protection for buildings and infrastructure against flooding. Once a dune achieves its protective function, it could be allowed to evolve as a naturally functioning environment.

3.8 Concluding Statement

Most projects that build dunes for shore protection are not designed as restoration projects, but they can accomplish restoration goals with minor changes in practice. Complex planting plans are not usually necessary after the initial stabilizer is planted, and there is often no need to stabilize bare sand patches or gaps in the foredune if landward portions of the dune remain to provide protection to coastal facilities. Natural foredunes are dynamic and fragmented, and the dune should contain patches that each bear a temporary vegetation facies according to the previous history of environmental events (García Novo et al. 2004). Portions of natural dunes are always in an incipient state. Species composition may change from year to year in places, although species richness and coverage may be nearly constant (Snyder and Boss 2002). Accordingly, attempts to immediately create a mature community structure in restored environments and to categorically stabilize bare areas may not be necessary. Placing restoration sites near habitat occupied by target species can offset lags in species recoveries.

4

Restoring Processes, Structure, and Functions

4.1 Increasing Complexity and Dynamism

Many actions can set human-modified beaches and dunes back on a natural
trajectory, including restoring the sediment, the basic shapes of landforms, the
characteristics of microhabitats, the natural process that rework the landforms, and
the space for this reworking to occur. Considerations for restoring sediment and
the basic shapes of landforms were presented in Chapters 2 and 3. This chapter
identifies actions to make landforms already degraded and landforms created or
managed to achieve a limited number of shore protection or recreational values
function more naturally. This can be accomplished through human efforts to
remove invasive species, reestablish physical habitat characteristics, reinitiate
cycles of growth and decay, and favor reworking by waves and winds. Managing
landforms to achieve their full restoration potential involves evaluating them on
multiple criteria related to ecological, geomorphological, and social indicators
(Table 4.1). Management actions can vary greatly depending on site and situation.
The range of options for overly stabilized dunes (Table 4.2) may be limited by
geographical constraints, by the need to keep projects small because of cost or
spatial restrictions, and by local human needs and preferences (including the need
to prevent migration of landforms into human use areas). Despite the small size of
projects, they can collectively accomplish broad regional or national restoration
goals. Regardless of scale, a key to long-term success of environmental
restoration projects is how much freedom is given to natural processes. Many
of the actions identified in this chapter are feasible because they are designed for
landscapes removed from human infrastructure, often within natural preserves,
and consistent with the regulations governing their management. Restoring
natural processes in areas where restoration actions are subject to alternative
human uses and needs and have spatial constraints is the subject of subsequent
chapters.

Table 4.1. *Physical, ecological, and social indicators that can be used as a checklist to establish target conditions for restoration plans, assess outcomes, or provide the basis for adaptive management*

Physical

Processes
Wind direction, speed, seasonality
Rainfall amount, seasonality, tempering effect on aeolian transport
Wave height, periodicity, seasonality, storm characteristics
Water levels (tidal, storm surge)
Longshore currents and sediment budgets
Barriers to sediment exchange (protection structures, fences)

Landform characteristics
Beach width, height, scarps
Surface characteristics (transportable grain sizes, lag surfaces, wrack/litter)
Dune height, width, number of ridges, slacks, age, slopes, overwash sites
Locations of sand mobility in dune

Soil characteristics
Soil profile
Nutrients
Presence of seeds, infauna

Ecological

Composition
Distribution, patterns
Proportions and numbers of species and patch types
Proportions of endemics and exotics
Numbers of threatened or endangered species

Structure
Heterogeneity
Connectivity
Dispersion
Fragmentation
Patch size and frequency distribution
Foliage height, density, and layering

Function
Patch dynamics
Energy flows
Trophic levels and links
Productivity/growth rates
Disturbances and persistence
Adaptations
Life histories and demographic processes

Table 4.1. (*cont.*)

Social

Relationship to human use trends
Recreational opportunities
Historical or archaeological interest
Health and safety
Aesthetic qualities
Stakeholder interests
Consistency with government plans and policies

(*Sources*: Westman 1991; Lubke and Avis 1998; Espejel et al. 2004).

Table 4.2. *Principal potential management considerations for restoring overly stabilized coastal dunes*

Increasing aeolian transport
Maintaining or restoring sand supply (sediment budgets)
Encouraging localized landform mobility
Enlarging active sand surface areas
Reducing impact of nutrient enrichment
Restoring dune slack morphology and hydrology
Optimizing groundwater supply
Creating lost native woodland habitat
Removing exotic species
Controlling shrub invasion
Establishing appropriate grazing levels
Enhancing geomorphic resilience to climate change or sea level rise
Ensuring shore protection aspects are not compromised
Addressing environmental regulations and directives
Addressing societal needs (recreation, nature appreciation)

(*Sources*: Rhind and Jones 2009; Darke et al. 2016; Arens et al. 2020).
The management considerations are not mutually exclusive in their application. The entries in the table can also be used as a checklist of indicators to address outcomes.

4.2 The Issue of Dynamism

The implementation of massive dune stabilization projects following periods of dunefield activation in past centuries established a strong precedent for stabilizing all unvegetated patches in dunes. Stability still is often the major goal in management of dunes built or modified for shore protection (Simeoni et al. 1999; Bossard and Nicolae Lerma 2020), and stable systems can even be favored by natural resource managers who wish to protect a key species or a specific environmental inventory. It is not always easy for managers to consider change as a positive factor in conservation (Doody 2001). Nevertheless, many scientists advocate dynamic systems to allow nature to undergo exchanges of sediment, nutrients, and biota; follow cycles of accretion, erosion, growth, and decay; and retain diversity, complexity, and the ability to deliver ecosystem services (De Raeve 1989; Wanders 1989; Doody 2001; Arens et al. 2013a; Kutiel 2013; Pardini et al. 2015; van der Biest et al. 2017, 2020). Greater diversity, in turn, results in greater resilience (García-Mora et al. 2000). Moderate levels of disturbance may have beneficial effects on species recruitment (Grime 1979) and overcome encroachment by less desirable species (Ketner-Oostra and Sýkora 2000). Tolerance of sand dune species to burial is one of the principal causes of zonation of plant species on coastal foredunes, and burial of dune building plants can have a stimulating positive effect on plant growth and prevent degeneration (Maun 2004). Many floral and faunal species can occur in sand dunes, but the species most dependent on dune habitat and the greatest diversity are usually associated with early successional stages that depend on bare sand and aeolian transport (Castillo and Moreno-Casasola 1996; Rhind and Jones 1999; Howe et al. 2010; Provoost et al. 2011; Brunbjerg et al. 2015; Van der Biest et al. 2017).

Storms introduce the greatest natural changes to beaches and dunes. Storms are an integral part of a natural cycle of change, and they represent the most universal, recurrent natural hazard faced by sandy shore animals (Brown et al. 2008). Storms may cause great mortality of organisms, but recovery can be rapid, even in environments landward of beaches and foredunes (Valiela et al. 1998). The ability of certain organisms to survive by behavioral means is a key feature of many sandy shore animals, and interspecific competition is minimized because few macrofaunal species can tolerate the harsh conditions (Brown 1996). Some fauna can take advantage of the daily changes that occur on beaches, whereas others take advantage of new environments created by infrequent storm events. The natural environments favored by piping plovers (*Charadrius melodus*) for nesting are the inlet spits and overwash platforms that are by-products of recent and rapid physical change (Melvin et al. 1991; Maslo et al. 2019). Thus storms create and define the ecosystem, not threaten it (Brown et al. 2008). Human actions to minimize the impact of storms may threaten the ecosystem by stabilizing or truncating the dynamic zone or separating its component parts using protection structures.

Rapid change is the norm for coastal dunes. They are dynamic and heterogeneous environments that often exhibit an aggregate pattern with unvegetated and vegetated sites with uneven plant composition because of site- and time-specific processes and responses that produce a patchy environment maintaining their diversity (Ritchie and Penland 1990; Doody 2001; García-Novo et al. 2004; Martínez et al. 2004; Drius et al. 2013). A dune maintained as a dynamic system can be more resistant to erosion, cheaper to maintain, have greater species diversity, and be more sustainable than a fixed dune (Heslenfeld et al. 2004). The sequence of succession in coastal dunes rarely occurs in a straightforward progression from one stage to the next but reveals a complexity in its spatial and temporal variation that imparts a diversity not found in more stable habitats such as woodland, permanent grassland, or heath (Doody 2001). The actions taken to restore dunes range from revegetating and stabilizing the mobile substrate to (alternatively) removing plant cover and increasing substrate mobility (Lithgow et al. 2013); a variety of these actions may be necessary within a single dune system. The optimum condition for dune environments may be one where different sections are evolving at different rates or stages, providing many alternative environments characterized by different degrees of sediment movement and ground covers. Different breeding bird species prefer different types and densities of vegetation (Verstrael and van Dijk 1996). The richest zones in dunes are where clumps of stabilized vegetation occur near unstable open sands and where species nesting in open soils can feed or find refuge in the denser vegetation (Doody 2001). Some faunal species depend on the juxtaposition of bare sand and vegetated patches (Maes et al. 2006).

Wind-eroded gaps that form in dunes (blowouts) were often targeted by managers for closure and revegetation. Gaps disrupt the vegetative cover, undermine substrate, contribute to sand inundation at their margins, accelerate winds locally, enhance salt aerosol transport farther inland, and reverse successional trends (Hesp 1991). Blowouts may also deflate down closer to the location of the water table where they can evolve into slacks (Doody 2001). Many of the effects of blowouts were considered undesirable into the 1980s (Nordstrom and Lotstein 1989), but opinions about them changed, especially in the Netherlands (Arens et al. 2013a; Riksen et al. 2016; Ruessink et al. 2018). The environment below slacks can be relatively hospitable for interstitial fauna (van der Merwe and McLachlan 1991), and slacks act as centers of diversity within either mobile or higher stabilized dunes (Grootjans et al. 2004; Lubke 2004). Poorly vegetated dry areas higher in the dune in both newly created and evolving environments are not as barren as often perceived. Invertebrate fauna thrive on the open and dry portions of dunes that are often warmer than surrounding areas (Doody 2001; Howe et al. 2010; Kutiel 2013). Relatively high numbers of species and abundances of soil microarthropods can occur in young dunes despite low levels of plant cover, plant species richness, and organic material (Koehler et al. 1995).

Often, the limit to tolerance of dynamism is proximity to the nearest human infrastructure. Where insufficient space occurs to allow coastal landforms and associated vegetation to develop, human efforts may be required to control and adapt to change rather than prevent it or allow it free reign, using a strategy that can be called "controlled dynamism" (Nordstrom et al. 2007c). Controls can be placed on the magnitude of change by providing protection against some storm effects or some human inputs but not all of them or controlling the location of change by protecting or abandoning natural habitat or human uses in some regions but not all of them. This chapter and the ones that follow identify many of the compromises that will have to be made to increase the dynamism and natural function of coastal landforms.

4.3 Restricting Beach Raking

4.3.1 The Value of Wrack

One of the most common environmentally damaging actions is eliminating natural beach wrack (litter or beach cast) by mechanically removing it, creating a beach with recreational value but little natural resource value (McLachlan 1985; Ochieng and Erftemeijer 1999; Nordstrom et al. 2000; Colombini and Chelazzi 2003; Dugan et al. 2003; Dugan and Hubbard 2010; Kelly 2014, 2016; Zielinski et al. 2019). Wrack lines consist of natural litter containing macrophytes, such as algae and seagrass, driftwood, fruits, seeds, and carrion, along with cultural litter consisting mostly of inorganic anthropogenic materials (marine debris) (Colombini and Chelazzi 2003; Balestri et al. 2006). The literature on cultural litter on the beach is large and increasing (reviewed in Zielinski et al. 2019). Much is known about the value of natural wrack as habitat and food source (Colombini and Chelazzi 2003) and the rationale for restrictions on removing it (Zielinski et al. 2019), but less is known about the best way to manage wrack on developed shores. Suggestions exist (Nordstrom et al. 2000; Kelly 2016) and are identified in Chapter 8, but the success of alternatives has yet to be evaluated. Cleaning beaches of cultural litter is a worldwide issue that must be addressed (Williams and Tudor 2001; Cunningham and Wilson 2003; Nachite et al. 2018), but finding ways to retain natural wrack while removing cultural litter is critical to conserving or restoring beach and dune habitats in developed areas.

The uppermost wrack line is especially important in providing a topographic obstacle to aeolian transport and contains the seeds and rhizomes of coastal vegetation and nutrients that form the core for formation of incipient dunes on the backshore and aid growth of new vegetation (Godfrey 1977; Ranwell and Boar 1986; Hesp 1989; Gerlach 1992; Dugan and Hubbard 2010; Del Vecchio et al. 2013). Retaining wrack allows the cross-shore environmental gradient and topographic diversity to be extended seaward, creating more microhabitats and increasing biodiversity. Species composition of freshly deposited wrack can differ from aging wrack, probably due to wrack decomposition (Orr et al. 2005). Some species have evolved biological cycles linked to wrack; other species have adapted their abundance, spatial distribution, and feeding habits to the spatial and temporal patterns of the wrack (Colombini and Chelazzi 2003). The lack of in situ primary production on beaches by macroscopic plants places great importance on the decomposing organic matter delivered from sources outside the beach environment (Brown and McLachlan 1992; Ochieng and Erftemeijer 1999). Beach wrack can also provide a physical obstacle to wave runup (Innocenti et al. 2018), and beaches with high levels of accumulated wrack can have greater vegetation cover and species richness (Del Vecchio et al. 2017).

Many terrestrial taxa not typically categorized as intertidal, such as beetles, spiders, and ants, occur in the wrack (Fairweather and Henry 2003). Natural litter provides food and shelter for macroinvertebrates that inhabit the dunes, which then provide a food source for higher trophic levels. Fruits and seeds provide a vital genetic link, a primary dispersal agent for plants, and a food source; carrion (commonly jellyfish, bivalves, fishes, and occasionally birds and other animals) provides a food source of special value for scavenging birds and mammals (Colombini and Chelazzi 2003).

4.3.2 The Problem of Raking

Removal of natural wrack prevents formation of incipient dunes that could grow into naturally evolving foredunes at that location (Nolet and Riksen 2019), and eliminates the habitat and the large quantities of invertebrates and a broad spectrum of other organisms within it, resulting in decreased biodiversity (Colombini and Chelazzi 2003; Kowalewski et al. 2014; Schooler et al. 2019). Mechanical raking also removes vegetation growing on the beach Kelly (2014) and shells from the surface of the backshore, which can negatively affect bird nesting (Maslo et al. 2011). Backshore habitat may contain a smaller number of species than dune habitat farther landward, but the species may be more rare, with a greater proportion of endangered ones (Acosta et al. 2009). Numerous beach species are currently listed as endangered, threatened, or vulnerable or have other conservation status, often as a result of raking as well as scraping (Chapter 3) or driving vehicles on the backshore (Kelly 2014). The flat, uniform appearance of a backshore groomed and used for recreation conveys the impression that the entire beach surface is a potential highway, and vehicle tracks may be found on many backshores in urbanized environments.

Litter of all types has a strong influence on beach site selection by humans, and beach quality awards and beach certification place considerable value on beaches cleared of litter (Somerville et al. 2003; Davenport and Davenport 2006; Gilburn

2012; Krelling et al. 2017; Zielinski et al. 2019). Beach raking can be done intensively by hand, as is commonly practiced by concessionaires in Italy, or extensively by municipalities using machinery. Mechanical beach raking became a common and widespread practice in the USA after medical waste was encountered on beaches (Nordstrom and Mauriello 2001; Fairweather and Henry 2003). In New Jersey, USA, large amounts of floating debris in the summers of 1987 and 1988 led to beach closings and a major loss of revenue (Ofiara and Brown 1999). Stakeholders in shorefront municipalities found that raking the beach to create a clean (but sterile) environment was good for public relations. By the early 1990s, many municipalities in New Jersey raked their beaches, and many were able to obtain loans to purchase equipment from the state Clean Beach Program. The added funding increased the budget and staff in public works departments and increased their ability to modify the beaches and dunes using their own equipment and labor (Nordstrom and Mauriello 2001). As in many locations throughout the world, raked beaches are now perceived to be the norm.

Sediment and biogenic material skimmed from the beach in cleaning operations may be deposited in the dunes (Nordstrom and Arens 1998). This process changes the grain size characteristics in the dune and adds nutrients, seeds, and parts of plants. Deposition of litter and sediment in the dunes is considered by managers to be a way of preserving the sediment balance in the beach/dune system, but the problem of creating visually unattractive and exotic microhabitats in the dune must be balanced against the advantages for shore protection.

Beach raking is, at present, a local action, so suggestions for managing litter and wrack compatible with restoration goals are identified in Chapter 8. The significance of raking is great enough to make it worth reevaluating at higher levels of management and providing better guidance to local communities.

4.4 Restricting Driving on Beaches and Dunes

Vehicles used on beaches can reduce populations of diatoms, microbes, and macrofauna such as ghost crabs, and can pulverize and disperse organic matter in drift lines, thereby destroying young dune vegetation and losing nutrients (Godfrey and Godfrey 1981; Moss and McPhee 2006; Foster-Smith et al. 2007). Vehicles driving in and just seaward of dunes can severely damage vegetation and prevent extension of the dunes onto the backshore (Anders and Leatherman 1987; Godfrey and Godfrey 1981; Kelly 2014, 2016).

Not all damage by off-road vehicles is due to wanton use. Drivers can think that they are not causing damage, even when they are, because they only consider the vegetated backshore to be vulnerable. Damage can also occur in the wrack lines (where much of the life on the beach is concentrated) or on underground rhizomes or on dormant vegetation that appears to be dead (Godfrey and Godfrey 1981; Anders and Leatherman 1987; Hoogeboom 1989). Restricting vehicle use to portions of the beach between wrack lines would leave the most species-rich microhabitats and incipient dunes intact. Restricting shore perpendicular vehicle access to a few prescribed crossings would reduce the number of impacted areas within the incipient dune zone and the spatial extent of the visual impact of tire tracks (Nordstrom 2003).

Regrowth of vegetation after termination of vehicle use indicates that rapid recovery is possible (Godfrey and Godfrey 1981; Anders and Leatherman 1987; Judd et al. 1989; Kelly 2016), so suspension of use in as many areas as possible is a first step in restoring degraded landforms. Godfrey and Godfrey (1981) and Priskin (2003) concluded that it is better to have a few well-managed and well-patrolled high-use trails than many dispersed low-use trails, so steps should be taken to reduce the number of trails where off-road vehicles are allowed.

4.5 Remobilizing Dunes

Dunes may be remobilized as an unintended or unwanted by-product of removing invasive species, but intentional managed remobilization is desirable in places to restore a broader range of successional stages ranging from pioneer mobile dune vegetation and embryonic dune slack to fixed dune grassland and scrub/woodland on the perimeters (Rhind et al. 2013). Removal of vegetation to create or rejuvenate blowouts may be more feasible over a smaller area and be more tolerant of regrowth of plants than removal of larger areas to create nesting ground for birds. Usually, the hope is to create patches of bare ground that can be sustained but do not evolve into massive bare areas that migrate out of the targeted zone. Some actions to eliminate exotic vegetation or intentionally remobilize sand dunes may result in burial of wetlands (Hesp and Hilton 2013) or the destruction of protected vegetated stable dunes (gray dunes). The latter would be in contravention of the 1992 European Commission Habitats Directive (Clarke and Rendell 2015). Managers may be initially hesitant to allow blowouts to occur if they see them as the beginning stage in the conversion of a stable landscape to a migrating sand sheet. However, there is evidence to suggest that isolated blowouts that form by natural processes and as a result of elimination of exotic species will eventually stabilize naturally without causing destabilization over a larger area (Pluis and de Winder 1990; Gares and Nordstrom 1995; Leege and Kilgore 2014; Castelle et al. 2019). Change in state from stable, vegetated dunes to bare, mobile dunes or from mobile to stable dunes often requires a significant change in environmental conditions (Arens et al. 2004; 2013a). These could include changes in temperature, precipitation, and storminess. It is likely that more bare ground can be tolerated in many vegetated dunes than is presently allowed without significantly increasing the risk of large-scale remobilization of dune fields.

Most studies of vegetation removal stress that one-time removal is not enough. Areas of removal can revegetate by undesirable species (Richards and Burningham 2011; Pye and Blott 2017). The heritage of remaining roots or rhizomes is an issue whether the project is to eliminate exotic species to encourage growth of native species or maintain unvegetated patches, requiring follow-up management (Hilton et al. 2009; Arens et al. 2013b; Hesp and Hilton 2013). Local activities such as increasing livestock grazing, reintroducing rabbits, clearing shrubs, and stripping turf are beneficial but may not themselves create self-sustaining dunes, implying that larger-scale vegetation removal projects and surface reprofiling would be required (Pye et al. 2014). Although full remobilization of dunes may not occur, the resulting landscape may contain habitat mosaics that enhance the survival of rare species and pioneer environments (Arens et al. 2013a).

Some options for vegetation removal may be more effective if used in combination with other options. Hand removal can minimize damage to the soil and allow for control to exacting standards (Kutiel et al. 2000; Pickart 2013), but it has high labor costs unless volunteers can be found. Emplacing a fabric may not kill rhizomes or keep seeds from dispersing or germinating. Bulldozing can remove rhizomes, but rhizomes and seeds will remain with the sediment when redeposited and may have to be removed by screening. Raking is effective on sparsely vegetated areas. Burning is fast and economical over large areas, but it can be dangerous to fauna; it does not kill rhizomes; and it may encourage subsequent growth of undesirable species. Herbicide is easy to apply and can be used on small areas but large-scale use can contaminate ecosystems and have unknown effects on human health. Salt is easy to apply but may contaminate the local environment (National Park Service 2005).

4.5.1 Case Studies

Artificial breaches have been cut in foredunes to reinitiate onshore aeolian transport in locations in the Netherlands (Riksen et al. 2016), Belgium (Van der Biest et al. 2017), and France (Castelle et al. 2019) where this would not increase risk of infrastructure to flooding (Figure 2.4). Box 7 identifies an example of one of these projects. Gaps have also been created in the foredune to allow sea water to periodically inundate the low ground landward. This occurred at Schoorl, the Netherlands, where a gap was created and vegetation and soil were removed from the swale behind the foredune (Arens et al. 2001; 2005). The project resulted in creation of several environmental gradients, including fresh to salt, dry to wet, vegetated to unvegetated, carbonate rich to carbonate poor. The wide dune zone

---

Box 7
Dune remobilization in the Netherlands: An example from Zuid-Kennemerland

Most of the Dutch foredunes were intensively managed because of their function as sea defense. There was no room for natural processes; erosion was suppressed; and the foredunes resembled sand dikes (Arens and Wiersma, 1994). The need for intense management decreased when the policy of "dynamic preservations" (de Ruig and Hillen 1997) was implemented to manage the coastline by means of beach nourishments. Also, awareness grew that influx of aeolian sand into the dune would benefit both nature and safety (Arens et al. 2005). Five notches were excavated into the foredune in the middle of the Dutch Mainland coast in Zuid-Kennemerland between the villages of Zandvoort and IJmuiden. The aim was to restore aeolian dynamics of the dune system, provide "windows" (notches) in the foredunes for sand to enter the dunes to landward, and create room for pioneer species by the formation of bare surfaces. After a few years, the notches became successful in transferring sand inland, creating a very dynamic landscape (Ruessink et al. 2018). Parts of the landscape are buried by sand, occasionally up to several meters, thereby destroying climax vegetation. At greater distance from the notches, sand deposition is mild, improving conditions there for the growth of species from dune grasslands.

   Contributor: Bas Arens, Bureau for Beach and Dune Research, the Netherlands.

Figure 4.1 Foredune notches, at Zuid-Kennemerland in March 2013, shortly after excavation.

Figure 4.2 Site in January 2020 after seven years of free development.

---

remaining landward of the gap provided adequate space for nature to evolve without threatening other resources landward (Arens et al. 2001).

   In some cases, simply ceasing management interventions that kept dunes stable can cause remobilization to occur spontaneously (Arens et al. 2013a). The deposition landward of small breaches (notches) can be limited to tens of meters where notches are <20 m wide – an effect considered disproportionately small relative to the effort required to construct and maintain the notches; in contrast, wide breaches can result in transport >300 m inland (Riksen et al 2016). Breaches with a width of 50–100 m at the top and excavation depths 9–12.5 m resulted in a depositional lobe up to 8 m thick and 150 m long (Ruessink et al. 2018). Sand supply from the beach to dune will be enhanced where reestablishment of blowouts is accompanied by beach nourishment (Arens et al. 2013a).

Reactivation has occurred at inland locations in dune fields by directly activating surfaces well landward of the contact between beach and dune. Arens et al. (2004) evaluated morphology, sand drift potential, and vegetation reestablishment on a reactivated parabolic dune 2 km from the sea. Erosion on the crest changed the form of the dune from a parabolic to a transgressive dome-shaped feature, with several blowouts in the crest. Deposition to the lee (landward) of the crest occurred with minor deposits up to 200 m away. Portions of the toe of the lee slope moved up to 12 m in 2 ¼ years. The dune remained largely unvegetated for at least 4 years. Results indicate that aeolian transport can be great enough to prevent establishment of vegetation in the short term, even if conditions are less windy and much wetter than average. Volume measurements revealed that the rate of sand transport at the inland location was about the rate of maximum foredune accretion rates in the Netherlands (Arens et al. 2004).

The most dramatic way of reinitiating succession in the beach/dune environment is to go back to the very earliest stage in dune evolution by favoring overwash or by grading dunes flat and removing vegetation to recreate conditions representing an overwash platform. This condition would favor some threatened species, including piping plover (*C. melodus*) (Maslo et al. 2011; Schupp et al. 2013). Mechanical removal of portions of foredune and vegetation has been employed in the USA where dunes have evolved to a condition where they no longer provide suitable nesting grounds for endangered shorebirds (Bocamazo et al. 2011; McIntyre and Heath 2011). Enhancing overwash was tried at Assateague Island, Maryland, USA, where the foredune acted as a nearly total barrier to overwash and wind action, and the habitat had changed from an early succession beach habitat to herbaceous and shrub communities, thereby diminishing foraging habitat and reducing productivity of piping plovers (Schupp et al. 2013). Multiple notches were cut through the foredune at an elevation expected to result in at least one overwash event per year. All of the notches allowed overwash to occur, and none of them widened or deepened enough to increase the risk of island breaching. Habitat was improved, and the resulting new overwash fans were viewed as increasing barrier island stability by increasing interior island elevation (Schupp et al. 2013). It may be too early to assess long-term success, but the result is of significance in demonstrating proof of concept for projects reinitiating a process normally prevented in shore protection programs. Raking may be needed as a follow-up action to control new growth in locations where maintenance of an unvegetated surface is important for species that nest on bare sand areas. Frequent spot raking can maintain an evolving beach/dune system at an early stage by targeting hummocks and vegetated areas on the verge of their conversion from incipient dunes to foredunes. Provisions can be made to leave some vegetation because some species select areas with more vegetation cover around nests (National Park Service 2005).

Removal of vegetation and topography is most appropriate where the resulting surfaces support multiple threatened species, not just a single species. Rejuvenation of stabilized dunes is not appropriate in many locations. Factors considered in selecting appropriate sites include physiographic setting, wind exposure, sediment availability, dune morphology, hinterland characteristics, importance of dunes for addressing coastal erosion and flooding, and ecological benefits that might arise (Pye and Blott 2020). The need for protection against inundation of landward features by waves and winds limits attempts to make dunes more mobile where human infrastructure or valued upland habitat is close to the beach.

4.6 Removing or Altering Sand-Trapping Fences

Sand fences can help construct dunes quickly (Chapter 3), but they create linear landforms and prevent the free exchange of sediment and biota across them. Nordstrom et al. (2000) suggest using sand-trapping fences only to create the first dune ridge that functions as the core around which the natural dune can evolve. The resulting dune would have more natural contours, greater topographical variability, and greater species diversity than the type of dune dike typically associated with development.

The use of sand fences to create a sacrificial dune seaward of an existing well-vegetated dune is a common practice, but sand-trapping fences may be unnecessary where the topography of the dune and the vegetation are available to trap sand. Vegetation is especially critical in causing deposition on the seaward slope or crest of the foredune (Brodie et al. 2019). A less expensive symbolic fence (e.g., foreground of Figure 3.6) could be used to prevent users from trampling the dune but would allow the sand to be blown landward into the existing dune to build up its volume (Nordstrom 2003). In some cases, sand-trapping fences are deployed seaward of foredunes to retain volume in the dune without increasing its height or mobility, in response to demands of users and shorefront residents for views of the sea. The advantages and disadvantages of this practice have not been fully evaluated.

Fences remaining landward of the seaward active fence rows may remain conspicuous in the dune (Figure 3.8) and may function as microhabitats within otherwise homogenous environments. They may also interfere with burrowing animals. Miller et al. (2001) suggest using biodegradable materials rather than wood slats and synthetic fibers to present less of a hazard to animals. Jute fabric that naturally degrades reduces the problem of remnant fences, and it is less expensive than wood slat fencing. Miller et al. (2001) found that the jute fabric they used for fencing deteriorated within a year to 18 months of deployment, but accumulation was equal to wood fences over that time. An interval of 18 months would allow fences to remain in place for two seasons of the strongest winds. Jute fabric loses its sand-trapping ability rapidly and the dune may eventually retain less sand and have a flatter profile than one created using wooden slat fences (Miller et al. 2001), but a flatter profile would be less of an obstacle to fauna.

Interference of wooden slat fences with movement of fauna can be reduced by employing fences in configurations that create corridors, either by leaving short gaps in longer sections of fence or by deploying fences as short sections oriented transverse to shore and to prevailing winds (Figure 3.2d$_3$). These gaps can be used by turtles or birds to gain access from the beach to the dune. Fences oriented at angles to prevailing winds and approach of storm waves reduce the likelihood of aeolian transport and overwash inland while providing access at the level of the backshore, but gaps would not be as desirable as paths over the dune crest if the dune is built for protection of human facilities and access is primarily for people.

The same fence types used to trap sand are often used to control pedestrian traffic and demarcate property lines (Nordstrom et al. 2000; Matias et al. 2005; Grafals-Soto and Nordstrom 2009), creating new depositional features that do not mimic natural landforms in location or orientation. Side spurs also create unnatural-looking shore perpendicular forms if they are overly long. Sand-trapping fences constructed along access paths to the beach create unnatural-looking shore perpendicular dunes. These fences visually isolate visitors from the adjacent dune environment, diminishing their nature experience. Symbolic fences or post and string fences that provide cues to visitors but do not interfere with sediment transport or movement of fauna would be better for controlling visitor movement.

Sand-trapping fences are often deployed in undeveloped parks and conservation areas where they are not necessary, converting what would otherwise be a dynamic dune environment into a stable one. One example is Island Beach State Park, New Jersey, USA, where dunes in an undeveloped portion of the park had been allowed to evolve by wave erosion, overwash, and aeolian transport for several years, providing one of the few examples of natural coastal dynamics along an intensively human-altered coast (Gares and Nordstrom 1995). A decision was subsequently made to seal all gaps in the foredune by deploying fences. The stabilized dune provided little insight into the way a natural environment would evolve or how the dunes in nearby developed areas would evolve if allowed to be more dynamic. Managers later changed their approach by restricting use of fences and allowing the dunes to become more mobile.

4.7 Protecting Endangered Species

Recent initiatives for protecting endangered species by controlling active human uses reveal great potential for restoring naturally functioning beaches and dunes (Breton et al. 2000; Nordstrom et al. 2000). In the USA, protection of endangered and threatened species is built into state coastal zone management regulations and also is the responsibility of the US Fish and Wildlife Service. Regulations require municipalities to ensure that nesting birds or turtles are not adversely affected by human activities. Identification of nests on beaches can lead to establishment of protected enclaves where activities such as nourishment, raking, bulldozing, scraping, and backpassing sand are restricted during the nesting season. Elimination of these disturbances leads to accumulation of wrack lines, colonization by plants, and growth of incipient dunes. The dimensions of the protected enclave should extend far enough across or along the shore to provide foraging or refuge areas or allow the foredune, beach, and wrack lines to provide a continuous cross-shore gradient of processes and habitats (Figure 3.1).

Once policies for endangered species are in place, they can be applied to other species. Protection of shorebirds in New Jersey has been restricted to species listed as endangered and threatened by the US Fish and Wildlife Service, but amendments were proposed by the state to address feeding habitat for other shorebirds. This would be done by changing the standards for beach and dune maintenance in the coastal zone management rules to confine raking to within 100 yards (91 m) of areas formally designated for swimming and watched by lifeguards. This expansion of no-rake zones would convert a substantial portion of the upper beach into a naturally functioning environment (Nordstrom and Mauriello 2001; Kelly 2016).

Protection of an endangered species can have negative effects when the landscape is modified specifically for it and the needs of other species are subservient to the endangered ones. One of the biggest problems in programs to enhance conditions for shorebirds is the natural succession of topography and vegetation, which can lead to demands for artificial clearing of backshores and dunes to keep them from evolving into later stages. Plovers with access to mudflat, bay beach, and ephemeral pool habitats appear to have a higher survival rate (Goldin and Regosin 1998), leading some managers to suggest that artificial cuts through dunes and pools in the backshore should be made. Some target species will nest on artificial substrates (Krogh and Schweitzer 1999), and there is great potential for enhancing use of coastal environments by these species through human actions, but ways must be found to make these environments naturally functioning and compatible with other species dependent on a coastal setting.

4.8 Altering Growing Conditions for Vegetation

4.8.1 The Issue

The history of dunes is characterized by phases of sand drift alternating with landform stability (Provoost et al 2011; Pye et al. 2014). In many environments, mobilization of dunes landward of the foredune appears to be initiated by natural or human-induced disturbance of the foredune, whereas disturbance within the dune field appears to be commonly caused by human actions (Delgado-Fernandez et al. 2019). The landscape in many interior dunes prior to intensive human disturbance likely had dense vegetation, including trees and shrubs (Sharp and Hawk 1977; van Aarde et al. 2004). Much of this vegetation was destroyed by land clearance for farming, burning, grazing, and other human-induced activities, resulting in open grasslands and sand drift that created hazardous conditions for human development (Sherman and Nordstrom 1994). This period was followed by efforts to combat dune movement and sand drift (Cullen and Bird 1980; Skarregaard 1989; Klijn 1990; Avis 1995; Hilton 2006). In Europe, planting vegetation, and restricting grazing and trampling by humans, combined with decrease in rabbit populations and eutrophication by deposition of atmospheric nitrogen, increased encroachment of grasses and shrubs, resulting in decreased diversity of habitat (Boorman 1989; Kooijman and de Haan 1995; de Bonte et al. 1999; Rhind and Jones 1999; Kutiel et al. 2000; Provoost et al. 2011; Pye et al. 2014; Clarke and Rendell 2015; Kooijman et al. 2017). A general increase in vegetation of coastal dunes also appears to be driven by the cumulative effects of global-scale increases in temperature and precipitation (vegetation growth stimulants), change in farming practices, and widespread reduction in windiness, which reduces sediment activity and promotes the spread of vegetation (Jackson et al. 2019; Gao et al. 2020). Of the 176 dunes in different parts of the world reviewed by Gao et al. (2020), stability increased at 93% of the sites over the period 1870–2018.

Many of the stabilizing species are exotic – introduced or aided by intentional or unintentional human actions. Attempts are now being made to remove many exotics to reestablish more dynamic conditions favorable for growth of native species (addressed later in this chapter). The concept of remobilization of stable dunes as a by-product of removal of exotics or other purposes is a recent phenomenon that goes against a long tradition of dune stabilization that still existed in the 1960s and 1970s (Provoost et al. 2011). Reinitiating early stages of dune mobility and succession is of great current interest in Europe and other locations (Hilton et al. 2009; Arens et al. 2013a; Walker et al. 2013; Brunbjerg et al. 2015; Pye and Blott 2017; Ruessink et al. 2018; Osswald et al. 2019; Arens et al. 2020; Creer et al. 2020). Box 8 provides an example of attempts to reinitiate activity in the United Kingdom.

Box 8
Dune restoration at Kenfig Burrows, South Wales, UK

Kenfig Burrows is a site of European nature conservation importance located on the southeast shore of Swansea Bay. Designated features of importance include *humid dune slacks, dunes with Salix repens* ssp. *argentea* (creeping willow), *fixed coastal dunes with herbaceous vegetation*, and *hard oligotrophic waters with benthic vegetation of Chara* spp. (Kenfig Pool). Two species are of European conservation importance – *Liparis loeselii* (Fen Orchid) and *Petaphyllum ralfsii* (Petalwort) – and the dunes and slacks contain a wide variety of invertebrates, including dune specialist species. Early in the twentieth century, the dune system was thinly vegetated, with extensive mobile parabolic dunes and transgressive sand sheets extending up to 3.2 km inland. Bare sand covered about 40% of the system in the early 1940s; by 2009, progressive spread of vegetation reduced bare sand cover to <4% (Pye et al. 2014; Pye and Blott 2016, 2017). Dune stabilization resulted from reduced stock and rabbit grazing, reduced coastal sand supply, cessation of sand and gravel extraction, enhanced deposition of atmospheric nitrogen, and climate change (including higher temperatures, reduced wind speeds, slightly higher rainfall, and longer growing season). Concerns about changes to biodiversity led the Countryside Council for Wales (later Natural Resources Wales) and Plantlife to instigate measures to increase sand mobility and extent of early successional stage dune and dune slack habitats. Management actions included turf stripping and creation of notches through the foredune ridge to increase flow of wind and sand from the beach. Interventions occurred in three main phases from 2012 to 2014. Monitoring revealed vegetation regrowth in some areas, but extent of bare, mobile sand and early successional stage vegetation in 2020 was much higher than before the interventions, and numbers of dune specialist invertebrates and plant species increased significantly.
Contributor: Ken Pye, Kenneth Pye Associates Ltd, Wokingham, UK

Box 8 (cont.)

Figure 4.3 Restored parabolic dune slack at Kenfig Burrows, Phase 1, March 2019, 7 years after initial turf stripping.
Source: Ken Pye.

Figure 4.4 Aerial image portraying Phases I–III at Kenfig Burrows, April 2015.
Source: Google Earth.

4.8.2 Techniques for Control of Vegetation

Rejuvenation of vegetation has been attempted by (1) reducing the height and biomass of grasses by controlled grazing or mowing (Jungerius et al. 1995; Kooijman and de Haan 1995: Kooijman 2004; Pye et al. 2014; Lundberg et al. 2017); (2) eliminating the undesirable vegetation by pulling (Pickart 2013), cutting, bulldozing or plowing (Jones et al. 2010; Darke et al. 2013), burning (Rhind et al. 2013) or applying herbicides (Hilton et al. 2009; Hesp and Hilton 2013), emplacing fabric over it to eliminate sunlight, mechanically cutting roots and rhizomes, or applying salt (National Park Service 2005); (3) changing landform characteristics by creating blowouts, removing dune ridges or building low-angle ramps to favor sand drift from the beach (Arens 2013a; Pye and Blott 2017); and (4) changing hydraulic conditions for slack development (Murphy et al. 2019). These techniques can have different effects that can be tailored to suit restoration needs. Removing vegetation and topsoil can reinitiate aeolian activity (van Boxel et al. 1997; Arens et al. 2004; Hilton 2006), whereas grazing and mowing can allow the existing vegetation to grow but at a different rate or stage.

Several different techniques for eliminating vegetation are often applied or suggested for use in the same dune field (Jones et al. 2010; Pye et al. 2014; Brunbjerg et al. 2015; Blindow et al. 2017; Pye and Blott 2017). Some techniques may be more effective in some dune fields than in others, depending on soil characteristics (Kooijman et al. 2017).

Ecological indicators (Table 4.1) can be used to assess the condition of the environment, provide warning signs of change, or diagnose the cause of the problem (Westman 1991; Lubke and Avis 1998; Espejel et al. 2004). They can also be used to establish target conditions for restoration and assess outcomes to provide the basis for adaptive management. Changes in vegetation composition and structure will continue to occur in restored dunes, resulting in changes in fauna (van Aarde et al. 2004), so evaluation of success of restoration efforts is an ongoing exercise that must be placed in the context of abiotic and human variables.

4.8.2.1 Establishing Levels of Grazing

Overgrazing led to widespread destabilization and sand movement, so grazing of domestic herds was eliminated in many locations throughout the world. In contrast, grazing at low levels (Figure 4.5) can locally control the development of thick grassy swards and scrub and result in a diverse pattern of open sand, grassland, and

Figure 4.5 Controlled sheep grazing introduced landward of the Sand Motor (Figure 2.4), south of The Hague, the Netherlands.

shrub in formerly stabilized landscapes (van der Meulen and Salman 1995; de Bonte et al. 1999). Establishing appropriate levels of grazing and monitoring is critical (Doody 2001; Baeyens and Martínez 2004). Effects on plants can differ spatially and in the short and long term depending on animal species type, number, and behavior (Millett and Edmondson 2013).

Controlled grazing can increase availability of light to species close to the ground, while trampling by the animals can open moss layers and topsoil, providing the opportunity for additional plant and animal species of conservation interest (Boorman 1989; Kooijman and de Haan 1995; Bonte et al. 2000; Kooijman 2004; Plassmann et al. 2010). Grazing will not necessarily meet the demands of every specialized species of open dune habitat (Maes et al. 2006), and optimum conditions for one group of organisms will rarely overlap with optimum conditions for another group (Rhind and Jones 2009). The effect of grazing will differ according to the types of grazing species, the number of animals, and the presence or absence of palatable species, and the methods should differ depending on whether the environment is grassland, slack, or heath (Boorman 1989; Bonte et al. 2000). Stocking rates may be high initially in scrubland and then reduced. Dense scrub and woods may require cutting before grazing can be effective. Grazing may alter the character and composition of species assemblages within a habitat but only marginally alter abiotic conditions, including water availability and sediment characteristics, and grazing may have a greater beneficial effect in dry dune habitat than moist slacks (Plassmann et al. 2010).

Advantages and disadvantages of grazing on dunes are identified in Boorman (1989) and van der Laan et al. (1997). Adverse impacts can be associated with large local concentrations of dung; soil compaction or loosening; trampling of lichens and other sensitive species (van der Laan et al. 1997), as well as collateral damage to trees and shrubs. Managers may be reluctant to use grazing because of (1) negative association with uncontrolled sand drift in the past; (2) hesitation to condone any agricultural practice in dunes; (3) negative views of the fences required to control the animals; (4) lack of detailed knowledge of the association of the grazing species and their ecological effects; (5) skepticism of long-term results; (6) fear of negative reactions between animals and visitors; and (7) fear of microbiological pollution of the groundwater (van der Laan et al. 1997). Like many other modifications to the landscape, timing is important, and grazing may be restricted during months when plants flower and set seed (Doody 2001). The conclusions of grazing experiments are that species diversity is improved by grazing despite the many different reactions of specific vegetation to grazing and unknown differences in preferences of grazing animals. Compared to mowing or cutting, grazing can be less expensive for widespread areas, more suitable for steep slopes, more conducive to creation of natural borders between plant communities, and more compatible with cultural heritage (van der Laan et al. 1997). The interactions between grazing, burning, mowing, and turf stripping are not well understood, and there is a need to evaluate the effect of different management methods both alone and in combination (Damgaard et al. 2013).

4.8.2.2 Mowing

Mowing, like grazing, is effective in counteracting grass encroachment and increasing species richness by opening the vegetation canopy to allow the lower vegetation types to compete for light, space, nutrients, and water (Anderson and Romeril 1992; Jungerius et al. 1995; Kooijman 2004). Mowing can also contribute to increased effectiveness of grazing by rabbits and domestic animals because the animals prefer to graze in open areas and find the new plant growth more palatable than old material (Anderson and Romeril 1992). Mowing may be necessary as an additional measure after some other treatments, such as fire (Bossuyt et al. 2007), to prevent expansion of competitive species and aid establishment of target species. In some cases, mowing should be accompanied by removal of the cuttings, so light can reach the surface. The frequency of mowing is related to the regional climate (that affects growth rate of the vegetation), and the number and kind of grazing animals that take advantage of the new vegetation growth. If mowing is not accompanied by grazing or other methods, annual mowing appears better than mowing at longer intervals (Lundberg et al. 2017).

4.8.2.3 Pulling

Manual methods of removing vegetation can be initially expensive but can have benefits that result in longer-term efficiency (Pickart 2013). Species can be precisely targeted, preventing elimination of desired species as an unwanted side effect. These species can provide a source for natural colonization and reduce the need for revegetation programs. Manual removal is labor intensive and is most suitable where (1) the vegetation has not become extensively established (Wiedemann and Pickart 2004); (2) other methods of removal have occurred and follow-up action is needed; or (3) stakeholders object to use of other methods. Volunteer labor can be forthcoming to reduce costs and provide a catalyst for community support for other restoration efforts (Pickart 2013).

4.8.2.4 Applying Chemicals

Hesp and Hilton (2013) identify some of the issues involved in using chemicals to eliminate unwanted *Ammophila arenaria* in New Zealand. Chemicals can be distributed by aircraft, which facilitates application over wide areas, although desirable species in proximity to the target species may be collateral damage in aerial application. Avoiding this collateral damage could leave adjacent portions of target species unsprayed. Aerial application may be less appropriate than other methods for treatment of widely scattered plants. Residual chemicals remaining in sand following application of herbicide does not appear to be a problem if an appropriate herbicide is selected, so colonization by desired species would not be restricted by this factor. Seed banks remaining in treated areas and seeds and rhizomes delivered alongshore from untreated areas will be an issue. Buried *Ammophila* seed banks appear to be viable for a decade and more and can lead to regeneration of vegetation as dunes erode (Hilton et al. 2019), and seeds can be dispersed by wind in large quantities (Hesp and Hilton 2013).

4.9 Replacing Lost or Exotic Vegetation

Vegetation may be planted on dunes to stabilize surfaces subject to sand drift; restore surfaces eliminated by mining or burial of pipelines; or restore surfaces previously planted with exotic species. In a restoration context, a reasonable working assumption is that revegetation efforts should attempt to replicate the pre-disturbance landscape. If that is not possible, species that are native to the region and require beaches or dunes for at least part of their life cycle could be used. There may be cases where species not naturally dependent on dunes may be targets of restoration because their preferred habitat is unavailable.

4.9.1 Stabilizing Sand Drift Areas in Dune Fields

A substantial literature on problems of blowing sand and methods of stabilization exists, including many reports dating back a century or more (e.g., Marsh 1885; Lamb 1898; Gerhardt 1900). Many examples of stabilization using exotic species exist, including the many afforestation projects common in Europe from the eighteenth to the twentieth centuries (Clarke and Rendell 2015). Detailed methods of stabilization and key species used for that purpose are presented in Schwendiman (1977) and Sharp and Hawk (1977). Stabilization of large areas characterized by much sand movement may be best done in phases, using an initial dune builder, such as *Ammophila* spp., followed in a few years by grasses and legumes after sand movement slows and the initial grasses lose vigor, then followed by woody plants (first shrubs, then trees) in landward portions of the stabilized area (Schwendiman 1977). These projects are not evaluated here because they are rarely true restoration projects, and current interest is in making many stabilized dune fields more dynamic or replacing the introduced exotic species with native species, which can also increase sand transport (Arens et al. 2001; Buckley et al. 2016; Van der Biest et al. 2017). It is doubtful if much thought was given to mimicking the original dunes in species types in past stabilization efforts, although some past sand stabilization operations became de facto restoration projects when native dune species were used and restrictions on use of the replanted area (usually by controlling grazing) allowed the landscape to evolve in a fairly natural trajectory. The emphasis on stabilization rather than restoration continued until the last few decades, when attention focused on using native species (Avis 1995; Hertling and Lubke 1999).

4.9.2 Restoring Mined and Excavated Dunes

Mining of beaches and dunes has been conducted all over the world where there is a use for the sediment for construction aggregate, minerals, new substrate for crops, and landfills (Mather and Ritchie 1977; Nordstrom 2000). In some cases, mined sites are left as degraded landscapes, without any restoration efforts (Sheik Mujabar and Chandrasekar 2013). Restoration actions are better documented for the large-scale commercial projects in South Africa and Australia (Lubke and Avis 1998; Dodkin and McDonald 2019; Dlamini and Xulu 2019). Excavation of beaches and dunes also occurs for construction of pipelines and storm outfall pipes. Excavation destroys the existing topography, vegetation, and soil profile, and increases the likelihood of aeolian transport. Fauna using the habitats will also be displaced but probably only temporarily if conditions are restored following active uses (Lubke and Avis 1998).

Suggestions for rehabilitating mined sites are presented in Table 4.3. A priori knowledge about the purpose and use of the post-restoration landscape is useful for making stakeholders aware of the significance of changes that will be made and for

Table 4.3. *Suggestions for rehabilitating dune mining sites*

Prepare a rehabilitation plan prior to initial disturbance
Agree on the long-term post-mining land use
Minimize the area cleared
Maintain pockets of representative plants in adjacent areas
Rehabilitate at a pace that keeps up with rate of mining
Prevent the introduction of weeds and pests
Reshape the disturbed land to make it suitable for the desired long-term use
Make landforms compatible with the surrounding landscape
Minimize the potential for erosion during and after disturbance
Remove hazardous or unnatural materials
Reuse topsoil (without long times in storage) or a substitute soil with similar seed bank
Loosen compacted surfaces and ensure that the soil can support plant growth
Revegetate with species consistent with preexisting or surrounding types (unless they are exotic)
Monitor and make adjustments until vegetation is self-sustaining or meets management need

(*Sources*: Environmental Protection Agency 1995; Lubke and Avis 1998).

ensuring that relevant data can be gathered before the mining project begins. Pre-construction maps of topographic contours and ground cover and baseline data on characteristics of soil and fauna are important to provide a standard for subsequent reconstruction of the landscape if the landscape prior to construction is the target condition. Data will have to be gathered on cultural variables where there is human use of the dune (Lubke and Avis 1998).

Restricting the amount of land that is modified and restoring it in sections at a pace commensurate with excavation will help minimize disturbance and the potential for wind-blown sand. In some cases, the vegetation from the removal site can be stored and replanted on the newly reshaped surface (Portz et al. 2015). Bare sand surfaces and newly planted surfaces that might be subject to deflation can be protected by spraying them with a binding material or by covering them with thatch, bark, vegetation cuttings, or geotextile fabric (including cuttings removed during excavation and not viable for replanting) in addition to sand-trapping fences.

Earth-moving equipment is used to refill the excavated area and reshape the dune. A decision can be made to mimic the original dune or reshape it to perform an alternative function. Restoring the pre-disturbance form seems intuitively obvious, but the pre-disturbance form may have already been altered by humans and not representative of a natural dune. Some departure from both the pre-disturbance form and a natural dune form also may be required if machines are used for replanting. Slopes may have to be gentler to allow revegetation using mechanical techniques, requiring a more landward setback of the dune crest (Ritchie and Gimingham 1989). This alternative may not be feasible if the reshaped dune does not mimic the adjacent dunes. The greater compatibility of planting by hand may make this method preferable to mechanical planting. It may be desirable to allow some dunes to assume their own form and pattern before stabilizing them with vegetation (Lubke and Avis 1998), but this alternative is not easy to put into operation, given the uncertainties in the effect of blowing sand and the short time frame of most contracts for restoring excavated land.

Procedures for interior dunes may differ from foredunes in that surface topsoil removed during excavation of interior dunes should be stored and eventually replaced on the restored dune surface to allow for growth of species typical of a later stage of succession (van Aarde et al. 2004). Sediment should be stored on ground where important species will not be destroyed by burial (Ritchie and Gimingham 1989). Topsoil should be replaced as soon as possible, so it does not lose nutrients, microfauna, and flora (Lubke and Avis 1998). Surfaces compacted by earth-moving machinery may have to be loosened as well. Use of imported topsoil should not be used because it can lead to growth of weeds from seeds in the fill (Ritchie and Gimingham 1989). Active foredunes should require no addition of soil to allow the vegetation typical of them to survive.

Decisions have to be made about whether planting should mimic the original cover in species composition or simply be visually acceptable but environmentally compatible by using native species in their approximate niche. In the latter case, species from the surrounding vegetation may colonize the area, but time will be required for establishment. The reconstructed landscape may be stable and relatively species rich after several years, but it is likely to appear somewhat different from the surrounding landscape because the growth trajectories of planted vegetation will be different even if the distribution of plants is the same (Ritchie and Gimingham 1989). Monitoring and adaptive management can be conducted to ensure that the restored site evolves as a self-sustaining landscape but also to reassure stakeholders that any departures from the preexisting landscape or differences from the adjacent dune are acceptable.

4.9.3 *Controlling Exotic Species*

Exotic vegetation may have been planted because it is more effective for stabilizing dunes, more valuable economically, or more attractive. Exotics may also invade a site from alongshore dispersal of seeds and rhizomes (Konlechner and Hilton 2009), or from inland agricultural land (Doody 2001), residential and commercial properties (Mitteager et al. 2006), or undeveloped uplands. Examples of exotic vegetation introduced to dune systems and targets for removal include *Casuarina equisetifolia*, *Acacia cyclops*, and *Acacia saligna* from Australia to Africa (Lubke 2004; Lück-Vogel and Mbolambi 2018) and Israel (Lehrer et al. 2013); *C. equisetifolia* to Mexico (Espejel 1993; Moreno-Casasola et al. 2013); *Casuarina* and *Acacia* to India (Ghate et al. 2014); *Chrysanthemoides monilifera* to Australia from Africa (Chapman 1989; Mason and French 2007; Dodkin and McDonald 2019); *Ammophila arenaria* from Europe to the USA (Cooper 1958; Pinto et al. 1972; Wiedemann and Pickart 1996, 2004; Pickart 2013), South Africa (McLachlan and Burns 1992), and Australia and New Zealand (Webb et al. 2000; Hilton et al. 2006; Hesp and Hilton 2013); *Carpobrotus edulis* from South Africa to Southern Europe, the USA, and Australia (French 2012; Novoa et al. 2013); *Carex kobomugi* from Asia to the USA (Burkitt and Wootton 2011); *Eucalyptus* spp. in the Mediterranean basin (van der Meulen and Salman 1996); and pine trees (e.g., *Pinus nigra* ssp. laricio and *P. contorta*) used throughout the world (Doody 1989; Sturgess 1992; Leege and Murphy 2000). Box 9 provides an example of removal of *Ammophila arenaria* in New Zealand.

Exotic vegetation can form monospecific stands of vegetation and outcompete native species. By eliminating sand movement, exotics may accelerate plant succession, resulting in rapid local extinction rates with long-term negative consequences for the reestablishment of early successional species. Removal of

Box 9
Removing invasive *Ammophila arenaria* at Stewart Island, New Zealand

Doughboy Bay is the southernmost in a series of active dune systems located on the west coast of Stewart Island/Rakiura in southern New Zealand/Aotearoa. This site has been the focus of a sustained "dynamic" restoration program led by the New Zealand Department of Conservation and the University of Otago. Restoration at Doughboy Bay commenced in 1999 with the goal of restoring and safeguarding geomorphic processes for the protection and conservation of native dune flora. The short-term aim was to increase rates of sedimentation, dune mobility, and dune dynamism. This involved eradication of the invasive marram grass (*Ammophila arenaria*), which had formed a dense and almost continuous canopy (Figure 4.6). Dune mobility had ceased, allowing mid to late successional species to establish. Marram was eliminated by initial applications of a grass-selective herbicide spread by helicopter, followed by annual ground-based applications in selected areas of regrowth. Remobilization of the sand surface followed, and by 2002, the dunes were devegetated and geomorphically active. The native sedge *Ficinia spiralis* (pīngao) was planted to facilitate its return. These plants thrived and now form large shadow dunes in the relatively exposed sections of bay. Other native dune plants reestablished a semi-continuous cover in sheltered sections of the bay (Figure 4.7). Overall, the program resulted in a system shift from a stable dune system to a geomorphically dynamic one that is similar in physical and ecological character to nearby dune systems where marram was never present. The system continues to evolve, demonstrating the need for a long-term perspective of dune restoration and careful consideration of the ecological requirements of native species to achieve restoration goals. See Konlechner et al. (2014, 2015a) for more information.

Contributor: Teresa Konlechner, National Center for Coasts and Climate, University of Melbourne

Figure 4.6 Doughboy Bay, New Zealand in 1999, prior to removal of Ammophila.
Source: Mike Hilton.

Figure 4.7 Doughboy Bay in 2017, showing varied vegetation cover.
Source: Mike Hilton.

dense stands of exotics can return dunes to earlier stages of succession, creating a less stable surface that favors species that rely on disturbance and openings in the vegetation (Wiedemann and Pickart 2004).

Afforestation (planting trees) was common in Europe in the past to protect crops from winds from the sea, establish a forest industry, and stabilize dunes (Blackstock 1985; Sturgess 1992). The practice is now out of favor in most countries because it is associated with loss of flora and fauna, changes to soil characteristics, lower water tables, seeding into adjacent unforested areas, and unfavorable conditions for native species (Sturgess 1992; Janssen 1995; Muñoz-Reinoso 2003). Restoring a diverse flora for areas forested for many years is not possible by simply clear-cutting the trees because of major changes to the soil, including low pH and high organic content that retains more water than bare sand, but conditions favoring many original dune species may be improved by removing the litter layer with the trees (Sturgess 1992; Sturgess and Atkinson 1993). Whether the soil is acidic or calcareous appears to be important for the types of vegetation that develop in pine clearings, and the vigor of remaining forests adjacent to clearings and exposure to coastal influences determine whether closed thicket (in more sheltered areas) or vegetation similar to evolving heathlands or grasslands (more exposed areas) develops (Lemauviel and Roze 2000). If resources to remove plantations are limited, a decision should be made on which new type of environment has the greatest value for the funds available. Restoration of the original dune flora following clear-felling a dune plantation may not be possible, but it may be feasible to create a vegetation type with greater value for conservation than the preexisting pine plantation (Sturgess and Atkinson 1993). One example is the substitution of endangered *Juniperus oxycedrus* ssp. *macrocarpa* woodlands for abandoned pine plantations in Spain (Muñoz-Reinoso 2004; Muñoz-Reinoso et al. 2013).

The advantages and disadvantages of exotics vary. Some exotic species allow for maintenance of populations without excluding other species (*A. arenaria* in South Africa); others totally replace natives and change the morphology of the dunes (*A. arenaria* in the USA) or involve complex ecosystem changes by changing soil characteristics that can trigger invasions of other species or prevent colonization of native species (Lubke and Hertling 2001; Knevel et al. 2002; Lubke 2004; Wiedemann and Pickart 2004; Hilton et al. 2005; Parsons et al. 2020). Some exotic species may add to the attractiveness of the dune. The appeal of *Oenothera biennis* in Europe (Doody 2001) and *Rosa rugosa* in Europe and the USA (Christensen and Johnsen 2001; Mitteager et al. 2006) can diffuse the perceived need to remove them. Invasives can be nurse plants, help in seed recruitment of natives, contribute nutrients, prevent further invasions, and be more likely to thrive on marginal lands. Accordingly, Rodrigues et al. (2011) and Rai and Kim (2020) indicate that exclusion of exotic plants in restoration efforts may be counterproductive unless their full ecological and socioeconomic payoffs are assessed. Understanding the positive values of invasive species can be important in designing effective control measures, especially where governing bodies must respond to multiple competing agendas and where control of invasives becomes a politicized issue (Hanley and Roberts 2019).

Exotics are often preferred over native species for landscaping on private lots on dunes just landward of the beach. Many species on private lots are not adapted to dune environments and are problematic because they require considerable efforts to maintain (Mitteager et al. 2006). The incursion of these species into publicly managed portions of the beach and dune is not likely, but the exotics that are adapted to dune environments are problematic. An example of the latter is *Carex kobomugi* that was used in place of the native *Ammophila breviligulata* on some private lots on the northeast coast of the USA and is invading dunes in natural areas. Much more can be done to control land use on private lots to achieve restoration goals, as identified in Chapter 8.

Early efforts to remove exotic vegetation and replace it with native species targeted only the most obvious invaders. More recent advances in understanding of dune ecosystems led to early detection of problems and more system-based approaches focusing on reestablishing dune processes and managing multiple taxa in an integrated way (Wiedemann and Pickart 2004). Removal of exotics, like other vegetation, can be accomplished by manual removal, excavation, burial, burning, and application of herbicides. Manual removal has the advantage of allowing relict native plants, soil characteristics, nutrients, and moisture to remain in place and speed recovery (Wiedemann and Pickart 2004). Seeds of exotics may remain after manual removal of growing plants, so a maintenance program may be necessary to remove returning invasives as they germinate (Lubke 2004).

Simply removing plants does not necessarily eliminate the legacy effects that the species had on native ecosystems, and additional steps may be needed to remediate some of the remaining physical, chemical, and biological effects (Corbin and D'Antonio 2012; Konlechner et al. 2015b; Parsons et al. 2020). These effects can limit colonization of native dune plants and enhance the potential for secondary invaders to take advantage of the decrease in abundance of primary invaders (Parsons et al. 2020). The magnitude of the problem of invasive species that are already well established can exceed the resources available to address it (Wiedemann and Pickart 2004), making detection and management of initial invasions of critical importance. The difficulty of eliminating legacy effects may warrant focusing on low-risk / high-yield restoration efforts, such as removal of invasives in lightly or recently invaded areas or in foredunes, where overwash is expected to create new substrate free of legacy effects (Parsons et al. 2020), or prioritizing the prevention of spread of invasives into uninvaded areas (Konlechner et al. 2015b).

4.10 Restoring Slacks

Dune slacks are low-lying, nutrient-poor, species-rich, periodically flooded interdunal wetlands. Many of the uncommon plant species in dune slacks tend to prosper during the early successional stages of slack development and are no longer generated by natural processes, requiring creation of new slacks or regeneration of existing older slacks (Rhind and Jones 1999). Slack management is underway in many countries (Stratford and Rooney 2017), and restoration projects are relatively frequent in Europe, where biodiversity was reduced due to lowering of water levels, reclamation for agricultural use, and afforestation with pines (Grootjans et al. 2004). Fresh dune slacks can be created by deflation, whether naturally or artificially induced (Arens et al. 2013a). Slacks can also be created by digging pits to create ponds, with the level of the bottom of each pond determining in which period of the year and for how long water will be present (Vandenbohede et al. 2010; Murphy et al. 2019), or by constructing artificial dunes on the seaward side of low washover flats (Grootjans et al. 2002). Slacks may be incorporated into beach and dune building designs by creating low-lying troughs between new dune ridges or between dune ridges and seawalls, which can also be buried forming new dune ridges (Figure 4.8). Creating the appropriate abiotic environment from the start and during the development period is important in artificially building new slacks; factors include sediment size, CaCO content, exposure to wind and rainfall, fluctuating and fresh groundwater, and nutrient-poor sediment (van der Meulen et al. 2014). Water table elevation and its temporal variation is critical to dune slack ecology and biodiversity (Stratford et al. 2013). Restoring groundwater levels in existing slacks is a fundamental issue where levels are too low (van der Hagen et al. 2008; Geelen et al. 2017). Artificial groundwater recharge is a key component, but the chemical content of the water is also critical for success (Geelen et al. 2017). Groundwater flow modeling can combine available data on hydrogeology, geology, water quality, drainage level, and water extraction, and provide a tool for planning groundwater management in restoration interventions (Vandenbohede et al. 2010). Actions such as sod cutting, mowing, and grazing are important to allow relicts of target species in the nearby vegetation to colonize the slack (Jungerius et al. 1995; Rhind and Jones 1999; Grootjans et al. 2004, 2013; van der Hagen et al. 2008; Geelen et al. 2017).

The construction of a slack is often unintentional, by forming landward of sand dikes or beach ridges or dunes accreting seaward of a low surface (Grootjans et al. 2002, 2004; van der Meulen et al. 2014). This kind of slack developed at Avalon,

Figure 4.8 Artificially created dune slack at Spanjaardsduin, the Netherlands. The project was designed to provide habitat characteristics of gray dunes and dune slack as compensation for losses to dunes in association with the extension of Rotterdam Harbor (van der Meulen et al. 2014, 2015).

New Jersey (Figure 4.9), where new sand-trapping fences were placed well seaward of the landward limit of the wide backshore created during a mid-latitude storm in March 1962. The slack in Avalon was initially perceived as a potential breeding area for mosquitoes and was treated for larvae sterilization by the state Mosquito Control Commission. The slack was eventually considered an environmental resource rather than a detriment by shorefront residents, in part because of the diverse vegetation community (Nordstrom et al. 2002).

Slacks are not common in developed areas where restrictions to width of dune zones or attempts to increase the volume of protective dunes reduce the likelihood that slacks will form and survive, but creation of new slacks is possible as a component of beach nourishment operations. This option would work if the fill creates a platform that is wide enough to accommodate both the foredune and the slack seaward of human infrastructure in a situation analogous to the naturally functioning slack identified in Figure 4.9. Where space is more limited, creative use of sand fences or earth-moving equipment could help build a foredune with

Figure 4.9 Dune slack at Avalon, NJ, USA, that formed on a beach overwashed by a storm in March 1962. The slack is protected from overwash and wind-blown sand by a dune created by placing sand fences on the seaward side of the overwashed beach. The photo was taken 45 years after the protective dune was created.

ridges and swales between, creating a moister sheltered environment characterized by species different from the more exposed ridges (Grafals-Soto 2012).

4.11 Allowing Time for Naturalization

Time is an important element in achieving a desired target state. Restoring habitats and landscapes using fences or earth-moving machinery is simply a way of compressing the time dimension to achieve a target state sooner than possible by natural processes. These projects should be considered emergency solutions rather than standard solutions. It is likely that some restoration outcomes that are perceived as bad result from a failure to allow the landscape sufficient time to evolve. The eagerness for immediate results can be questioned in situations where nature can accomplish the goals, given time (Arens et al. 2001).

Nourishment operations can make use of fill sediment that differs from native materials in size, shape or roundness. Time is required for natural processes to selectively grade and sort the materials and make them more suitable for aeolian transport. Wave energy and beach morphodynamics are critical to this process. Reworking of sediment in the turbulent breaker, surf, and swash zones will be faster on beaches with a high wave energy regime, and buried sediments will be exposed more frequently on beaches that undergo cycles of erosion and deposition. Beaches that are nourished to an elevation where wave uprush cannot reach the upper portion of the backshore may require grading to facilitate wave action. It is understandable why human interventions are sought to create desirable characteristics in beaches and dunes modified primarily for human use, but interventions to truncate the time dimension appear less justifiable in locations where the attempt is to achieve a natural state. The new landforms should be as sustainable as possible through natural processes alone.

4.12 Determining Appropriate Levels of Dynamism

Dynamism and complexity of topographic relief are important natural characteristics favoring biodiversity, but we do not know how much dynamism and complexity are critical to maintain long-term viability. The natural interplay of sediments, landforms, and biota does not have to be limitless to retain species. Cycles of erosion and deposition may be truncated in space or altered in magnitude or periodicity by humans, thus changing the natural history and areal extent of landforms and biota, but this does not mean that landforms and biota will be eliminated.

The formation of gaps in foredunes, accompanied by cycles of erosion and accretion at the gap and adjacent dune, is a common occurrence in natural dunes (Gares and Nordstrom 1995). These gaps can be replicated by human processes by eliminating surface cover. In some cases, simply not using fences to repair notches in the dune crest will accomplish the same goal. The vigor of many dune forming species, such as *Ammophila* spp., and their potential value in facilitating growth of other species makes planting vegetation a natural alternative to fencing. In many cases, the optimum action for rejuvenating landscapes is to do nothing and allow natural gaps in foredune crests to form and evolve. The variety of local relief in the topography would create small-scale differences in sheltering and proximity to the water table, which would enhance the variety of habitats over short distances across and along the shore and provide a more realistic image of the kind of naturally functioning landscape that evolves when native species dominate.

It is easier to design for maximum levels of protection than for optimum levels of dynamism. Suitably erosional conditions are difficult to predict and vary for different species; there is a risk of providing too much or too little protection from natural stresses; and designs may have to be by trial and error with little foresight into how the final product will behave (Barton 1998; Lee 1998). The capability to quantitatively predict and simulate overwash and the resulting deposits is emerging (Donnelly et al. 2006; Schupp et al. 2013), but it is still difficult to restore beaches for nesting by some species. The optimum elevation to create plover habitat is lower than the elevation required for protection of human structures (Maslo et al. 2011), so establishing a target state is difficult.

Dune management will always be complicated because the systems vary considerably in time and space (Sherman 1995). Their dynamism should be accepted and written into management plans, but dunes still should be monitored so they do not reach an excessively eroded condition through neglect or become overly managed and maintained as static features (Davies et al. 1995). Developed coasts can retain their natural dynamism if development is set back at a sufficient distance from the shore to retain a natural buffer against wave damage (Janssen and Salman 1995). One of the reasons portions of the dunes in the Netherlands are being reactivated is that development is so far landward that no facilities are threatened. Human efforts can create a more stable form of protection landward of the dynamic zone, thereby partly compensating for lack of space. Adding dynamism and other values and functions once levels of protection have been achieved can be accomplished at a relatively large scale (van Koningsveld and Mulder 2004). Most projects are likely to be experimental and small scale initially, but these projects are required to document feasibility for application to larger areas.

4.13 Offsite Activities

Comprehensive restoration programs should include controlling offsite activities that influence sediment budgets, biotic changes, or pollutant levels to the extent possible (Morand and Merceron 2005). For example, more could be done to attempt to control the litter that enters the beach from the outside by tracking the sources and modifying practices there (Williams and Tudor 2001; Cunningham and Wilson 2003). Reducing human litter would not only reduce the hazards associated with non-biodegradable pollutants and the exotic habitats and undesirable species attracted to them, it would also reduce the need for beach raking. Dam removal in streams that empty onto the coast (Willis and Griggs 2003) or dredging sediment trapped in dams for beach use (De Vincenzo et al. 2019) are other offsite actions that would help restore coastal environments, in this case by supplying the sediment needed to provide space for natural features to form seaward of human facilities. Influencing some activities in the basins of rivers discharging into coastal waters may be easier to accomplish if the river basins are controlled by the same regional authorities that control coastal activities or if river basins are included in integrated coastal zone management plans. Establishment of river basin authorities in Italy in 1994 has resulted in a more sustainable management of rivers, with a requirement to consider the sediment budget to the coast due to stream flow (Autorita' di Bacino Del Fiume Arno 1994; Cammelli et al. 2006).

Some overarching problems, such as sea level rise (Stocker et al. 2013), atmospheric nitrogen deposition (Bobbink et al. 1998; Jones et al. 2004), and production of materials outside the coast that become litter on beaches (Borja and Elliott 2019), are beyond the control of local managers but can be mitigated by restoration actions in the field. Excess atmospheric nitrogen inputs can be compensated by reducing inputs of nitrogen from vegetation litter through grazing (Kooijman and Smit 2001). Erosion associated with sea level rise can be offset by beach nourishment in some areas, at least for the near future (Houston 2017). Broader regional actions to address external threats are not discussed in detail here because offsite adjustments are beyond the capabilities of most beach managers, and political and economic constraints may prevent linkage of offsite activities to onsite restoration efforts.

4.14 Concluding Statement

Increasing the dynamism of stable landforms enhances exchanges of sediment, nutrients, and biota, and cycles of accretion and erosion and growth and decay that will retain diversity, complexity, and resilience. The initial mobilization may occur by direct human actions or by suspending existing stabilization programs. Thereafter, landforms should be allowed to evolve with as little human input as possible. Restoration efforts in dune areas cannot necessarily compensate for the ecological function of natural processes, including intensive sand blowing, natural erosion/accretion cycles, regular supply of ground water, and clean air with minimum human interference (Grootjans et al. 2013; Delgado-Fernandez et al. 2019). Sufficient space and time are critical for equilibrium forms to develop, and equilibrium states can take decades (Konlechner et al. 2015a). If space and time are restricted, ongoing human efforts may be required to adjust the level of dynamism or reestablish cycles of change.

Leaving wrack lines in place is an easy and cost-effective way to retain natural conditions. Greater dynamism can be accommodated by removing or modifying shore protection structures or allowing them to deteriorate or become buried, as indicated in Chapter 5. Programs for protecting endangered species have great potential for restoring naturally functioning beaches and dunes by restricting human uses, but they can have negative effects if the needs of other species are subservient.

Exotic vegetation and vegetation and soil that represent later stages of succession can be removed to support greater variety of native species.

Rejuvenation of surface processes and vegetation can also be accomplished by reducing the height and biomass of grasses. Vegetation may be planted on dunes to restore surfaces that have been eliminated or are subject to sand drift, but stabilization over large areas, as has occurred in the past, appears counter-productive, given the current emphasis on diversity and complexity of landforms and habitats.

The amount of dynamism and complexity critical to maintain long-term resilience of habitats and species is site specific and poorly known. Many of the methods taken to increase dynamism represent departures from past practices, so it is likely that most projects will be experimental and small scale until the feasibility of large-scale future projects is demonstrated. The many experiments in the Netherlands to mobilize dunes were feasible because of (1) the relatively undeveloped and extensive dune landscape resulting from large-scale beach nourishment and dune building projects; and (2) restrictions on human use because of the national importance of dunes for flood protection and provision of drinking water (Grootjans et al. 2002; van der Hagen et al., 2008; Arens et al. 2013a). For example, the shoreline where the Spanjaards dune (Figure 4.8) was constructed was artificially increased from a width of 100–150 m to 400–450 m (van der Meulen et al. 2014). We highlight these experiments (and the Sand Motor discussed in Chapter 2) because they provide much useful information on feasibility of innovative strategies, but the context in other countries may vary markedly (Delgado-Fernandez et al. 2019). Application of dynamic management strategies elsewhere would require careful tailoring to local spatial, social, and political constraints that may be severe. Chapters 6–8 identify approaches that can be used to restore environments under these constraints.

5

Altering or Removing Shore Protection Structures

5.1 Rethinking Erosion

A major deterrent to management solutions that attempt to work within the dynamic natural system is the perception that erosion is categorically bad. Removing shore protection structures would result in increased erosion, but it would also contribute sediment volumes to the beach, permit sediment exchanges between beach and dune or upland, and help restore landforms and associated habitats. Availability of sediment and space are key elements in sustaining natural features (Orford and Pethick, 2006; Cooper and McKenna, 2008; Psuty and Silveira 2010; Morris 2012; Myers et al. 2019). Coastal retreat and lack of sediment are highly correlated.

Natural coastal landforms and habitats are connected by sediment transfers offshore, onshore, and alongshore. Cross-shore sediment transfers and changes in topography associated with coastal storms are often cyclic, with offshore transport of sediment removed from beaches and dunes during storms followed by onshore transport after storms that replenishes the backshore. Intense storms can erode the seaward portion of the dune or wash over the dune, but recovery of the beach after storms provides a source for wind-blown sand for dune building and creates a wider buffer against erosion of incipient dunes by small storms, allowing dunes to grow. Waves from intense storms deliver sand farther landward, but the sand remains within the coastal system to build up the substrate and provide a source for aeolian transport to new dunes that can develop farther inland (Godfrey et al. 1979). Coastal dunes can also migrate landward by wind erosion of the seaward side, with deposition on the landward side (Ollerhead et al. 2013). Marshes are coupled to environments seaward of them by overwash (Zinnert et al. 2020), inlet breaching (Miselis et al. 2016), and aeolian transport (Rodriguez et al. 2013). Onshore migration of topographic features via sediment transfers maintains their elevation and provides the substrate for ecological productivity, but cross-shore and longshore sediment transfers are often prevented by human actions to avoid erosion or inundation of buildings and infrastructure.

Erosion of coastal bluffs can be a source of sediment to beaches and nearshores fronting them and downdrift of them (Dawson et al. 2009; Brooks and Spencer 2010; Rees et al. 2015). The value of bluff erosion in enhancing the beach sediment budget was acknowledged by public authorities in the United Kingdom over a century ago (Owens 1911). The need to reestablish sediment budgets and limit public expenditures for shore protection places increasing attention on the desirability of abandoning protection structures or providing only partial defense (Brampton 1998; Bray and Hooke 1998; Lee 1998; Nordstrom et al. 2007c; Moore and Davis 2015). Bluff erosion does not play a significant role as a source of sand for beaches in some locations because of resistance of the substrate or overly fine sediment (Runyan and Griggs 2003; Pye and Blott 2015), but natural bluff evolution has other benefits, including ecological, aesthetic, paleontological, and geodiversity values (Brampton 1998; Howe 2015; May 2015; Rees et al. 2015).

Adoption of any system of working with natural processes will require a change in approach from the prevailing view of coastal erosion as a problem (Cooper and McKenna 2008). Vast amounts of sediment are needed in many locations for restoration of beaches and wetlands under present conditions, and the need will only increase with acceleration of sea level rise (Orford and Pethick, 2006; Morris, 2012). Under natural conditions, the characteristic features of eroding landforms and their resulting habitat and aesthetic values are self-generating, in that erosion may displace the system farther landward or alongshore, but the active processes will proceed unimpeded, restoring landforms and habitats in new locations. The need for erosion to supply this sediment should be highlighted, but designs that work with natural processes are likely to be viewed as acceptable to stakeholders only if the result (new sediment source or rejuvenated habitat) is specified in advance as a positive product (Nordstrom et al. 2016).

People have used structures to protect eroding coasts for centuries, establishing a precedent that is not easy to break (Harman et al. 2015). Shore parallel walls (seawalls, revetments, bulkheads), in particular, restrict movement of sediment and biota; truncate or eliminate beaches, dunes, bluffs, and marshes; and foreclose future nature management options (Pilkey and Wright 1988; Jolicoeur and O'Carroll 2007; Defeo et al. 2009; Dugan et al. 2011, 2018; National Research Council 2014). Walls eliminate the native riparian vegetation that provides important functions in the nearshore and foreshore, including habitat structure, microclimate regulation, fish prey habitat, recruitment of large wood and organic debris, habitat for riparian dependent species, and corridors for wildlife movement (Brennan 2011). The recent emphasis on using beach fill as the principal shore protection option has greatly reduced the number of new structures built and has added sediment seaward of existing structures, but existing structures can also be modified to allow landforms and habitats to be more dynamic.

Managed retreat, involving removal of protection structures, is a viable option for restoring a geomorphically sustainable (i.e., self-regulating) coastal system on developed coasts (Cooper and Pethick 2005). Buyout and relocation programs exist as mitigation (Binder et al. 2015), but application of these programs has been limited by reluctance of homeowners to relocate, reluctance of local governments to reduce their tax base, and expectation that public funds will prevent the market from discounting the value of properties at risk (Nordstrom and Mauriello 2001; Blott and Pye 2004; Kousky 2014). These conditions may change in the future. Existing shore protection structures will decrease in effectiveness through time. Their structural integrity will diminish because of aging; their elevations relative to future sea levels will make them susceptible to overtopping; and the expected future increase in frequency/magnitude of coastal storms will increase their susceptibility to structural damage. Abandonment of coastal structures may be a more promising option in anticipation of these future problems. Small-scale projects where human investment is low, such as removing shore protection structures to allow farmland or campsites to undergo wind and wave reworking and relocating the intensive uses farther landward (Isermann and Krisch 1995), reveal possibilities for allowing erosion in more developed locations.

5.2 Breaching Dikes to Allow Inundation by Sea Water

Dikes are being relocated landward or breached in several locations throughout the world but usually in estuarine environments (Holz et al. 1996; Abraham 2000; Warren et al. 2002; Myatt et al. 2003; Balletto et al. 2005; Garbutt et al. 2006). These alterations may be driven by the need to restore lost salt grass meadows, improve conditions for endangered species that use wetlands, and concentrate protection efforts for more densely developed areas by building higher, shorter dikes that provide more protection than the old dikes (Jeschke 1983; Lutz 1996).

Part of the appeal of relocating dikes and allowing some structures to deteriorate or be removed is the savings in the cost of protecting a longer coastline. Relocation can also accomplish many of the coastal management goals in Table 1.3, including creating ecological corridors, adding portions of the coast to the public domain by converting them from farmland to nature areas, adding habitat for endangered species, and providing a wider buffer against sea level rise for developed lands farther landward. Costs for projects that have a restoration component can also be defrayed by using funds for compensation or mitigation for actions elsewhere (Nordstrom et al. 2007c).

Natural breaches in dikes on the open coast can have a positive effect on biodiversity (van der Veen et al. 1997). Managed realignment on the Baltic Coast of Germany is considered an appropriate strategy because the coastal plain is sparsely inhabited, the tidal range is negligible, many dikes are reaching the end of their useful lives, flood-prone areas are relatively small, and coastal surges are infrequent (de la Vega-Leinert et al. 2018). A new dike will be built in the center of a spit at the national park near Zingst in the state of Mecklenburg-Vorpommern. This action will be followed by cutting breaches in the old dike on the seaward side and removing an old dike on the bayside (de la Vega-Leinert et al. 2018). The old dike system protects land that was reclaimed to provide pasture and protect the mainland from flooding. The changes will allow wave overwash from the Baltic Sea to occur over a larger area, diminishing wave energies before reaching the new dike and creating a mosaic of sand flat, dune, slack, heath, and shrub communities on the former pasture. Overwash will occur across more of the spit but not across its entire width, as was possible during large storms prior to construction of the old dikes. The new dike will still protect the mainland from full storm surges in the Baltic Sea and prevent formation of an inlet. Removal of the bayside dike will allow for periodic flooding of the spit from that side, where wave heights and surge levels are lower than on the Baltic Sea side. This flooding will cause the low surface to evolve as salt grass meadow. Periodic floods in the former polders will deliver more fine-grained sediment, reinitiate peat growth, and allow the surface to keep pace with sea level rise. The surface will be unavailable for human use during periodic inundations but will provide pasture during the summer, when floods do not occur. Virtually the entire spit will evolve by natural processes while flood protection for the mainland will be retained (de la Vega-Leinert et al. 2018). The difference between the restored landscape and a truly natural landscape will be the decoupling of the upland and marsh contact at the new dike location, the lack of direct input of ocean processes on the marsh, and less inundation and reworking of the mainland during large storms.

Dunes originally built or maintained to function as dikes can be mechanically breached or allowed to erode to the point where they are breached by storm waves. State planners in Mecklenburg-Vorpommern are allowing another artificially constructed protective dune northeast of the developed community of Markgra-fenheide (in Rostock) to erode naturally, reducing its protective value and allowing the low upland landward of it to flood and increase the area of wetland habitat. To protect the developed community from flooding, a new ring dike was constructed around it (producing a ring levee enclave), providing a higher barrier than previously existed. As at Zingst, the length of upland protected by the dike and dune system is reduced, but the level of protection to the human infrastructure is increased (Nordstrom et al. 2007c; Schernewski et al. 2018). Erosion of the formerly protected upland provides sediment to a coastal town 4 km downdrift (northeast), reducing erosion there and extending the benefits of the project outside the immediate planning area (Schernewski et al. 2018).

5.3 Altering Hard Structures

Recent studies call for the need to make protection structures more compatible with natural values as well as human recreational needs (Escudero et al. 2014; Nordstrom 2014; Kochnower et al. 2015; Manno et al. 2016; Alves et al. 2017; Pranzini et al. 2018b). Hard structures that already exist can be modified to accommodate greater dynamism, e.g., by retaining only the lower portions of seawalls and breakwaters to serve as sills for beaches landward (Zelo et al. 2000; Aminti et al. 2003; Cammelli et al. 2006). Protection structures can be built or altered to enhance habitat by reducing their slopes or increasing surface complexity (Moschella et al. 2005; Borsje et al. 2011; Browne and Chapman, 2011; Chapman and Underwood 2011; Evans et al. 2016). Damaged structures can be allowed to deteriorate without being repaired (French 2006; Morris 2012) or structures can be allowed to become buried beneath new foredunes and remain to provide backup protection (Nordstrom 2019).

Groins are often allowed to deteriorate if money for repair or replacement is lacking or if there is a need to restore natural sediment transfers. Groins are not being replaced in some locations where sediment is needed on adjacent beaches or where the dunes and dikes landward of them are allowed to breach (Nordstrom et al. 2007c). Existing groins can be removed (Williams et al. 2016); they may be notched (Donohue et al. 2004; Rankin et al. 2004); their length or height can be reduced (Basco and Pope 2004; Bocamazo et al. 2011); and wood pile groins can be altered to provide greater spacing between pilings to allow more sediment to pass them. Alternatives for groin design that allow for sediment bypass while slowing erosion rates are identified in Kraus and Rankin (2004).

Design of groins should consider many aspects, including technical, environmental, constructional, economical, navigational, safety, recreational, and aesthetic, with all of these aspects considered together (Gómez-Pina 2004). Several aspects can be improved by decreasing the interference of groins with natural processes. More attention could be paid to aesthetic aspects to enhance appreciation for nature. Beach users prefer sites without prominent human structures (Morgan and Williams 1999), and acceptance of (and demand for) more naturally functioning coasts may be increased as more natural features dominate the landscape.

5.3.1 Examples of Modifications of Protection Structures

Recent policy changes in Italy to decentralize responsibilities from national to regional and local authorities, combined with European legislation requiring new integrated coastal zone management policies, have encouraged local authorities to implement softer options for shore protection. Coastal management in Italy now includes using submerged protection structures or converting existing high-profile structures to lower ones (Lamberti and Mancinelli 1996; Tomasicchio 1996; Lamberti et al. 2005; Polomé et al. 2005; Archetti 2009; Pranzini 2013; Pranzini et al. 2018b). Examples include the detached breakwaters at Marina di Pisa and Senzuno in Tuscany that were lowered to 0.5 m below sea level (Cammelli et al. 2006; Pranzini et al. 2018b). These conversions can improve aesthetics and safety and make beach widths more similar alongshore. Conversion of other breakwaters to low-profile structures is planned. Willingness to modify protection structures indicates that they are not a permanent solution but a solution that can change based on stakeholder needs and changes in beach processes. Initial construction can be just one stage in a series of adaptive adjustments that can override the original goal of creating a more stable beach for shore protection (Pranzini et al. 2018b).

Projects to abandon or reduce levels of protection on the open coast can be considered feasible where (1) the new natural environments have enough value for the project to be considered cost effective; (2) the human infrastructure that will be exposed to natural processes has so little value that protection efforts can be discontinued; and (3) existing protection structures are insufficient to provide future protection and would have to be reconstructed anyway. Several factors make removal of structures in estuarine environments feasible, as indicated in the following paragraphs.

Many examples of successful projects to enhance the formation of natural environments exist in Puget Sound, Washington, USA, where fetch distances for wave generation are relatively short. Restoration there is aided by the acknowledged value of beaches as habitat for salmon prey, which provides political and economic incentive (Shipman et al. 2010). Nearly all of the projects involve adding beach fill for habitat and shore protection, and many projects involve adding driftwood logs on the upper beach. Projects include (1) burying rock revetments in the beach; (2) completely removing sections of bulkhead, revetment, or seawall; (3) removing the upper portions of shore parallel walls and reestablishing the natural coastal bluffs landward while using the subaqueous remnant of the wall as a sill to hold the beach in place; and (4) using anchored logs instead of concrete or rubble riprap for shore protection structures (Zelo et al. 2000; Toft et al. 2010). Many projects are designed for private properties where erosion control was the major concern of the owner, but where public agencies expressed early concerns about the potential environmental impacts of standard armoring solutions.

Puget Sound has a convoluted and heterogeneous shore, with numerous pocket beaches and small isolated drift cells. The small beach compartments make localized projects feasible, and projects occur at the scale of private residential and commercial/industrial lots and along short lengths of shore in public recreation areas (Shipman 2001; Hummel et al. 2005; Brennan 2011; Toft et al. 2013). The sites in Figure 5.1a–d illustrate how (1) a natural beach can be inserted into an intensively developed and armored urban shore to provide public access to the water (Olympic Sculpture Park); (2) how a preexisting suburban park can be altered to provide for intensive recreational use as well as natural value (Seahurst Park); (3) how restored natural habitats can be mixed with remnant human structures in a meaningful way (Dickman Mill Park); and (4) how private lands can be modified to achieve goals compatible with residents and public interests (Powel Site).

The sites on Puget Sound indicate that removal of protection structures can occur where the structures are in good condition or where they are deteriorating, and this removal can be accompanied by actions to enhance natural features and protect against erosion, or the sites can be left to evolve without human alterations. The artificially created landforms and habitats may not reflect the former natural history of the sites, and they may not be in an overall natural context. Nevertheless, the many projects have increased the visibility of restoration projects and added to the expertise of the participating stakeholders, while expanding the information base needed to design future projects.

(a)

Figure 5.1a Olympic Sculpture Park, Seattle, WA. A portion of a seawall was removed; sand and gravel were placed in the intertidal and supratidal zones; driftwood logs were added at the upper limit of swash; and sediment was placed on the fronting nearshore terrace to create new habitat as refuge and foraging grounds for migrating juvenile salmon. The enhancements increased densities of larval fishes and juvenile salmon feeding and provided habitat for invertebrate assemblages different from armored shorelines and with a high taxa richness (Toft et al. 2013).

(b)

Figure 5.1b Seahurst Park, Burien, WA. Groins were removed in the north to create a naturally functioning feeder bluff to nourish the developed shore (in the far distance). Rock from the groins was crushed to supply gravel fill to the beach or used to build a new groin to protect a bulkhead and building south of the bluff. Artificial tide pool microhabitats were built into the new groin. A wall was removed (in the middle distance); the beach was nourished there; and fixed access steps, gazebos, and picnic benches were added. Protection structures were removed in the remainder of the park and fill was placed to provide a better representation of a naturally functioning shore.

(c)

Figure 5.1c Dickman Mill Park, Tacoma, WA, at the site of a former lumber mill. Many pilings were removed from a former pier; a beach was built with artificial fill; and a wetland and feeder tidal stream were constructed landward. Some deteriorating structures were left in place to reduce project costs. The site resembles neither the preexisting natural landscape nor the developed commercial site, but the juxtaposition of natural and deteriorated cultural elements does not seem out of place in an urban coastal environment. Didactic panels at the site identify the rich cultural history and significance of the remaining human structures.

(d)

Figure 5.1d Powel Site, a private beach in West Port Madison, WA. The owners wanted to rebuild failing shore protection structures and asked the local city and land trust about a conservation-oriented solution. Site-specific solutions include removing concrete and riprap bulkheads, retaining a rock revetment to protect a house (center of photo), and constructing return walls (far left) to prevent flanking from unprotected segments. Costs were defrayed by donated time of stakeholder groups, which was considered critical to the process (Brennan 2011).

Many projects to remove protection structures are on sites where waves are of relatively low energy. The range of alternatives is more restricted on ocean sites, where public interest in beach recreation is great and the land has greater human use value (Nordstrom et al. 2015), but not all creative alternatives are precluded (Zelo et al. 2000). Removing structures and using more flexible construction methods is occurring on some high-energy sites. In many cases, the main impediment to removing or relocating structures is a shorefront road, whether the road is protected by a seawall or revetment or is itself the obstruction. Examples of road removal exist to provide templates for future actions. A road was removed on the Atlantic coast of Osório, Rio Grande do Sul, Brazil, and backfilled using sand similar in size to the existing foredune, increasing the width of the foredune up to 20 m (Portz et al. 2018; Box 10). Managers of Assateague Island National Seashore on the Atlantic coast of the USA relocated one threatened shorefront road landward, using shell and clay as construction material for the new road. Other roads and parking areas will be moved in the future and built with erodible shell, gravel, and clay instead of asphalt to reduce cost of construction, facilitate the decision to abandon or relocate roads when threatened, and eliminate problems of exotic debris when roads are damaged (National Park Service 2011; Nordstrom et al. 2016). A project conducted at Surfer's Point Park on the exposed Pacific coast of Ventura, California, USA (Kochnower et al. 2015), involved removal of concrete barriers, a concrete bike path, asphalt parking area with fill underneath it, and a riprap revetment on the beach. A new parking area and bike path were built

Box 10
Atlantida Sul Beach, Brazil: Adaptation to state legislation

The maintenance and recovery of the coastal dune system in the state of Rio Grande do Sul, Brazil, are carried out according to the Dune Management Plan, which mandates a Permanent Preservation Area of 60 m minimum width in urbanized zones. Publication of the plan in 2000 changed actions of the municipalities regarding recovery of this environment. The construction of a condominium behind the foredune at Atlantida Sul Beach required reconstruction of the dune system as compensation. Portz et al. (2018) concluded that the installation of sand fences alone would not attain the required 60 m wide preservation area because the shorefront road limited dune growth inland. Accordingly, the road was removed and the area was filled with sediments with the same granulometric properties as the current dune field. Tree branches or bark were then deposited to reduce sand removal by wind in the newly unvegetated areas, and sand fences were installed on the foredune. Native vegetation naturally colonized the restored area. An outfall channel through the dune was contained in a structure that was subsequently buried under the foredune. The gain in the dune width from these actions was 1–20 m.

Contributor: Luana Portz, Universidad de la Costa, Colombia

Figure 5.2 Atlantida Sul Beach in January 2012.

Figure 5.3 Atlantida Sul Beach in August 2020.

farther landward, and a cobble berm was placed in the excavated area and buried under reconstructed sand dunes. These adjustments are compromises that attempt to maintain traditional recreational uses and provide facilities close to the water while providing for restoration of natural processes. Not surprisingly, further beach retreat occurred, eroding parts of the new bike path, necessitating planning for the next phase of the project (Griggs and Patsch 2019).

In some cases, safety can provide the rationale for eliminating deteriorated structures. A deteriorating steel bulkhead and a deteriorating wood bulkhead in

Gateway National Recreation Area, New York, USA, were modified to reduce the threat to beach users and reestablish the connection between the beach and upland. These structures were built to protect facilities landward of the beach that no longer are in use. Rehabilitation funds made available following damage from Hurricane Sandy in October 2012 provided the opportunity to address the safety and erosion problems. The management alternative selected was to cut the bulkheads to an elevation below the surface of the active beach. This alternative was considered cost effective and compatible with the need to respond to future climate change. Cutting the bulkheads to a lower elevation offered the best chance of improving natural coastal processes and sediment pathways and maintaining habitat connectivity between the beach and upland while also conserving the historic fabric of the bulkheads (considered a cultural resource because of their age) by leaving intact portions in place below the beach surface.

5.4 The Case for Not Stabilizing Bluffs

Coastal bluffs can provide a unique habitat for plants tolerant of physical stresses such as salt spray, poor soils, and strong winds. Bluffs can support rich assemblages of invertebrates, including species restricted to that environment (Howe 2015; Rees et al. 2015). Bluffs are a source of groundwater, nutrients, and organic debris and provide critical roosting and nesting habitat for seabirds and birds that dwell in cavities and ledges (National Research Council 2007). Eroding bluffs often reveal a diverse mosaic of microhabitats in close proximity to each other. Bare surfaces exist in the vertical scarps created by slope failures and the fresh cones of debris below these scarps. The portions of the bluff that remain stable for several seasons can establish a vegetation cover, contributing to surface variety. The vegetation cover and root mass at the top of the bluff can create an overhanging canopy that provides shelter to the upper portion of the bluff and allows access to roots and subsurface characteristics that would not be exposed otherwise. Water table outcrops in the bluff face provide sources of ground water that would otherwise not be available. Digging by fauna creates excavations that provide local patches of moist surface juxtaposed against adjacent dry surfaces. The texture and color of the patchy microhabitats give eroding bluffs aesthetic appeal (Figure 5.4).

Shore parallel protection structures at the base of bluffs decouple beaches from their sediment sources and, as in other coastal locations, place a physical barrier between what would otherwise be two mutually evolving natural habitats. Risk to upland habitats should be viewed in terms of loss of the naturally functioning ecotone and natural scenic beauty caused by placement of a structure at the beach–upland boundary and not just in terms of loss of area of upland that is the

Figure 5.4 Eroding bluff at Monomoy National Wildlife Refuge, USA, showing habitat variety on an active bluff face.

traditional rationale for shore protection. It is doubtful that most bluff tops have habitat that is as rare as the eroding face of the bluff itself.

The first step in deciding whether to stabilize a bluff should be to determine the potential for moving threatened infrastructure, although this option is rarely implemented. Siddle et al. (2016) provide an interesting case study of a site in North Yorkshire, England, where houses threatened by bluff failure were demolished and residents were relocated to housing plots farther landward. Getting residents to accept relocation was facilitated by the availability of land nearby and the consideration that structural protection would be less cost effective and constrained by the designation of the seaward face of the vegetated bluff as a Site of Special Scientific Interest (SSSI) by Natural England (Siddle et al. 2016).

If relocation of threatened structures is not an option, alternative designs for bluff protection may be developed. Alternatives are presented in Nature Conservancy Council (1991) and include lower rock sills and permeable timber palisades on the beaches fronting bluffs, and creating separate strong points at some locations, resulting in an irregular plan shape. Some methods have the added advantage of being considerably cheaper than protecting the entire length of cliff (Brampton 1998) and can create greater variety of habitats.

Groins have been used at the base of bluffs for decades to provide partial protection (Brampton 1998), but there is increasing interest in abandoning their use. In Mecklenburg-Vorpommern, groin fields fronting some eroding bluffs not backed by human structures are being allowed to deteriorate to supply sediment to adjacent beaches (Nordstrom et al. 2007c). Less mobile bluffs are believed to act as headlands (hinge points) that help stabilize adjacent eroding lowlands and prevent a change in orientation of the coast, so they are not allowed to erode. Breakwaters are often used at these locations as well as groins and beach fill. The breakwaters are designed to allow enough wave energy to pass them to facilitate transport of sediment to downdrift beaches. The sediment bypassing breakwaters is not derived from bluff erosion; beach fill is used to protect the toe of the bluff and provide material for transport. Sediment transfers are maintained, but these bluffs do not have a natural form characterized by active cliffs, slumps, debris falls, and patchy surfaces with vegetation mosaics that occur on actively eroding bluffs. The bluffs could remain in a more natural state if the only means of shore protection is artificial nourishment applied at a rate that corresponds to natural beach losses. The base of the bluff would then be subject to erosion during extreme events but not as much as under the preexisting conditions (Nordstrom et al. 2007c).

5.5 Managed Realignment for Beach and Dune Environments

5.5.1 Feasibility

Human adaptation to coastal hazards by retreating from the coast is limited, despite the increased interest in this management option (Abel et al. 2011; Morris 2012; Roca and Villares 2012; Berry et al. 2013; Niven and Bardsley 2013; Cooper and Pile 2014; Fouqueray et al. 2018; Gesing 2019). Retreat can be reactive rather than proactive (Ledoux et al. 2005). Removal or abandonment of shore protection structures now occurs mostly on low-energy coasts, where salt marshes are the dominant natural environment (French 2006; Garbutt et al. 2006; Rupp-Armstrong and Nicholls 2007), although managed realignment may be economically efficient only at selected sites along these coasts (Luisetti et al. 2011). Plans for retreat on more exposed coasts have been suggested (Pranzini 2013; Kousky 2014; Nordstrom et al. 2015) but often not implemented. Demonstration sites that document the value of deconstructing or abandoning massive coastal defenses on exposed coasts are lacking (Cooper and Pile 2014), and decisions for proactive retreat may have to be made on feasibility assessments (e.g., Nordstrom and Jackson 2013; Nordstrom et al. 2015) rather than assessments of actual retreat.

Determining the potential for managed retreat involves (1) identifying the general rationale for retreat; (2) using demonstration sites to document the feasibility of accommodating retreat; and (3) identifying the kinds of geomorphic and ecological changes that will occur and the advantages of allowing those changes to occur (Nordstrom et al. 2015).

The loss of private properties and income from commercial establishments and difficulties of relocating existing infrastructure may make plans for retreat unacceptable to stakeholders (Abel et al. 2011; Niven and Bardsley 2013). The retreat option should be most feasible where few structures or stakeholders are directly affected and the costs of compensating owners are minimized (Rupp-Armstrong and Nicholls, 2007). As a result, most examples of the potential to create beach environments through retreat are in locations where houses and commercial establishments are not threatened and the beach is managed for public use, especially where land use change has eliminated the original need for protection (Nordstrom et al. 2016).

Determining priority for removal of a shore protection structure or allowing it to deteriorate is assisted by using consistent decision-making criteria (Table 5.1). Reasons for recommending removal include (1) infrastructure landward is movable or has no clear historical value; (2) the resulting sediment sources and space will be sufficient for new landforms and habitats to form and evolve; and (3) permission by regulatory agencies and stakeholders will be forthcoming. Reasons for not recommending removal include (1) threat to key infrastructure in use (e.g., houses, commercial structures, bridges, or roads) or to a historic feature, even if that feature is unusable; (2) limited potential for new habitat if the structure is removed; (3) threat to important feature just outside the jurisdiction; (4) potential for pollutant release; (5) difficulty of obtaining permits from regulatory departments; (6) lack of sufficient data to make informed decisions; (7) lack of funds; and (8) simple reluctance to allow any erosion or changes to the status quo to occur (Nordstrom et al. 2016).

5.5.2 Addressing Uncertainty and Stakeholder Concern

Uncertainty about long-term outcomes can play a major role in resisting managed realignment (Abel et al. 2011; Esteves and Thomas 2014). Lack of data is a common reason for forestalling management decisions (Jantarasami et al. 2010). Uncertainty is increased when climate change and natural dynamics are included in solutions (Cooper and Lemckert 2012; van den Hoek et al. 2012). Future changes can be modeled, but the new processes and end stage may not resemble target characteristics, especially where the starting point is not natural (French 2006; Mossman et al. 2012). The concept of an "end" stage may not even be applicable in a dynamic coastal environment (Nordstrom et al. 2016). Nevertheless,

Table 5.1. *Decision support criteria to retain hard structures or remove/allow to degrade*

Factor	Not favoring removal	Favoring removal/degradation
Buildings/ infrastructure landward are in use.	Buildings/infrastructure are outside local jurisdiction, not easily movable, or place- or water-dependent; space for habitat is limited.	Buildings/infrastructure movable; space available for new natural habitat.
Buildings/ infrastructure landward are not in use.	Structure has historic value; permits for removal difficult to obtain.	Habitat or historic value unknown or unclear or not constrained by site-specific location.
Availability of space landward.	Little ability of new habitat to migrate farther landward through time.	Space for restoration and further evolution of water-dependent habitat is available.
Type of coastal formation.	Toxic fill that is difficult to remove; steep slopes or bedrock that limit evolution of landforms; erodible substrate fronting key infrastructure.	Low land with space and unconsolidated sediment favoring ready reformation and migration of new landforms and habitats.
Construction material of protection structure.	Environmentally benign chemically; not a safety hazard; massive and difficult to move or porous and does not restrict water/sediment flows (e.g., pier).	Material toxic to biota or favored by invasive species; safety hazard for humans; material of high quality that can be reused elsewhere.
Significance of existing landforms and habitats.	Existing landforms and habitats are similar to what would occur if structure is removed.	Existing landforms and habitats are truncated, unable to migrate, or not native.
Ease of obtaining permission and removing structures.	Economic cost assumed high; overcoming adverse environmental impacts challenging.	Small, easily moved components; clearly demonstrated environmental advantages.
Stakeholder concerns.	Stakeholders have commercial/ legal interests that would be threatened and cannot be easily compensated.	Concerns are based on convenience or past precedent and not on current/future needs, commercial or legal interests, or water-dependent uses.
Institutional capability.	Lack of data, funds, or human resources; fear of unknown consequences; lack of appreciation of the benefits of erosion.	Forward vision; ability to obtain institutional resources; acceptance of outcome, even if unexpected.

(*Source*: Nordstrom et al. 2016).

uncertainties should not deter efforts to provide space to accommodate the unknown changes that will occur (Barbour and Kueppers 2012; Bell et al. 2014). Adaptive management will be important, given the uncertainty in forecasting future conditions (Harris et al. 2012).

Changes to shore protection practices can be resisted by stakeholders (Morris 2012; Roca and Villares 2012), even those outside the boundaries of the managed zones (Nordstrom et al. (2016). Stakeholders can simply prefer their familiar cultural landscape to a primordial natural landscape (de la Vega-Leinert et al. 2018). The human preference for the status quo will always be a deterrent to allowing change, whether that change restricts beach dimensions (Ardeshiri et al. 2019) or accommodates freer interplay of natural processes by removing or restricting structures (Leafe et al. 1998; Tunstall and Penning-Rowsell 1998; Nordstrom et al. 2007c, 2015), but the status quo cannot be easily maintained in a coastal environment. Uncertainty about how changes related to climate will affect the coastal landscape and its use can lead to inaction, but providing scenarios can make future changes more meaningful and acceptable (Lorenzoni and Hulme 2009). Good demonstration sites are suggested to aid stakeholder acceptance of strategies to allow the coast to evolve naturally (Parrott and Burningham 2008; Nordstrom et al. 2015, 2016). The initial sites for removing protection structures may have to be in locations where change will be most easily predicted or most clearly demonstrated after a major storm has established natural features farther landward.

Rejection of the retreat option is largely based on the assumption that the functions and services provided by the existing situation are better than those that will be provided in the future. Change is not the same as loss, although change will be perceived as loss until the values of the new features are appreciated (Nordstrom et al. 2015). Seeing what actually happens under managed retreat will provide more useful information than hypothesizing unknown future values. In any case, adaptive management could be used if retreat is allowed and the benefits are not realized (Nordstrom et al. 2015). Removing structures in the USA in Puget Sound and at Gateway National Recreation Area and relocating paths, roads, and parking areas at Assateague Island and Surfer's Point, identified in Section 5.3.1, are promising ways to demonstrate adaptation, and they provide excellent templates for other locations to follow.

Opposition to coastal retreat can come from natural resource managers as well as from managers of developed property. Loss of terrestrial habitat, particularly habitat protected by environmental regulations, can be a barrier to managed realignment (Goeldner-Gianella 2007; Rupp-Armstrong and Nicholls 2007; Nordstrom et al. 2015). Many freshwater bird sanctuaries and planted pine forest are human artifacts. These habitats are not always coast dependent and it can be argued that this habitat is less critical than the habitat that would replace it. Pine plantations, in particular, could be replaced by a vegetation type with greater diversity and value for conservation (Doody 1989; Sturgess and Atkinson 1993).

Enhanced funding and popularity of managed realignment is likely to occur if fostered by government authorities (French 2006). These authorities can take a leadership role in adapting to sea level rise and demonstrating how the impact of climate change can be addressed by interpreting their own sustainable practices. The National Trust in England and Wales decided to abandon sea defenses at some of its ocean coast sites, consistent with its policy of no intervention in natural coastal processes (Cooper and Pile 2014). Recent memoranda by the US National Park Service underscore the need to provide better data defining the importance of park assets relative to new hazards and allowing some built and natural resources to be abandoned rather than restored following storms (Nordstrom et al. 2016).

Storms are opportunities for passing legislative initiatives favoring adaptation (Fouqueray et al. 2018). Environmentally friendly landscape actions can result if prior planning and compliance are in place to take advantage of post-storm funding to remove damaged protection structures and facilities. Storm damage led to the projects at Gateway and Surfer's Point, mentioned in Section 5.3.1. The cost of removing structures and difficulties of conducting environmental assessments, obtaining permits, and addressing the concerns of stakeholders will take time and effort. Having examples of successful actions will aid implementation of future projects.

5.5.3 Allowing Protection Structures to Deteriorate versus Removing Them

Allowing structures to deteriorate may be perceived as benign neglect, but its formal adoption as a "Do Nothing Alternative" is actually a proactive approach to accommodating sea level rise where removal is difficult and there are no safety issues with structures left in the water (Nordstrom et al. 2016). Complete removal may not be required where safety is the primary concern, e.g., cutting down the bulkheads at Gateway, mentioned in Section 5.3.1. Deteriorating buildings, asphalt roads, and toxic fill landward of walls could be more problematic than stone or wood components of shore protection structures that are designed to be compatible with an aquatic environment. Irregular and sharp surfaces on deteriorating sheet piles, steel fittings, and cables used to construct bulkheads would be problematic because they would be difficult to see when submerged and could be hazardous to beach users and boats. Sediments used as backfill landward of many bulkheads and seawalls may contain pollutants if built prior to adoption of strict environmental regulations. Safety considerations may require removal of backfill material, even if the deteriorated structures are not removed. Removing structures that are already in the water may be difficult where strict environmental regulations require a comprehensive impact assessment.

5.6 Burying Hard Structures

Burial of hard shore protection structures can reestablish the connection between the beach and landward habitats, restore some of the natural process-response relationships between waves and currents and faunal interactions, and increase the aesthetic and recreational value of the shore. Burial of shore protection structures can occur as a by-product of beach nourishment operations, either by direct burial during the fill operation or later by aeolian action off the widened beach (Kana et al. 2013; Tresca et al. 2014; Kana and Kaczkowski 2019; Nordstrom 2019). Burial may also occur by natural cycles of accretion. Construction of new protection structures often occurs soon after intensive storms and after migration of tidal inlets exposes new portions of coast to erosion. In both cases, wave action may restore beach width during the depositional phase of these natural cycles, providing a greater source width for aeolian transport (Nordstrom 2019). Even large seawalls can be covered after the local sediment budget is enhanced (Irish et al. 2013; Smallegan et al. 2016; Nordstrom 2019). The likelihood of burial after nourishment projects is decreased if the new beaches are built too high for wave action to rework the new backshore and eliminate the coarse surface sediments that resist aeolian transport (as identified in Chapter 2). The absence of dunes seaward and landward of low walls on a widened beach is often more a function of human actions to eliminate the incipient dunes than the incompatibility of the walls with aeolian processes (Nordstrom 2014). Sand may even accumulate on top of walls (Figure 5.5).

Beach nourishment projects are increasingly deployed seaward of seawalls to provide extra protection to the wall and restore recreational benefits (Kana and Kaczkowski 2019, Box 1, Figure 5.6). Aeolian transport off these nourished beaches, and the resulting location and form of the dunes, can be controlled by human actions to keep the dune seaward of structures or expand the natural environmental gradient across them. Allowing slow natural accretion to occur is feasible if nourishment occurs seaward of protection structures designed as primary protection, where the time required to build a dune is not critical. Direct burial using earth-moving equipment may be required to reestablish dunes where beaches are too narrow to provide a sufficient source of wind-blown sand or the hard structure is too small to provide primary protection. US Army Corps of Engineers beach and dune nourishment projects conducted in New Jersey include dunes built to heights that exceed the height of most existing protection structures. Construction of these dunes will increase the potential for subsequent burial of the protection structures landward of them. In some municipalities, the Corps design calls for burial of the existing protection structures (Kana and Kaczkowski 2019; Nordstrom 2019).

Figure 5.5 Wind-blown sand accumulating on the revetment at Leffrinckoucke, France. The beach here is narrow at high tide, but delivery of sand is accentuated by a dominant wind that is nearly parallel to the shoreline and a large beach nourishment operation about 4 km upwind.

Figure 5.6 Foredune on nourished beach fronting seawall and promenade at Knokke-Heist, Belgium.

In many circumstances, the choice of a coastal risk reduction measure may not be an either/or decision between a traditional hard structure and a nature-based approach but a hybrid of both (Pontee and Tarrant 2017; Hofstede 2019; Almarshed et al. 2020). These projects can take advantage of the best characteristics of built and natural elements, allow for innovative designs, provide benefits besides coastal protection, provide a greater level of confidence than natural approaches alone, and be used where little space for a purely natural approach exists (Sutton-Grier et al. 2015). Many hybrid projects are of small scale and appropriate to provide temporary protection until larger or costlier protection projects are initiated (d'Angremond et al. 1992; Feagin 2005, 2013; Antunes do Carmo et al. 2010).

Placing resistant cores as backup protection in dune building projects is an example of a hybrid strategy appropriate to eroding sandy beaches where space is restricted. The cores can include geotextiles (Gibeaut et al. 2003; Feagin 2005, 2013; Antunes do Carmo et al. 2010; Harris and Ellis 2020), gabions – which are wire mesh baskets or mattresses filled with cobble-sized stone (d'Angremond et al. 1992; Furmanczyk 2013; Nordstrom 2019) – clay (Wamsley et al. 2011), and rock or concrete units (Basco 1998). The growth of geotextile container technology has been rapid (Saathoff et al. 2007), with increased use as shore protection systems (Hornsey et al. 2011). Geotextiles are often considered a soft solution that is more environmentally and user friendly than rock or concrete and more temporary, making them appealing to local stakeholders and environmental regulatory agencies (Saathoff et al. 2007; Hornsey et al. 2011; Corbella and Stretch 2012). Definition of hybrid structures as a soft solution is not strictly correct when cores are employed to resist erosion (Corbella and Stretch 2012), so reburial is critical where designation as a soft solution was a criterion for project acceptance (Nordstrom 2019). Burial is also necessary to maintain the integrity of backup structures. Geotextiles can be damaged or degraded by ultraviolet radiation, abrasion by coarse sediment and driftwood moved by waves and currents, vandalism, shifting of sand, or escape of sand from the containers (Saathoff et al. 2007; Antunes do Carmo et al. 2010; Corbella and Stretch 2012).

The sand cover should be designed to maximize its protective, environmental, and social value and support vegetation to stabilize the surface and provide natural habitat (Feagin 2005; Nordstrom 2019). The sand layer should not be so thin that it prevents animals from burrowing into it (Nourisson et al. 2014) or allows dune vegetation to root into the core and reduce its structural integrity (Corbella and Stretch 2012). Geotubes have been buried to depths of 0.5–1.5 m (Feagin 2005; Kessler 2008; Schreck Reis et al. 2008; Antunes do Carmo et al. 2010). The state of Florida requires maintaining a minimum depth of 1 m of beach quality sand, planting stabilizing vegetation, and requiring that the reconstructed dune does not cause significant impact to state or nationally listed threatened or endangered species (Hirsch et al. 2019).

Plantings can be vegetation types that reflect the cross-shore zonation of environmental stresses, e.g., *Panicum amarum*, *Sporobolus virginicus*, and *Spartina patens*, placed from seaward to landward, respectively (Feagin 2005). The widths of the vegetation zones are likely to be much narrower than on natural dunes, and the seaward slope of the artificial dune is likely to be steeper (Feagin 2005). Attempts to achieve species diversity in initial planting programs may not be needed or recommended where wave attack periodically removes the sand cover. Few ecological studies have been conducted on hybrid projects, but initial results on turtle nesting indicate that well-designed geocore projects may minimally affect nesting success (Hirsch et al. 2019). It should be emphasized that hybrid solutions may look like natural dunes, but they cannot evolve like natural dunes in the long term because of spatial restrictions and lack of sediment; episodic beach nourishment is likely to be required, especially after storm impacts (Almarshed et al. 2020).

5.7 Nature-Based Shore Protection Alternatives

The expressions "nature based," "building with nature," "green infrastructure," "ecological engineering," and "living shorelines" are often used to describe shore protection strategies that combine ecological and morphological elements to provide erosion or flood protection as an alternative to purely structural solutions. The generic term "nature based" may be used for these kinds of projects, although definitions can differ between countries and disciplines, and case studies of solutions reveal a great range of approaches that can be used (Narayan et al. 2016a). For example, traditional beach nourishment and dune building projects are sometimes called nature based, although vegetation enhancements may not be part of the design. The expression "nature based" may be more useful as a check on the extent to which ecological features are addressed, either initially or in expected long-term outcomes.

Nature-based projects are becoming popular because they appear to offer multiple social, economic, and environmental benefits not provided by static hard structures (Narayan et al. 2016a,b; Conger and Chang 2019; Morris et al. 2019; Powell et al. 2019). The natural habitats that provide this protection can include mangroves, salt marshes, coral reefs, and sea grass/kelp beds (Narayan et al. 2016b). Even vegetation on the backshore and dune (Feagin et al. 2015) and environmentally friendly structural engineering, such as adding surface complexity to construction materials (Moschella et al. 2005; Chapman and Underwood 2011), can be considered nature based. Nature-based protection strategies that use only natural features require large amounts of space (Pinsky et al. 2013; Salgado and Martínez 2017), which may not be available in developed areas. As a result, many projects designed for shore protection are hybrid projects that combine structures with natural features and are employed in low-energy environments.

The protection standards provided by a nature-based solution are less well understood than an engineering structure because of the variability and dynamism of natural habitats and the lack of detailed comparative evaluations of successful projects against alternatives (Narayan et al. 2016a). Information on projects is rapidly accumulating, and descriptions of many projects can now be found online. An overview of natural and nature-based solutions for coastal implementation is available, classifying solutions, developing performance metrics, monitoring and managing projects, and addressing policy changes (Bridges et al. 2015).

The terms "green infrastructure" and "living shorelines" appear to be appropriate where vegetation or faunal built reefs are the primary component of the protection. Using bulldozed or fenced dunes to provide protection is also a nature-based solution (as is beach nourishment), although the terms "living" or "green" appear to be less appropriate for solutions that are primarily geomorphically based. The more traditional and large-scale beach and dune building projects are rarely termed "nature based" and are addressed in Chapters 2 and 3. This section is devoted to projects where vegetation provides the primary benefit against wave erosion, which is only appropriate in low-energy environments. Designs that favor habitat development on hard structures are described in Sections 5.3 and 5.6.

A "living shoreline" is a type of estuarine shoreline erosion control popularized by many environmental groups and regulatory agencies in the USA as an attempt to incorporate natural habitats into stabilization designs (Currin et al. 2010). Living shorelines are not appropriate as primary protection against the effects of storm waves, high tides, or long-term sea level rise in high-energy environments (Griggs and Patsch 2019). Even in low-energy environments, living shoreline projects often require low-crested breakwaters and sills to protect vegetation plantings that, in turn, stabilize the upper intertidal zones (Currin et al. 2010; Scyphers et al. 2011). Use of the expression "living shorelines" conveys the impression of a natural solution, which may be appropriate if natural reef building organisms are used to provide the stabilization, e.g., oyster reefs (Scyphers et al. 2011). The concept is misleading where artificial hard structures not found in the substrate are added to provide the shelter to favor vegetation growth (Pilkey and Cooper 2012). Many living shoreline projects replace an eroding sandy beach habitat with a marsh, but many estuarine shores are sufficiently energetic that vegetation cannot survive at the waterline (Gedan et al. 2011). These locations are usually fronted by sandy beaches. If permanent structures are needed to allow vegetation to become established, it can be argued that a living shoreline solution is inappropriate because it could never resemble a previous condition undisturbed by humans, and it would not be sustainable without ongoing human efforts. Even where the eroding shore is a marsh, it is not axiomatic that background conditions will be similar to those that allowed a marsh to form in the first place. The seed sources and sediment inputs under urbanized conditions cannot be expected to match those prior to urbanization. In these instances, it is important to provide the rationale for eliminating the beach habitat that exists on the eroding shore prior to project implementation and explain how the new project will be more resilient than the previous beach habitat (Nordstrom 2014). Sandy beaches are also a threatened habitat (Schlacher et al. 2006; Defeo et al. 2009), particularly in estuarine environments, where the uplands capable of supplying sediment are favored for human development (Nordstrom 1992). The advantages of implementing a living shoreline solution for marsh construction are documented elsewhere (Davis et al. 2015; Bilkovic et al. 2016; O'Donnell 2017) and are not elaborated in this book, which is concerned with beach and dune environments.

5.8 Concluding Statement

Protection structures built in the past will become less effective as they deteriorate or their design elevations become exceeded because of sea level rise. Many landforms and habitats that are now stabilized will be subjected to increased erosion or flooding whether shore protection structures remain in place or are removed. Protective walls could be removed or allowed to deteriorate through time to allow natural processes to prevail. Erosion and displacement of landforms and habitats may be fundamental to the future sustainability of coastal ecosystems. Reevaluating past actions that used shore protection structures to restrict natural processes and habitats is important to allow for future actions unfettered by past decisions. This reevaluation and consideration of coastal retreat is facilitated by an objective decision framework (Nordstrom et al. 2016).

Local morphological or hydrodynamic conditions may make it impossible to use natural species for coastal protection, but protection structures can be designed to enhance ecological value by burying them or enhancing surface complexity; these approaches may not reduce costs, but they may mitigate adverse ecological impacts while helping to obtain permits and gain acceptance by stakeholders (Borsje et al. 2011). The need for any nature-based solution in a developed area is, in itself, admission that the erosion condition is problematic. The lack of space seaward and landward of the location where protection is most needed necessitates a compromise solution that should be considered temporary. At some point, storm damage can be expected to exceed the design specifications of small-scale projects, so plans should include subsequent actions, including options for retreat from the coast as well as providing a greater level of protection in place. Designs cannot be considered resilient without considering long-term needs in addition to short-term advantages. Performance standards and evaluation protocols (including rigorous monitoring) and specific contingencies for future repairs or modifications (adaptive management) should be built into the original project, especially for innovative projects with untested outcomes (Nordstrom 2019).

6

Options in Spatially Restricted Environments

6.1 Overview

Coastal environments should be managed as large units, taking adjacent areas into account (van der Meulen and Salman 1995; Gann et al. 2019), but restoration of full geomorphological and ecological gradients across the shore and lengthy corridors along the shore may be impossible because of spatial restrictions and fragmentation of habitats caused by humans (Rhind and Jones 1999; Doody 2001; Pethick 2001). Abiotic thresholds, physical barriers, and disturbances create boundaries in nature (Gosz 1991; Johnstone et al. 1992; Risser 1995). Human impacts often sharpen these boundaries (Correll 1991), segment them by land use or ownership (Forman 1995), halt natural flows across them (Harris and Scheck 1991), and create new flows (Bennett 1991). Every ecological restoration project can have beneficial outcomes regardless of size (Gann et al. 2019), and even projects at the scale of private properties can overcome some of the local boundaries caused by structures.

This chapter identifies the way cross-shore gradients are altered where houses are close to the water and the way natural habitats can be accommodated in environmental gradients that are truncated, compressed, decoupled, or fragmented. The ongoing conversion of natural landscapes to human use increases the need for maximizing use of remaining undeveloped enclaves and converting developed holdings to naturally functioning environments. Phased restoration of small tracts is possible and can be advantageous in the short term to address local problems with sediment budgets and flood risks, provide a sense of nature to homeowners and tourists, and institute long-term changes in thinking about the value of natural habitats and adaptive policies (Pethick 2001; Feagin 2013).

The degree to which restored landforms and habitats in developed areas can approximate natural ones depends on the space available and how natural processes and landforms are allowed to function. Ideally, nature should emerge from management efforts rather than be created by them (Simpson 2005), but active human inputs may be required where the temporal and spatial frameworks are compressed or fragmented into parcels managed by stakeholders with competing goals (Hesp and Hilton 2013). Proactive human efforts are required to maintain sediment budgets, provide space, and generate appreciation of the intrinsic, recreational, and educational aspects of nature (Brown et al. 2008).

Developed coasts often have a publicly managed (usually municipal) zone to seaward and a conspicuously different privately managed zone to landward. These zones are often separated by a shore parallel structure, such as a seawall, bulkhead, promenade, barrier against wind or blowing sand, or a trough landward of the dune crest, where sand is removed to minimize deposition on private lots (Figure 6.1). In some cases, publicly managed protective foredunes extend onto private lots (Figure 6.2), either because no shore parallel structure interferes with sediment transfers or because managers make no attempt to prevent natural aeolian accretion onto or over the structure. Attempts can be made to convert artificially differentiated management zones to a cross-shore gradient that contains the

Figure 6.1 Ocean City, New Jersey, USA, showing separation of the dune and private lots by three lines of cultural features – a landward sand fence, an unvegetated trough, and a bulkhead.

Figure 6.2 Ship Bottom, New Jersey, USA, showing integration of the dune across municipally managed and privately managed lots.

microhabitats found in undeveloped segments of the shore, from pioneer species and incipient dunes on the backshore to trees and woody shrubs in the more stable landward environments. The potential for initiating return to habitats that have been eliminated is illustrated in this chapter by examining the kinds of environmental gradients that are achievable on a developed shorefront where space is restricted. Some strategies accommodate natural processes and others restrict them to achieve specific restoration goals. Allowing nature to take its course is not always possible on developed coasts.

6.2 Natural Gradient

Natural gradients (Figure 6.3a) can occur on developed coasts where ample sediment is available between the water and structures and management strategies are compatible with evolution of landforms and habitats. The sediment can be derived from beach nourishment, but raking of the nourished beach must be prevented to allow the full suite of landforms and habitats to evolve. Natural disturbance is critical to the composition and richness of vegetation on beaches and dunes (Keddy 1981; Moreno-Casasola 1986; Barbour 1990; Ehrenfeld 1990), and the kinds and levels of disturbance differ with distance from the water on both sand and gravel beaches (Doing 1985; Randall 1996; Walmsley and Davey 1997b; Dech and Maun 2005;

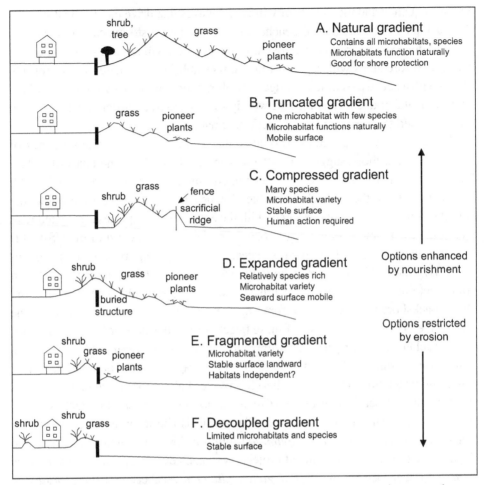

Figure 6.3 Alternative types of dunes on developed shores, representing examples of target states for restoration, modified from Freestone and Nordstrom (2001) and Nordstrom and Jackson (2003).

Lortie and Cushman 2007). Richness is diminished near the beach, where few species can tolerate the stresses of sand mobility and salt spray (Moreno-Casasola 1986). Pioneer plants (e.g., *Cakile edentula*), which are tolerant of salt spray and sand blasting, form embryo dunes on the backshore, and grasses (e.g., *Ammophila* spp. and *Spinifex* spp.) form foredune ridges (Hesp 1989; Seabloom and Wiedemann 1994). Landward of the foredune, protection from salt spray and sand inundation favors growth of woody shrubs in the seaward portions and trees and upland species in the landward portions. On natural dunes of coasts with a relatively balanced sediment budget, the transition from pioneer beach plants to fully mature forests can extend over environmental gradients of hundreds to thousands of meters, depending on the frequency and magnitude of winds and waves that drive the physical stresses (McLachlan 1990). Erosion of beaches and dunes during storms can eliminate the seaward dune subenvironments. Post-storm deposition will replace sediment on the beach, and foredunes will begin to reform, reestablishing the natural gradient. On shores with a negative sediment budget, the elimination of the seaward portion of the environmental gradient can place the woody shrubs and trees closer to the beach and the associated stresses than would occur on a natural shore.

Ideally, restoration programs should replicate the natural cross-shore zones of vegetation (e.g., those suggested by Rodrigues et al. 2011 for the coast of India). Full expression of zones can only happen in eroding developed areas if beach width is restored. It is theoretically possible for beach nourishment to provide the conditions for formation of a gradient with the entire suite of species associated with a natural beach/dune system (Figure 6.3a). On the northeast coast of the USA, this scenario would favor threatened species, such as piping plovers (*Charadrius melodus*), black skimmers (*Rynchops niger*), least terns (*Sterna antillarum*), and common terns (*Sterna hirundo*), that nest in or use the sparsely vegetated sand and shelly backshore surface and use foraging areas along the intertidal zone or in the wrack line. The dune landward of these beaches provides habitat for the American oystercatcher (*Haemotopus palliatus*) and horned lark (*Eremophila alpestris*), which use coastal beaches and sand dunes and forage on seeds and insects within low-lying vegetation. The unraked beach provides habitat for invertebrates, such as the northeastern beach tiger beetle (*Cicindela dorsalis dorsalis*) that makes use of the upper intertidal to high-wrack zone. Plants can also be accommodated, including seabeach amaranth (*Amaranthis pumilus*) found in interdune areas, and bare sand deposited from beach nourishment projects; seabeach evening primrose (*Oenothera humifusa*), seabeach knotweed (*Polygonum glaucum*), seabeach purslane (*Sesuvium maritimum*), and seabeach sandwort or sea chickweed (*Honckenya peploides*) can grow on the backshore or dunes (National Park Service 2005). Some of these animals and plants are sensitive to the presence of humans or direct trampling, but those impacts can be controlled by restrictions on beach use.

The height, width, seaward slope, and vegetation cover of the seaward foredune ridge reflects the interplay between storm wave erosion and aeolian transport. If this ridge evolves by natural processes, it will be low and hummocky in initial stages of evolution, such as soon after nourishment or after a severe storm has removed any earlier ridge. The foredune ridge may become relatively high and wide if several years pass without severe storms (Figure 3.1). A ridge that forms without the aid of a sand fence is likely to be broader and lower than a ridge formed using a sand fence, and the area of bare ground may be wider and patchier, but the ridge will provide a degree of shelter from wind and wave action in areas landward of it and increase the likelihood of colonization by new species.

Cost constraints on renourishment projects and competing human demands for the space created by the initial fill may not allow for construction of a beach of sufficient width for a natural dune and full vegetation gradient to form and survive. Dune restoration on a developed coast usually will occur across a much shorter cross-shore distance than a natural dune occupies. The issue then becomes how many of the human and natural values can be accommodated in the space available. Where space is critical and restricted by human structures landward of the beach, the dune may be managed to provide a dynamic and naturally functioning incipient dune microhabitat (truncated environmental gradient) or a spatially restricted sampler of a wider transect of a natural dune (compressed environmental gradient) that must be enhanced by human efforts (Freestone and Nordstrom 2001).

6.3 Truncated Gradient

Naturally evolving gradients truncated by human structures (Figure 6.3b) may consist of only the subenvironment of the dune characterized by pioneer plants, but that subenvironment provides habitat for nesting birds, seed sources for pioneer species that in turn provide food for fauna, and examples of the cycles of growth and destruction that underscore the dynamic nature of natural coasts (Nordstrom et al. 2000). Wave attack of the incipient dunes that form on the backshore is frequent, and these landforms do not survive long. Their small size and proximity to the water provide little shelter from salt spray and blowing sand. Plants characteristic of stable backdunes cannot thrive, and vegetation is characterized by species only found on the active beach and seaward portions of natural dunes. Buildings and grounds landward of the beach are subject to inundation by blowing sand, but residents and tourists retain their views of the sea.

A truncated gradient can be viewed as a no-action (do nothing) management approach that is also a viable restoration outcome. It represents an achievable initial attempt at restoration on coasts where managers now grade and rake the beach to maintain it as a flat, litter-free, vegetation-free recreation platform. To make this conversion, managers must first acknowledge that natural values, including the human values associated with nature appreciation, can compare favorably with previous recreational uses, and stakeholders must accept natural beaches as suitable recreation sites (Nordstrom and Jackson 2003).

6.4 Compressed Gradient

The cross-shore sequence of vegetation species found in natural areas can be represented in spatially restricted environments (Figure 6.3c), even if the distances cannot be representative (Nordstrom et al. 2002; Feagin 2005), but the species typical of the backdune environment can only exist close to the beach if growing conditions are enhanced by providing a relatively stable surface that is protected from inundation by sand, water, and salt spray. In compressed space, greater topographic variation can act as a substitute for greater distance from the shoreline by providing extra protection against coastal stressors (Bissett et al. 2014). Artificially maintaining a protective, sacrificial barrier on the seaward side of the foredune using sand fences can provide greater protection against wave overwash, flooding, and wind-blown sand and result in greater species richness for a given space than on truncated gradients.

A dune in Avalon, New Jersey, USA (Figure 6.4), reveals how a landform built and used as a protection structure can evolve into a condition that appears natural, at least in terms of surface vegetation. A state/municipal nourishment project in 1987 used sediment dredged from a nearby inlet to create a protective beach and sand dike that was shaped by earth-moving equipment to have a flat top at an elevation of 3.7 m above mean low water. The initial shape of the structure and the large amount of coarse shell and gravel within the fill revealed that it was not created by aeolian processes. Subsequent use of sand fences on the seaward side to

Figure 6.4 Avalon, New Jersey, USA, showing diversity of vegetation favored by high protective foredune and narrow eroding beach that prevent sand inundation landward.

build a high foredune to protect against storm flooding resulted in greater topographic diversity. The foredune is higher than a natural dune would be this close to the backshore, providing considerable shelter landward of it. The cross-shore zonation of vegetation (Figure 6.4) is roughly similar to that of a natural dune, but the environmental gradient is compressed into a much narrower zone, and backdune species are within only a few meters of the backshore. Woody shrubs, such as bayberry (*Myrica pennsylvanica*), break up the surface of the graded dike, increasing its aesthetic value and its value for fauna (Nordstrom and Mauriello 2001; Nordstrom et al. 2002). The dune in Figure 6.4 does not convey a truly natural image because of its engineered shape and altered spatial context, but it has great resource potential in its species diversity. The diverse vegetation on a compressed gradient provides what many local stakeholders would consider an aesthetically pleasing landscape. At times, a pleasing representative image may be preferable to an ecologically pure image as a means of involving more members of society and developing an appreciation for sustainability goals (Parsons and Daniel 2002; Özgüner and Kendle 2006). The compressed gradient provides a more reassuring image of geomorphic stability than is provided by a truncated gradient, giving it a utilitarian value that fosters acceptance of dunes where natural landforms are underappreciated. The increase in dune crest height required for stability may restrict resident views of the sea, which is likely to make the option unattractive to some stakeholders.

Active management can create a compressed gradient quickly, but this kind of dune must be maintained by ongoing human efforts on a narrow beach where wave erosion or aeolian transport contributes to dune instability. These efforts will have to include rebuilding dunes rapidly using bulldozers or replacing sand fences removed during wave erosion and using periodic (but often small-scale) beach nourishment operations to maintain at least a narrow fronting beach. The present attempts of many managers to maintain dunes as protection structures on eroding shores creates the kind of landform that would allow a compressed gradient to occur, but the full suite of species that is possible may not evolve and survive if the bulk of the dune (not just the sacrificial part seaward) is considered solely as a protection structure and thus is considered expendable (Nordstrom and Jackson 2003).

6.5 Expanded Gradient

Municipal managers who build or maintain dunes to provide habitat or storm protection for human infrastructure usually can only do so on the publicly owned portion of the beach, resulting in a dune that is small relative to its natural counterpart. Even if the privately owned landward portion of the dune is contiguous to the publicly managed portion, it can bear little resemblance to it or to a natural dune if managed according to suburban landscape tastes. Developers and property owners often replace native coastal vegetation with species that require human efforts to maintain them outside their natural range (Conway and Nordstrom 2003; Mitteager et al. 2006; Kumar et al. 2009). Studies of shores where private properties are just landward of the beach call attention to the value of residents accepting natural landscaping alternatives on their lots to provide wildlife habitat, enhance the image of a developed coast, influence future landscaping actions taken by residents, and increase the likelihood that natural features will be a positive factor in resale of coastal property (Conway and Nordstrom 2003; Mitteager et al. 2006). Humans benefit from a closer and reciprocal engagement with nature (Gann et al. 2019). Allowing the dune to expand onto private lots can expand the environmental gradient (Figure 6.3d) and engage owners in the restoration process.

Private coastal lots have received little attention from scientists and managers as potential restoration sites, but these lots are important because the cumulative effects of numerous individual management actions can be great. In some countries, individual residents have great freedom in selecting landscape options, and they do not have to wait for public funds or endure the many public meetings and reviews that government entities have to go through prior to their management actions. The division of land into many small lots results in a high manager to land ratio, so management can be more intensive than on municipally managed dunes. Watering, fertilizing, adding topsoil and mulches, planting alternative species, and constructing barriers of different sizes and configurations can be more practical.

Vegetation on private lots is often selected to recreate a suburban landscape. Sometimes it is used to demarcate a property line or to obtain privacy. Whether a lot owner will invest time and effort to plant or maintain natural vegetation is partly based on resident preference and partly on the constraints of natural processes. An approach that minimizes human alterations can result in the species typical of a dynamic environmental gradient (Figure 6.2), whereas, planting and watering can favor backdune species typical of locations much farther landward. These two options are somewhat analogous to the truncated and compressed gradients possible on municipal dunes and have related advantages and disadvantages.

Tree lines are sometimes used to demarcate the sides of properties. Viewed from the side, a shore perpendicular tree line may mimic the cross-shore gradient in height of vegetation, with lower vegetation close to the water, but trees of similar height planted in a line perpendicular to the shore can convey the image of a fence rather than a representative subenvironment. Many of the desirable changes to the landscape on private lots are more easily achievable if lot owners are made aware of the reasons. More detailed suggestions for managing private lots are made in Chapter 8.

6.6 Fragmented and Decoupled Gradients

Portions of dune may remain adjacent to buildings on cleared lots (Figure 6.3e and f), or longer segments of dune may remain in undeveloped lots as enclaves within developed coasts. If the patches of vegetation are allowed to evolve, they can have variety in size and age and contain nearly as many species as non-fragmented areas (Escofet and Espejel 1999). Dune remnants are a potential resource that should not be ignored in restoration efforts. The dunes may be separated from the beach by seawalls, revetments, bulkheads, promenades, boardwalks, buildings, lawns, roads, parking areas, or recreation surfaces (Figure 6.5), but this separation does not preclude establishment of native species that would occupy a stable backdune (Nordstrom and Jackson 2018; Nordstrom et al. 2018). A distinction is made here between fragmented dunes, where patches of dunes exist within a matrix of cultural features, and decoupled dunes, which stand as isolated landforms.

Decoupled dunes landward of shore parallel protection structures can survive long after beaches and seaward portions of these dunes are eliminated by wave erosion. Many remnants have vegetation species found on active foredunes (e.g., *Ammophila*), but reduction in the amount of sand that can be entrained on the narrow beaches or pass over structures can allow shrubs characteristic of static backdunes to thrive.

Figure 6.5 Decoupled dune at Leffrinckoucke, France.

Fragmented and decoupled dunes vary greatly in size, location, duration, and vegetation because they can occur under many land use scenarios and reflect many different landscaping preferences. In some cases, the human structures can appear to have little negative effect on the distribution of vegetation on either side of them (Figure 6.6). Both kinds of dunes may be remnants of the original dune that was modified to accommodate construction of human facilities, or they may be human-created forms, such as disposal areas for overwash or wind-blown sand cleared from lots or adjacent roads.

Dunes between properties and on the landward side of buildings are the most unnatural of all dune types in terms of vegetation and internal characteristics of the sediment (where shaped mechanically). The lack of dynamism on these dunes and their location relatively far landward make them suitable sites for growth of backdune species that require time to evolve, including trees such as American holly (*Ilex opaca* Ait.) (Nordstrom and Jackson 2003). Lot owners often view the seaward and landward sides of houses as two different management units, with the sides of the houses a transition between the two. Mitteager et al. (2006) found that owners considered the landward side of houses to be the front yard. Space for

Figure 6.6 The foredune at Rehoboth, Delaware, USA, showing a promenade with little effect on the cross-shore distribution of vegetation.

nature on this side is restricted by driveways and parking space for cars, and ornamental shrubs are often the only vegetation planted. There appears to be little incentive to adopt a natural appearance landward of houses if lots on the other side of the shore parallel road (and not in the dune zone) are maintained as suburban landscapes (Mitteager et al. 2006).

6.7 Implications

Some outcomes in Figure 6.3 are relatively easy to achieve, whereas others require changes in policy and practice. Natural gradients require space, and beach nourishment may be necessary to create this type of gradient in developed areas. A fully functioning dune is likely to be considered a luxury by managers who think a dune interferes with recreational use or is a sacrificial shore protection structure. Natural gradients fronting developed areas are now found only where environmental or safety regulations specify great setback distances for new construction or prevent construction or intensive recreational use on accreting areas, including nourished beaches.

Truncated gradients, with their natural dynamism, require no active management actions, and they are less expensive to maintain than their alternatives (a flat recreation platform or the protective dune of a compressed gradient). The dynamism that gives the truncated gradient its natural value is usually perceived as a negative characteristic by local managers. A public information program may be required to identify the advantages of letting wild nature occur or to allow the truncated gradient to evolve into an expanded gradient through landward migration (Nordstrom and Jackson 2003).

The compressed environmental gradient, with its species-rich inventory, its value for shore protection, and its location completely within the publicly managed part of the shore, is a common and preferred option in many locations. The cost of maintaining the dune as a primary barrier against continued storm wave attack using sand fences and bulldozing makes this option relatively expensive, but funding from national, state/provincial, or municipal sources is often available for maintaining the dune as a protection structure. The compressed gradient could theoretically evolve into an expanded (to landward) gradient if fences and bulldozing were no longer used to maintain the sacrificial ridge, but this action would signify abandonment of a public policy of providing shore protection to landward properties, which is not likely (Nordstrom and Jackson 2003). If the dune did migrate landward, the responsibility for maintaining the species-rich landward habitats would shift to private lot owners.

An expanded gradient represents an extension of a truncated gradient, with fragmented and decoupled gradients representing eroded forms of a former expanded gradient. A consensus among many stakeholders would be required to facilitate formation of expanded gradients along great lengths of coast. The cost of maintaining natural vegetation on the landward side of the dune crest could be spread among many stakeholders, and intensive management within each lot is possible, but each resident would have to be informed about ways to use compatible construction and landscaping practices, and there is no assurance that property owners will adopt them unless a municipality specifies this in its zoning ordinances. The expanded gradient has distinct advantages over the truncated gradient in achieving a more naturally appearing and functioning dune system, but the option is less easily achieved. Individual residents have exercised this option, demonstrating its feasibility over a longer segment of coast (Nordstrom and Jackson 2003). Buildings within dunes create localized scour and accretion zones and alter topography, migration rates, and vegetation assemblages from natural conditions (Nordstrom and McCluskey 1985; Hernández-Cordero et al. 2017; García-Romero et al. 2019). An expanded gradient that represents natural conditions may be more achievable where houses are elevated on pilings, and the threat of damage from inundation by swash and blowing sand are reduced. Elevation of houses on pilings is usually viewed in terms of protecting people against natural processes, but it can also be viewed as a way of accommodating natural processes and habitats. Full advantage of this option would require property managers to resist maintaining ground surfaces for uses more typical in suburban areas.

Fragmented and decoupled gradients provide the opportunity for nature to occur landward of protection structures, where it would not occur otherwise. Stakeholders often consider the first cultural feature to be the boundary between nature and human habitation. Maintenance of natural dune enclaves on private properties would help eliminate this distinction and provide a more compatible image of the coast as a natural environment. Fragmented and decoupled dunes may be the only vestiges of former dunes, which makes their preservation and enhancement critical, despite their small size and limited dynamism. Lot line dunes and dunes on the landward side of buildings that owe their existence to disposal of sediment are not remnants of natural dunes, and their value as targets for restoring nature is unclear.

The scenarios described in this chapter are short-term targets that may not occur without change in human perceptions and practices. The amount of space required to maintain coastal habitats in the future will have to be much wider across the shore–land transition than single lots (Burger et al. 2017), but acceptance of change to a more natural system in many locations will not be immediate and will depend in large part on precedents established for the first row of developed lots (Nordstrom and Jackson 2018). The ways programs can be implemented to achieve these spatially restricted restoration states, given stakeholder interests, are presented in Chapters 7 and 8.

7

Stakeholder Interests, Conflicts, and Cooperation

7.1 Obtaining Public Support

Public support and accountability are becoming increasingly important in restoring natural environments and adapting to climate change (Hickman and Cocklin 1992; Higgs 2003; van der Meulen et al. 2004; Wiens and Hobbs 2015; Gann et al. 2019; Jellinek et al. 2019; O'Donnell 2019), and the politics of sustainability may rest more on human perceptions and values than on the intrinsic worth of natural systems (Doody 2001). Many of the most critical issues are at the intersection of natural sciences, social sciences, engineering, decision science, and political economy (Kopp et al. 2019). Perceptions of the acceptability of coastal management actions can be polarized into ecocentric and anthropocentric views or along disciplinary lines (e.g., ecology, geology, engineering), and the different values, interests, and goals among stakeholders require considerable dialog and compromise before solutions are acceptable (Ariza et al. 2014; Lucrezi and van der Walt 2016; Prati et al. 2016; Roca and Villares 2018). Restoration is undertaken to satisfy personal, cultural, social, and economic values as well as ecological values, which can lead to improved resilience (Gann et al. 2019). Social justice and the concept of "who owns the coast" are important themes in coastal management (Stocker and Kennedy 2009). Protecting or restoring locations subject to intensive development pressure will be difficult without considering all stakeholders as part of the solution, especially where natural features are important for the livelihood of local residents. Stakeholders can help define the vision, targets, goals, objectives, and methods of implementing and monitoring restoration projects and can provide political and financial support for long-term project sustainability (Gann et al. 2019).

Incorporating stakeholder interests into restoration plans may be desirable and often necessary for project approval, but some stakeholders can thwart projects. Tourists and local residents can place greater value on human-altered elements of nature than on naturally functioning habitats and species, giving introduced species preference over the natural dune landscape (van der Meulen et al. 2004). Stakeholders may also place greater value on human-induced stability than natural dynamism, indicating that acceptance of sand movement as a natural process of dune evolution may require a fundamental change in attitude (Doody 2001; Charbonneau et al. 2019). The desire to maintain the status quo can override actions to improve natural environments. Remobilizing the protective dune in Mecklenburg-Vorpommern (Section 5.2) is an implementable option because it is acceptable to local stakeholders, but state managers were unsuccessful in overcoming resistance to a similar project on a nearby island, where those local stakeholders preferred a familiar landscape to an unknown dynamic one (Nordstrom et al. 2007c). This reluctance to accept change is noted elsewhere (Leafe et al. 1998; Tunstall and Penning-Rowsell 1998; Goeldner-Gianella et al. 2015).

Coastal managers tend to seek and accept technical advice based on precise predictions without caveats about uncertainties (Cowell et al. 2006), but restoration and adaptation to natural processes will have uncertainties (Wiens and Hobbs 2015; Nordstrom et al. 2016). Communication and collaboration are important when uncertainties are involved and are facilitated if both specialists and local stakeholders have a shared frame of reference with a common operational objective for multiple goals (van Koningsveld and Mulder 2004). Effective communication and information dissemination between governments and local interests can be seriously limited in some countries (De Ruyck et al. 2001; Barragán Muñoz 2003), but even if communication is relatively effective, restoration may only be achievable if it coincides with local interests and perceptions (Swart et al. 2001; Clarke and Rendell 2015; de la Vega-Leinert et al. 2018). Carro et al. (2018) present an example of capacity building of staff and stakeholders in a municipality in Santa Catarina State, Brazil, resulting from knowledge exchanges with decision-makers at the national level and scientists, followed by ecosystem-based adaptation by the local government. These actions occurred after innovations were made in the coastal management institutional structure.

Treating coastal problems with holistic and integral approaches greatly complicates the decision process (van Koningsveld et al. 2003). For nourishment projects, approval often comes after lengthy discussions and debates among competing interest groups and alterations to original project designs. Final designs are often compromises between engineering practicality, shore protection, recreational needs and ecological considerations. Conflict is always likely where decisions and choices are available (Myatt et al. 2003). Conflicts are not just confined to a simple dichotomy, such as nature versus economic or recreational interests. Disputes can occur between two pro-environment stakeholders. Bulldozing vegetation on gravel beaches to encourage breeding terns (Randall 1996) is an action that may appeal to some environmental interests but not others.

7.2 The Need for Compromise Solutions

Environmental debates need to be expressed in terms that will permit compromise, flexibility and pluralism of values with more pragmatic and policy-based approaches to human–nature relationships (Barrett and Grizzle 1999; Minteer and Manning 1999; Katz 1999; Norton and Steinemann 2001). Public participation is most effective if the process is transparent and honest and there is early participation by all interest groups, with continuous interactive consultation and opportunity for feedback (Johnson and Dagg 2003). Consensus is especially important in locations where the character of the coast will change dramatically. Legal battles such as those resulting from use of exotic sediment for beach fill (e.g., Nordstrom et al. 2004) can be diffused if stakeholders know in advance that the nourished beach will depart dramatically from their perceived ideal. Discussions can stimulate initiatives to obtain financial support for more suitable fill materials or add a treatment phase to make the borrow materials more compatible and convert single-purpose projects to multiple-purpose projects with greater restoration potential.

The success of compromise solutions can be enhanced by combining the skills of a range of specialists to identify objectives, design appropriate options, and monitor the methods used to refine subsequent designs and protection methods (Brampton 1998; Brennan 2011; Martínez et al. 2019). Compromise solutions that combine the need for shore protection or adaptation to coastal hazards with environmental benefits included can work, although they may be initially resisted. The key to acceptance seems to be (1) identifying and involving key participants from start to finish; (2) identifying areas of disagreement and forums for resolution; (3) expanding the conception of public good and appreciating the needs of all stakeholders; (4) identifying a sufficient number of alternative solutions so at least one will be acceptable; and (5) accepting increased costs or inconveniences where necessary to achieve added values (Nordstrom 2003). Acceptance is enhanced where engineers, ecologists, regulatory personnel, and local stakeholders are willing to work together, and the desire for an expedient solution encourages consideration of alternatives beyond traditional solutions (Zelo et al. 2000). Better plans and programs may require participating professionals and stakeholders to broaden their disciplinary, practical, or personal horizons and become curious, open, and respectful of other interests and approaches (Martinez et al. 2018).

7.3 Contrasts in Stakeholder Perceptions and Values

Opinions about how coasts should be managed are often polarized and can even be reduced to stereotypes. It is difficult to articulate another person's goal, even if it is

Table 7.1. *Contrasting human and natural optimums*

Human optimums	Natural optimums
Safety	Free interplay of sediments and biota
Good views	Topographic relief
Familiarity/predictability	Diversity and complexity
Stability	Cyclic accretion/erosion, growth/decay
Cleanliness	Wrack (including biodegradation)
Property demarcation	Cross-shore gradients, habitat zonation
Access (paths, roads, parking)	Unfragmented landscapes

(*Source*: Nordstrom 2003).
Human optimums favor desires; natural optimums favor geomorphological and biological diversity.

done with good intentions. A serious problem occurs when municipal managers base their strategies on their perceptions of the expectations of tourists, who often have no common topical agenda. Data on the wants and expectations of tourists and residents are often gathered but not necessarily sufficient to definitively conclude whether they will collectively accept a given management goal or accept a more naturally appearing landscape. Local management authorities and private interests often ally against conservation bodies in debates over how natural features and how much of their dynamism should be tolerated (Barton 1998, Leafe et al. 1998), but enough exceptions exist to make options to increase dynamism feasible.

Compromise solutions are required to allow for some of the dynamism required to maintain elements of natural coastal systems while providing people enough stability to maintain infrastructure or retain property values (Powell 1992; Brampton 1998). Table 7.1 presents a few of the more significant contrasts in human and natural optimums arranged as sets of opposing attributes, although it is acknowledged that coastal systems and human needs are too complex for mutually exclusive categories (Nordstrom 2003). Managing coastal landforms and habitats to have both natural and human values lies in viewing differences in these characteristics along a continuum rather than as bipolar opposites and then separating the characteristics and needs that are critical from those that are simply desirable.

The concept of safety encompasses major threats to life and property from storm waves and flooding that must be addressed and minor nuisances resulting from blowing sand that can be tolerated. Increasing the natural dynamism of portions of dunes seaward of private properties may cause only a minor nuisance. A good view from a shorefront residence is important to property owners, but the view can be from a sitting position in a downstairs parlor, from a standing position there, or from an upstairs room or viewing platform. Views can be accommodated at higher elevations within a residence while allowing a protective dune seaward of it to increase in height and provide a stable backdune environment to form in restricted space (Figure 6.3c). Access to the beach through the dunes may be desired by every lot owner, but it is unnecessary to make access this convenient if it results in trampling and fragmenting natural habitat and jeopardizing the value of the dune as a protection structure. The concept of cleanliness can be viewed in terms of aesthetic preference or a true health hazard, which would greatly influence the timing, location, and method of beach cleaning. Inconveniences can be tolerated in management strategies (both psychologically and economically), whereas major losses cannot (Nordstrom 2003).

A key element in establishing beaches and dunes as naturally functioning environments is redirecting managers toward accepting native species, natural litter, incipient dune forms, larger and more dynamic foredunes with greater topographic and species diversity, and greater amounts of blowing sand. This change in attitude applies to municipal managers operating in the public interest and to developers and shorefront residents who are often viewed as operating in their own interests. Restoration goals may not be achieved unless there is also a change in the attitudes of scientists, restoration specialists, and environmental regulators. These stakeholders may have to accept a less-than-perfect definition of what is natural or how a natural trajectory should be defined. Specific suggestions for changes in attitudes, goals, and roles that stakeholders must play are difficult to articulate without a management context, so examples are provided of actions that can be taken to achieve more naturally functioning dunes on a developed coast (Table 7.2).

Table 7.2. *Actions that can be taken to enhance natural value of beach and dune*

Municipal managers
Establish unraked segments alongshore and shore parallel zones across shore
Remove cultural litter manually in these no-rake areas
Avoid mechanical grading except for hazard prevention
Minimize vehicle use on backshore
Travel on prescribed routes
Minimize extent of sand-trapping fences
Manage municipal parks for nature experiences
Change expectations of tourists and residents by education and involvement
Establish environmentally friendly zoning ordinances

Developers/Property owners
Tolerate or accommodate natural processes or vegetation
Leave space for natural environments
Do not demarcate seaward property line
Adopt a sense of coastal heritage and appreciation for natural value
Provide a compatible coastal landscape image

Table 7.2. (*cont.*)

Scientists
Consider humans intrinsic to landscape evolution
Identify reference and target states for developed systems
Define the meaning of "natural" in developed areas
Adapt research to include smaller temporal and spatial scales
Respond to research needs of property owners and municipal managers
Ensure that the advisory process is two way
Engineers
Favor designs that enhance multiple aspects of the environment
Mimic natural landforms and favor diversity and complexity of habitats
Create structures capable of being altered to address future changes
Address needs or physical constraints of the broader region in addition to local needs
Integrate ecological and geomorphological functions over the lifetime of a project
Address stakeholder inputs in maintaining projects as well as planning and designing them
Evaluate applicability of nature-based alternatives for exposed coasts
Environmental advocates and regulators
De-emphasize preserving a static environmental inventory
Accept alternative natural trajectories
Emphasize restoration of multiple-use habitats in target species programs
Use stakeholder preferences to help develop restoration
Avoid restrictions that penalize environmentally friendly actions

(*Source*: Nordstrom 2003).

7.4 Stakeholder Actions

7.4.1 Municipal Managers

Municipal inertia can stifle actions to promote strategies to address effective coastal management (De Ruyck et al. 2001; Colenbrander and Bavinck 2017), but municipalities can also be proactive in instituting programs to enhance natural landforms and habitats (Nordstrom et al. 2002; Carro et al. 2018). Municipal programs may require a phased incremental approach, requiring small-scale projects to document proof of concept and obtain data to support large-scale projects that may eventually be supported by higher levels of government. An example is the phased approach taken by Myrtle Beach, South Carolina, to protect infrastructure using beach nourishment and dune building in a location where seawalls had been preferred (Kana and Kaczkowski 2019). The first phase involved scraping the beach to provide a low-cost alternative for protecting the most vulnerable shorefront buildings and parking areas. The second phase was a municipally funded beach nourishment project, using a 2 percent tax on hotel rooms. The earlier phases were important in demonstrating to the community and the state that beach nourishment was a viable alternative to building hard protection structures, and it provided interim protection until a federal (national) beach nourishment project was conducted. The federal project that comprised the third phase was large enough to provide a beach for shore protection and recreation but also bury shore protection structures (Figures 2.2 and 2.3), providing a more natural image of the shore (Kana and Kaczkowski 2019). The municipality banned construction of new seawalls in the early 1980s before the state banned them in the Beach Management Act passed in 1988, revealing that municipalities can take a lead role in demonstrating and documenting the advantages of policy changes for implementation at higher levels of government (Kana and Kaczkowski 2019).

Beaches and dunes created by national and state/provincial governments are often subsequently managed by municipal authorities, who play a crucial role in ensuring long-term success in restoring coastal habitat. Restoration projects in local communities can make environmental actions an everyday reality (Moreno-Casasola 2004) and transform the tourism experience. One of the most important actions is establishing undeveloped natural enclaves to serve as demonstration areas, breeding sites, and seed banks. This action may require funding from higher levels of government to defray costs of purchase of shorefront real estate, but municipal officials may have to initiate the process of obtaining these funds. Many of the municipal actions elaborated in Chapter 8 can be taken at little or no added cost, but municipal officials may have to work more closely with stakeholders to obtain public support.

Municipal actions are critical in changing the culture of coastal resorts. Municipalities can create an environmentally friendly posture through reestablishing natural environments and instituting programs of education and involvement that will change the expectations of visitors accustomed to landscapes modified by humans. Visitors learn best when they are engaged in at least some conservation effort (Hose 1998), so attempts should be made to enhance the role of tourists in environmental management and stewardship. Requiring visitors to haul their own garbage off the beach, rather than placing trash barrels right on the beach, accomplishes part of this goal while eliminating the need for maintenance workers to drive vehicles onto the beach to empty trash containers (Nordstrom 2003).

Converting municipal coastal parks to respond to ecological needs where they were given low priority is another option. This conversion often requires educating park staff and developing new maintenance skills to restore and sustain park landscapes as living systems (Cranz and Boland 2004). Removing infrastructure that encourages active recreation on raked, paved, or mowed grass surfaces (that can be accommodated on inland parks) and not relying on cultural features for aesthetic appeal would help change visitor expectations and favor gradual acceptance of municipal parks as a nature experience.

Post-storm improvements are often piecemeal and the work of individuals, not the whole community (Andrews 2016). Actions can be taken by municipalities to help private property owners contribute to a broader restoration program. Beaches and dunes on private lands are not accessible to the public, but restoration of ecosystem functions and services on these private lands can be considered a common good worthy of public action by local governments or environmental groups. Initiatives include (1) extending planting programs on municipal beaches farther landward, onto private properties; (2) educating residents and the professional landscapers they hire about the advantages of planting natural species; (3) providing tax credits or permit exemptions for natural landscaping; or (4) requiring use of native coastal species in municipal ordinances. Ordinances based on safety could highlight the value of native species in trapping and stabilizing sand, resisting erosion, and being more tolerant of salt spray than exotics (Nordstrom and Jackson 2018). Municipalities often have beautification funds that could support initial efforts, and local environmental commissions can provide encouragement and expertise to establish initial demonstration sites.

Municipal master plans provide a long-term vision for environmental protection, and lead to development of ordinances and a means of regulating uses on a frequent basis through permitting and enforcement. These mechanisms for control work best if management criteria recognize and protect ecological functions and processes and incorporate restoration programs and ongoing environmental monitoring, assessment, and adaptive management in accordance with new scientific findings (Best 2003). Maintenance of self-sustaining ecosystems can be incorporated into municipal plans, but effective programs are not readily achievable without higher levels of government providing state-of-the-art requirements, updated science-based guidelines, and funding (Best 2003; Portz et al. 2018).

7.4.2 Developers and Property Owners

Natural environments are often identified in advertisements used as marketing strategies for coastal real estate, but many of these environments do not survive the development process. The names given to streets and subdivisions in coastal municipalities, such as Bayberry and Dune Vista, often refer to natural features that are eliminated by the time the new properties are sold. Ways must be found to translate the value of natural features into real estate values, so the perceived loss of income associated with restrictions to access or views can be offset or diffused. Decisions affecting restoration include economic and political criteria (Jackson et al. 1995; Higgs 1997), so there will be little economic justification to restore natural environments unless they can be valued in those terms (Clewell and Rieger 1997).

Property owners often stall efforts to reinstate natural processes (Abel et al. 2011; Kousky 2014). A key element in a restoration plan is to change the attitudes of owners to make them accept more naturally functioning landscapes (Mitteager et al. 2006; Cerra 2017; Charbonneau et al. 2019). Regulations based on safety considerations or incentives based on aesthetic appeal, appreciation of natural heritage, or economic benefits may be required to get owners to take action (Nordstrom and Jackson 2018). Employing natural landscaping is most critical on the seaward edge of private shorefront lots, which can be one of the most conspicuous boundaries between wild nature and human habitation (Conway and Nordstrom 2003). Plantings of natural vegetation across the dune can mask the line separating private property from the public beach (Feagin 2005) and change the expectations of tourists who may currently think of the shore only in terms of human use. This change in perception may increase the likelihood of adopting future strategies to create sustainable landscapes palatable to tourists.

Restructuring the attitudes and actions of residents to accept natural landscaping can be difficult (Feagan and Ripmeester 2001) but is crucial considering the amount of coastal land they manage (Feagin et al. 2015). Conversations with homeowners interested in contracting for landscaping services indicate that they are not aware of (1) the advantages of using natural species for landscaping; (2) the great differences in viability of alternative species in the harsh dune environment; or (3) the high levels of maintenance required for species that are not adapted to coastal stresses (Mitteager et al. 2006). Most residents acknowledge that there is a difference between natural and suburban landscapes. Many want a landscape that looks natural, but what they think is natural is actually a cultural landscape. Residents often do not realize that a natural landscape does not need constant pruning or watering or that it is cheaper to maintain, because it can survive with minimal input. They often opt against natural landscapes because of a perceived lack of control over vegetation plantings and a reduction of lawn or open space. They need sound advice on how to proceed, and they require visual evidence of the benefits of a natural landscape (Mitteager et al. 2006).

Advice for property owners and professional landscapers should come in the form of convenient guidelines that may first require collaboration between scientists and municipal managers. Establishing well-placed demonstration landscapes will provide the needed visual evidence. Property owners often get ideas from looking at the landscapes of their neighbors (Henderson et al. 1998; Zymslony and Gagnon 2000), so use of demonstration areas on private lots can result in diffusion of natural landscaping alongshore.

Shorefront residents frequently trust professional landscapers to provide the concept plan for their lots. Landscapers thus have great latitude in suggesting more natural options and designs that will fit into a broader neighborhood plan that will achieve a less fragmented landscape. Most landscapers now tend to use the same plants on each job, and many of these can be noncoastal species. Getting landscapers to change practices can be difficult, but once their practices change, inertia will work in favor of maintaining natural landscapes. Owners and landscaping professionals alike must be educated about what nature really is (Mitteager et al. 2006).

7.4.3 Scientists

Scientific research has highlighted the ways developed coasts differ from natural coasts but has only recently acknowledged that natural landscapes are a myth, that human agency is now a part of the coastal environment rather than an intrusion, and that human-altered landscapes can and should be evaluated as a generic system (Nordstrom 1994; Doody 2001; Jackson and Nordstrom 2020). We are now aware that reestablishment of natural environments will not occur in many locations unless aided by human actions and that the environments will evolve on trajectories that are human induced and influenced. A priority task must be to determine reference and target states for developed areas. This need is hampered by the lack of appropriate reference sites and the difficulty of adopting a static approach to a reference site or to a dynamic feature such as a coastal dune (Provoost et al. 2011).

The scale of shore protection and restoration projects often occurs within local jurisdictions and is confined to short lengths of coast over short time periods, while natural coastal systems are driven by processes that exceed these temporal and spatial scales (Pethick 2001; Lazarus et al. 2016). With the exception of issues associated with sea level rise, the large scales that often characterize studies of natural coastal systems have little relevance in assessing changes that now confront coastal managers and may have little relevance in developing workable solutions to current problems. Scientists often acknowledge problems of scale in alternative views of problems and solutions (Pilkey 1981; Schwarzer et al. 2001), but they may think that it is the responsibility of local stakeholders to make dramatic changes in their temporal and spatial frameworks. In the past, many local managers saw little use for scientists and may have actively opposed suggestions for management (Gares 1989). Scientists can have inadequate understanding of the needs of local managers and may not be prepared to listen to them (Schwarzer et al. 2001). A two-way advisory process is required to allow managers to educate scientists about the constraints of local politics and economics and allow scientists to then devise workable strategies tailored to these constraints (Schwarzer et al. 2001; Nordstrom 2003; Lazarus et al. 2016). It is in the interest of both coastal managers and scientists to monitor and bridge the relevance gap that occurs between them (van Koningsveld et al. 2003; Lazarus et al. 2016).

Human alterations at the scale of individual properties or municipalities have not benefited from much scientific study in the past because these alterations were considered too small, temporary, site specific, or artificial to be of interest.

Landforms and habitats at small temporal and spatial scales take on increased importance when they are ubiquitous or recurring, which argues for placing increased emphasis on scientific studies of them and development of relevant management guidelines (Nordstrom 2000; Nordstrom and Jackson 2018).

Models of future change are likely to have little impact on managers if human inputs are not incorporated as fundamental variables (Lazarus and Goldstein 2019; Jackson and Nordstrom 2020). Geomorphological investigations are required to identify the coastal landscapes that could be created under alternative scenarios for sea level rise with humans as active participants rather than passive victims. The option of human retreat from the coast is not being undertaken in most developed areas (Schmahl and Conklin 1991; Lee 1993; Hooke and Bray 1995; Abel et al. 2011; Nordstrom et al. 2016), so scenarios should be developed to predict how restoration goals could be achieved if managed according to the Dutch policy of no retreat, using beach nourishment and nature engineering. There is much to be learned about strategies appropriate to the no retreat policy, even in the Netherlands (Koster and Hillen 1995).

7.4.4 Engineers

Engineers face the problems geomorphologists and ecologists face in developing models of beach and dune change and management plans that incorporate human inputs as fundamental variables, but engineers often have the more challenging task of implementing projects on the ground. Coastal engineers are in a strong position to assess and construct risk and adaptation strategies, but the complexity of issues suggests the need for extensive collaboration and synergies across many disciplinary fields (Martínez et al. 2019; Toimil et al. 2020). One of the greatest challenges is to design projects to overcome erosion and inundation and lack of sediment sources while accommodating natural species and human development. Designs that accommodate multiple species but are also expected to be stable are especially challenging. Cycles of growth and decay of habitats are important in achieving the diverse and often complex habitats that can deliver a broad set of ecosystem services (Section 4.2). Engineering designs that enhance only certain aspects of the environment (Figure 2.11) may not convey the proper image or function of a natural environment that would occur if shaped by natural processes. Static, linear landforms can appear like structures even if the project is defined as nature-based. Designs should allow for the natural and ecological processes of a site, by using shapes that mimic natural landforms or by allowing the constructed features to evolve to the extent possible, given safety considerations.

Differences between the vision of a linear and static coast attributed to engineers in the past and the vision of a dynamic coast attributed to many scientists have been reduced, and scientists and engineers now appear to share a common vision, resulting in more innovative engineering solutions (Gueben-Venière 2016). Standard boilerplate solutions may not satisfy current needs. Nature-based solutions and hybrid structures (Chapter 5) are examples of more innovative solutions. Projects that retain levels of protection while accommodating some dynamism are being tested, such as modifying groins to enhance sediment transport (Donohue et al. 2004; Bocamazo et al. 2011), or altering the elevation, length, or orientation of breakwaters to reestablish a more naturally functioning and appearing shore (Polomé et al. 2005; Pranzini et al. 2018b). Creating new structures using elements that can be relocated provides managers with greater flexibility in the future. Protective features created using unconsolidated sediment can be more readily altered, e.g., creating gaps in dunes to allow overwash (Schupp et al. 2013) or aeolian transport (Riksen et al. 2016; Van der Biest et al. 2017). The onshore limit of the beach/dune system considered critical for storm protection had historically been the dune crest that functions as a barrier against flooding and overwash. Recognition of the natural value of the landscape landward of the dune crest has resulted in new designs for publicly managed shorelines, where nature protection or enhancement is one of the management goals. Solutions for allowing private properties landward of the dune crest to be more dynamic would add to the inventory of natural features in locations where these features have been eliminated. Structures are often designed to be less vulnerable to natural hazards (e.g., elevated on pilings), but the structures and their use could also be designed to have less interference with the landforms and habitats.

Engineers in private practice, who are called on to address local shore protection issues, often limit their designs to local needs, not the needs or physical constraints of the broader region. A broader approach may be of limited interest to the local client, who may only see the immediate problem and seek a remedy after other potentially viable options are already foreclosed. The broader regional perspective may not be translated into appropriate action unless the base of technical and economic support is increased and involves actions at higher levels of government. In this case, participants in the planning process should include representatives of adjacent regions and higher levels of government with expertise in many disciplines to evaluate options for more comprehensive programs with access to greater funding. The opportunity to integrate ecological and geomorphological functions over the lifetime of a project can enhance natural values beyond the problem at hand.

Stakeholder inputs in planning and design are critical, but their actions after projects are built are also important in achieving restoration goals. Beaches and dunes constructed and vegetated to initial engineering designs are often subsequently managed by local interests. Aspects of the subsequent evolution of the project could be incorporated into original designs and formally addressed in monitoring and adaptive management. Often, subsequent efforts include maintaining the template used in initial engineering designs. Care should be taken to ensure that replacement of lost sediment is done in a way that does not perpetuate the initial static, linear artifact created in the original project.

The current emphasis on nature-based solutions is leading to more creative designs to couple restoration goals with shore protection goals for new projects and conversion of structures already built. Projects on estuarine beaches can often be more creative than on exposed ocean beaches because estuarine beaches are often located in small isolated compartments, the magnitude of wave and current energies and scale of landform changes is smaller and solutions are less costly. The projects identified in Section 5.3.1 provide examples that could be developed for application at larger scale on coasts with higher wave energies. The highly creative projects on the ocean coast of the Netherlands were implemented, in part, because levels of protection had already been achieved to the point where public safety was no longer an overriding issue and the national emphasis on dunes made funding available. The scale of projects may be smaller elsewhere, but many of the principles should apply.

7.4.5 Environmental Regulatory Departments

The natural features of the coast and the social, political, and economic processes are dynamic, so static approaches to retain or enhance environmental functions of beaches and dunes are not likely to be successful in the long term (Pethick 1996). Lack of appreciation of the dynamism of the coast is apparent both in municipal management plans (Granja and Carvalho 1995) and plans for nature protection areas. Examples in nature protection areas include protecting landward areas against wave overwash by filling low parts of the crest of gravel barriers or using sand fences to maintain linear dunes (Orford and Jennings 1998; Nordstrom 2003). Change may be perceived as degradation by some environmental interest groups, but as indicated in Chapters 4 and 5, mobility itself is worthy of conservation, requiring more flexible management approaches (Pethick 1996; Bray and Hooke 1998; Doody 2001; Moore and Davis 2015).

Target species protection is a legal (but not especially welcome) way of overcoming public resistance to efforts to return parts of a severely altered system to a naturally functioning one (Breton et al. 2000). Prevention of human action in nesting areas on the backshore can result in colonization by plants, growth of new incipient dunes, and eventual reestablishment of new dunes (Breton et al. 2000; Nordstrom et al. 2000) (Figure 3.1). Problems can occur if the subsequent landscape evolves to the point where the site no longer appears desirable to target species, such as when vegetation becomes too dense (Breton et al. 2000; National Park Service 2005). Proactive attempts to reestablish conditions favorable to a single species through human action, such as removing vegetation and associated topography to create bare sand for bird nesting, can negate the broader advantages to other species that require a landscape at a later evolutionary stage. Single-species management may be best viewed as a last chance action than as a supplemental action. The optimum solution may be to obtain a new restricted area for target species, leaving the previous area to develop naturally (Breton et al. 2000). Reestablishing an appreciation for natural landforms and habitats in developed communities through a comprehensive demonstration and education program may facilitate the process of designating new restricted areas.

Existing environmental regulations may have the negative effect of discouraging environmentally friendly actions on the part of private residents. In the State of New Jersey, for example, private lots that occur within a dune fall within a regulatory zone under the New Jersey Coastal Area Facilities Review Act and are subject to more severe environmental restrictions than lots without dunes. Conversations with property owners reveal a reluctance to allow natural migration of a dune onto their properties because they do not want to be in the regulatory zone. An accord with state regulators may be required before this kind of environmentally friendly action could take place. It seems logical to assume that subsequent landscape options should be permitted if the natural value of the new dune landscape exceeds the old landscape. This precedent need not require an institutional change because it could be handled as a permitted exception, but it would require greater flexibility and perhaps an added workload on the part of environmental regulators (Nordstrom et al. 2011). The gain in natural habitat and in developing working partnerships with residents would be major benefits.

The regulatory requirement to address any and all potential adverse impacts can impede new projects to restore coastal landforms and habitats. Project planning and preparation of environmental impact statements in connection with permits can cost more than dune enhancements, such as installation of plants and sand fencing. Demand for paperwork can deny or stall projects in a way that prevents them from occurring, especially small projects with limited funding. The review process is important because it can identify inadequacies or unintended consequences, so the solution requires streamlining the permit process without overlooking serious environmental consequences. In some cases, different municipalities have to submit an independent environmental impact statement to regulatory authorities for virtually the same projects. Streamlining the process would free private and municipal resources for construction and dune enhancement activities (Nordstrom et al. 2011).

Beach and dune restoration projects conducted by private interest groups can be more innovative than those conducted by national and state agencies because there may be fewer constraints on how funds are used. These projects can provide templates for larger publicly funded projects, but the case must be made to these interest groups for incorporating nature into the designs, and initiatives must be made by regulatory groups toward overcoming the financial or administrative burdens involved (Nordstrom et al. 2011).

7.5 Implications

A more dynamic and naturally functioning landscape can be achieved in developed areas by adopting a management strategy that (1) tolerates change (including cycles of destruction and regrowth); (2) uses zones, not lines, for delineating the boundary between management units; (3) appreciates incipient features as proto stages in evolution of landforms and habitats; (4) tolerates inconveniences that are not true hazards; (5) rediscovers environmental heritage and the value of a natural landscape image with people in it; and (6) attempts to enhance the nature-based aspects of tourism and an appreciation for nature that can be reflected in real estate value. The new nature in developed municipalities may be small but more complex than in natural areas because it will include both human and natural processes. It may also require more frequent human participation where landforms and biota must be maintained in nonequilibrium states to survive. A newly restored landscape may initially appear to offer fewer opportunities to stakeholders who have grown accustomed to a static, cultural landscape modified to suit personal tastes. Many restored landscapes on developed coasts will be artifacts, but the added natural values and significance of getting these coastal locations off a human trajectory may be better than alternatives that are redundant with inland locations. Many of the concepts identified in this chapter have already been applied in site-specific cases, so their feasibility has been demonstrated. Some of the concepts have yet to be worked out in detail, but the opportunity exists to devise more creative plans for restoring natural environments on developed coasts in the future.

8

A Locally Based Program for Beach and Dune Restoration

8.1 The Need for Local Action

Addressing the main causes of decline in the health of ecosystems can be beyond the capability of individual management units, but local management measures can be used to restore individual components of the landscape (Verstrael and van Dijk 1996). Locally restored environments can be part of a more holistic regional approach to restoration, if the importance of connections between them is acknowledged and favored. The expression "think globally, act locally" is as appropriate to coastal restoration as it is to addressing problems of pollution or water management. Local actions are often critical in optimizing the value of projects and plans developed by higher authorities (De Ruyck et al. 2001). Large-scale beach and dune construction projects are usually funded, designed, and built by governments at the national and state/provincial levels and tend to be capital intensive, with minimal maintenance costs (Townend and Fleming 1991) and without follow-up monitoring or performance criteria based on natural habitats. These projects are often managed by local jurisdictions, who also have their own projects to implement and monitor. Comprehensive locally based (municipal) programs are needed to ensure that projects achieve their full restoration potential. Often, little more needs be done to restore basic beach and dune vegetation structure and function than to simply prevent adverse human activities (Dugan and Hubbard 2010; Kelly 2014), but this may require changes in policy and practice. Some communities may still eliminate evolving natural features to enhance active recreation rather than tailoring recreational activities to natural attributes (Gangaiya et al. 2017).

An integrated locally based program for increasing the number, size, and cumulative benefits of natural environments in developed areas involves (1) getting stakeholders to favor restoration initiatives that accept natural landforms and habitats as appropriate elements in a coastal landscape modified by humans;

(2) identifying environmental indicators and target reference conditions using characteristics of existing natural enclaves; (3) establishing demonstration sites to evaluate the positive and negative effects of return to a more dynamic system; (4) developing realistic guidelines and protocols for use in restoring and managing landforms and habitats; and (5) developing environmental education programs to establish an appreciation for new naturally functioning landscape components (Nordstrom 2003). This chapter is a synthesis of elements observed in local-scale restoration programs that can accomplish these goals (e.g., Breton et al. 2000; Nordstrom et al. 2002; Gesing 2019). The rationale for many of the suggestions is presented in previous chapters and is based on observations of local activities on the northeast coast of the USA. Most of the management suggestions are generic and apply to many locations throughout the world. The examples of species are more specific to the region, but the species appropriate to other areas can often be obtained from local guidelines on web pages.

Most of this chapter evaluates activities where retreat from the coast is not presently considered an option. The end of the chapter addresses ways local governments should be planning for the future, when management of infrastructure in place becomes less tenable.

8.2 Gaining Acceptance for Natural Landforms and Habitats

The appearance, function, and use of landforms and habitats will change considerably following restoration, so anticipating and addressing stakeholder concerns should be a major component of a restoration program. Surveys of beach users often reveal a preference for cleanliness over natural topography and vegetation (Cutter et al. 1979; Vandemark 2000), underscoring the difficulty of allowing nature to evolve on beaches that have been raked to eliminate litter and produce a neat, orderly surface. Convincing residents and municipal officials of the need to create natural beach and dune habitats where they no longer exist is one of the most challenging aspects of a program for restoration (Mauriello 1989). The difficulties of overcoming local inertia to change can be overcome by a long-term commitment to implementing incremental changes at increasing levels of funding, as demonstrated by the restoration of beach and dune at Myrtle Beach, South Carolina (Kana and Kaczkowski 2019, Box 1).

In attempts to build dunes in New Jersey, state regulators had to contend with local interests that resist building dunes or argue for making them as small as possible to retain views of the sea from boardwalks and residences (Nordstrom and Mauriello 2001). Case studies of Lavallette, New Jersey, USA (Mauriello and Halsey 1987; Mauriello 1989), provide perspective on the compromises required in constructing dunes where none existed. The dunes were built as a condition for

Figure 8.1 The dune at Lavallette, New Jersey, USA, showing the low, narrow, linear nature of a protective dune formed where there is little space for its evolution and little interest in increasing its size because of the desire to retain views of the sea.

accepting federal and state funds for post-storm repair of damaged facilities. They were constructed to existing Federal Emergency Management Agency guidelines that specified a dune with a minimum height 2.56 m above mean sea level and a cross-sectional area of >50.2 m^2 above the 100-year recurrence interval storm flood level. Concern about inundation of human facilities by blowing sand led to placement of sand fences on the landward side of the foredune and retention of a flat trough between the dune and boardwalk where earth-moving equipment could be used to mechanically remove sediment blown inland. These practices caused the dunes to be narrow, linear, and fixed in position (Figure 8.1).

The dune at Lavallette, like many others on developed coasts, was built where beach widths were insufficient to allow a dune of that size to be created naturally, and it could not evolve into a naturally functioning landform because it was not allowed to migrate landward. Dunes in space-restricted areas are usually a single narrow ridge close to the sea and provide little shelter from salt spray and blowing sand. Plants characteristic of a stable backdune cannot thrive, and vegetation is characterized by species commonly found on the seaward portions of natural dunes. In this case, *Ammophila breviligulata* dominates because it was planted, but subsequent growth of seaside goldenrod (*Solidago sempervirens*) provided at least some diversity of vegetation. The image that this dune conveys is not a natural one because of its linear, truncated look, but the dune was a step toward resident acceptance of a dune as part of the coastal system. The next step in this restoration process was to use beach nourishment to add more space to allow for growth of a more dynamic larger dune with greater diversity of topography and vegetation.

The dune in Lavallette is an example of how storms provide an opportunity for restoration of natural features by providing stable sources of funding for beach nourishment and dune building and for passage of laws that restrict development or favor retreat from the coast (Nordstrom and Mauriello 2001; National Research Council 2014; Kousky 2014). In some cases, storms provide incentive for property owners to sell their land, providing the opportunity for public purchase and conversion to larger conservation areas (Nordstrom et al. 2002). This opportunity is only occasionally realized because few people are willing to move, and there is little public funding to buy properties at fair market value. Public funding is more readily available for protecting remaining property owners against future storms. These funds can be put to good use for nourishing beaches and building dunes, even in areas where dunes did not exist prior to the storm.

National and state/provincial governments have leverage over local managers via aid agreements for shore protection, making money available if local jurisdictions make their activities and regulations consistent with coastal zone management rules on beaches, dunes, and erosion hazard areas. The state of New Jersey, for example, adopted a formal hazard mitigation plan that recommended creation and enhancement of dunes as one of the primary hazard mitigation efforts. Funds were made available to municipalities to provide vegetation for planting and materials for constructing sand fences. As a condition of funding, municipalities signed an agreement requiring them to adopt or amend municipal ordinances to conform to state coastal zone management rules.

Although shore protection can be a driving force for construction and maintenance of dunes in developed areas, their construction does not ensure restoration of habitat, or aesthetic or heritage values. Regulations often allow bulldozing, and vegetation (including exotics) can often be planted. The resulting dunes may bear little resemblance to natural dunes and provide few of the goods and services identified in Table 1.2, but stakeholders would be more likely to accept dunes and be aware of at least one of their most important utility functions.

Acceptance of dunes can evolve slowly, relying on positive actions of a few key people and programs in local municipalities, but preference for natural features can change as people are enlightened as a result of incremental improvements in nature quality (Arler 2000; Kana and Kaczkowski 2019). Conversations with managers in municipalities where dunes were created by national and state initiatives indicate that they now recognize dunes as an effective and inexpensive means of reducing storm damage. Many of them also think that the dunes are an aesthetic improvement because they restore a degree of natural beauty, although they still may not like the direct blockage of ocean views (Nordstrom and Mauriello 2001). The subsequent arrival of interesting species or charismatic fauna (such as some species of birds) can provide catalysts to new appreciation of nature within dunes (Baeyens and Martínez 2004).

8.3 Identifying Reference Conditions

Selection of a reference condition or state in a dynamic system subject to human alterations is a challenging task (Aronson et al. 1995). It is important to select reference sites that are not so naturally favored that those conditions are unachievable at restoration sites in developed areas (White and Walker 1997; Ehrenfeld 2000). An inventory of species typical of natural environments in the backdune of a stable or accreting shore cannot be used to determine species that should be planted in a narrow, eroding dune close to the sea, unless ongoing human actions are taken to ensure survival. Several reference sites that are generally similar to potential restoration sites in their juxtaposition to nearby natural and cultural features can provide the broad description of ecological variation and range of spatial and temporal contexts and better models than a single site (Clewell and Rieger 1997; White and Walker 1997).

Readily defensible reference sites for a municipality can be defined using data from other locations in that municipality or adjacent municipalities that have the greatest species diversity and richness in the most naturally functioning (but not necessarily naturally favored) dune habitat. Dunes must first be classified to identify the number of distinctly different types. Field data can then be gathered along representative transects to provide detail on differences in biota along environmental gradients from the litter lines to the landward limit of the dune system. Distinctive habitats resulting from the cross-shore gradient in physical processes include foreshore, backshore, incipient dune, seaward side of foredune, foredune ridge, lee side of ridge, sheltered backdune, and slack, if present. The landward limit for restored habitat is not necessarily the existing limit. Portions of the landward side of the foredune that are presently isolated from the beach and seaward portion of the foredune by human uses or structures (Figure 6.3 e,f) can be used as representative target states for restoration of backdune habitat where construction of a continuous cross-shore transect is not possible (Nordstrom and Jackson 2018).

Cross-shore surveys of topography and vegetation provide detail on effects of elevation, distance from the shoreline, and sheltering from wind and wave runup.

Species should be identified as exotic or native to assess potential invasions from private lots nearby. Data should reveal the relationship between biodiversity and management variables (e.g., raking, fences, planting) as well as natural variables (e.g., distance landward of high water, dune dimensions) to assess the significance of human actions to success of restoration outcomes. A follow-up study of a few growing seasons would determine if the dune vegetation increases in diversity and speed of colonization where wrack lines are allowed to persist, where natural processes are allowed greater freedom, or where management actions are taken to enable sediment exchanges, enhance topographic characteristics, or nurture the vegetation.

8.4 Establishing Demonstration Sites

Demonstration sites provide a way for managers to document the feasibility of implementing conversion to more naturally functioning systems (Breton et al. 2000; Morris et al. 2019). It is unrealistic to assume that the conversion of human-altered environments can occur quickly and at large spatial scales. There are too many unknown variables to convince all stakeholders that change is good or possible, and there is insufficient information on some issues to say that a planned restoration project will not create new problems. Demonstration sites can be used to (1) test whether projects function as designed; (2) identify unwanted side effects; (3) provide specific technical information to municipal managers and property owners; (4) reveal the value of naturally functioning landscape components to residents and tourists; and (5) provide evidence that restoration options are achievable with changes in policy or practice (Breton et al. 2000; Nordstrom 2003).

Some of the useful characteristics of local demonstration sites are presented in Table 8.1. The characteristics should exist within the same region as the potential restoration sites and be of the same spatial scale. Dunes on private lots or dunes on undeveloped lots remaining between two developed lots can be used as demonstration sites for private lots. Naturally functioning portions of municipal dunes (Figure 3.1) should be used as demonstration sites for adjacent municipalities where dunes were eliminated or restricted in size or number of utility functions.

Publicly managed access ways to the beach have great potential as demonstration sites. These access ways are where tourists and residents who live landward of the first row of houses obtain their first view of the dune on the way to the beach, and they are often the only way visitors can see all of the cross-shore subenvironments as they traverse the dune. Public access ways may be among the few locations where municipal managers have complete control over the entire width of the dune in developed areas. By changing the appearance of the

Table 8.1. *Useful characteristics of demonstration sites for restoration projects in developed municipalities*

Close to potential future restoration sites
Similar in space constraints and relationship to natural and cultural features
No mechanical removal of litter or use of vehicles on the beach
No impermeable barriers or sand-trapping fences used for crowd control
Dune dimensions sufficient to provide storm protection to landward human structures
Portions of dunes function as dynamic subenvironments without sacrificing shore protection
Free of exotic vegetation
Integrated with demonstration sites on properties landward
Sections available for nature-based experiments
Sections dedicated to specific environmental management goals (e.g., bird, turtle nesting)
Monitored in ongoing programs
Integrated with education programs

topography and vegetation adjacent to these cross-shore access ways, managers can influence the image that a municipality presents to its residents and visitors. The feasibility of planting native species and the visual appeal that these species can have can be displayed here.

Some municipalities have relatively large undeveloped segments within otherwise developed shorefront (Breton et al. 2000; Nordstrom et al. 2002). The 1.5 km undeveloped segment that the Borough of Avalon, New Jersey, USA, purchased following a storm in 1962 provided space for landforms to evolve naturally and a location for testing environmentally compatible management strategies. The site has been used to evaluate results of suspending use of sand fences and raking (Nordstrom et al. 2012). The US Department of Agriculture also initiated an experiment to test the growth potential of sea oats (*Uniola paniculata*) farther north than it is normally found. Suspension of use of sand fences and raking resulted in a foredune that compared favorably in volume (but not height) with dunes built with fences but having a gentler seaward slope and fewer restrictions to cross-shore movement of sediment and biota. The dune also provided greater species diversity (Nordstrom et al. 2012).

The New Jersey Division of Fish, Game and Wildlife requires municipalities to ensure that shore-nesting birds are not adversely affected by pedestrian and vehicle traffic, beach raking, and beach nourishment. The undeveloped segment in Avalon allows that municipality to accomplish this state management goal without seriously restricting uses of the beach in adjacent segments that are backed by houses. Accordingly, Avalon was the first municipality to agree to a bird management plan/agreement with the state. The undeveloped site was monitored on an ongoing basis and incorporated into the municipal education program identified in Section 8.6.

Actions to ensure compatibility between human use and natural processes in municipal demonstration areas are detailed in the following section and include confining litter removal to manual pickup of cultural material, using symbolic fences to control access while allowing for free exchange of sediments and biota, allowing portions of dunes to be dynamic once they have achieved dimensions required for shore protection, using only native vegetation in planting programs, and removing exotic species. Actions in dunes managed by individual property owners include removing barriers to transport of sand onto their properties, removing exotic vegetation, and planting (or simply not removing) natural vegetation species that would survive there given the degree of exposure to blowing sand and salt spray (Nordstrom 2003). Private property owners are unlikely to adopt the same type of dune or mix of vegetation as their neighbors. Even if consensus among them could be found, the great variety in site-specific conditions caused by buildings, roads, and protection structures would make a single approach unrealistic. Landscaping approaches have to be flexible and offer a spectrum of target states that suit varying degrees of commitment and different exposure to the beach and natural coastal processes.

8.5 Developing Guidelines and Protocols

Guidelines for restoring natural processes and system components should identify best and worst management practices and their impacts in changing the character and value of the shore. They also should be tailored to different levels and intensities of management and potential for success given physical constraints. Guidelines for nourished beaches will differ from unnourished beaches because the larger nourished beaches will provide greater amounts of sediment for building dunes, more space for dunes to grow, and more time for them to evolve before they are threatened by erosion. Guidelines for developed lots cut off from the active beach and foredune by shore parallel protection structures or promenades (truncated and decoupled gradients) may require more specific statements about the need for ongoing efforts to control exotic species. The following subsections identify factors that must be considered in developing guidelines and protocols for publicly and privately managed beaches and dunes. Management alternatives derived from these factors are presented in Table 8.2.

8.5.1 Litter and Wrack Management

Suggestions for managing beach litter (Table 8.2) should include ways to selectively remove it without using mechanical methods. Manual removal is feasible where the environmental ethic of the community is strong and residents

Table 8.2. *Alternatives for restoring and maintaining beaches and dunes in publicly and privately managed coastal segments*

Managing beach litter
Clean only some portions of the beach
Separate beach into management zones with different methods or frequency of cleaning
Clean beaches less frequently
Remove cultural litter and leave natural litter in place
Use nonmechanical removal methods
Retain the uppermost storm wrack line as core of incipient foredune
Retain the most recent wrack line as a foraging area
Establish no-rake zones alongshore
Restrict cleaning only to months when beaches are intensively used

Grading
Grade primarily for shore protection or return overwashed sediment from lots or streets
Restrict grading on naturally evolving surfaces
Mimic natural landforms
Do not exceed replenishment rates in borrow areas
Monitor borrow and fill areas to assess success

Driving vehicles on beach
Prevent driving on beaches and dunes where possible
Confine driving to a few heavily used sections
Restrict trails and shore perpendicular access to a few locations
Restrict both private and public vehicles
Restrict driving in wrack lines and vegetated areas, particularly on the landward backshore

Access paths
Confine visitors to a limited number of access paths through dunes
Clearly mark entrances, but make crossings as unobtrusive as possible
Use walkover structures only where intensive use is expected

Structures
Reduce the physical and visual impact of structures, including movable ones
Locate buildings as far landward as possible and reduce size and number of dwelling units
Elevate human use structures above active surfaces
Allow natural features to form at and on shore protection structures
Restrict use of impermeable barriers
Avoid demarcating property lines, especially with exotic materials

Landscaping on landward lots
Avoid "suburban" landscapes, such as grassy lawns and trimmed hedges
Replace exotic vegetation that is costly to maintain with natives
Plant vegetation in patches for a natural image and sense of topographic relief
Select plants on the basis of maintenance, visual interest (height, color, texture), or smell.

Implementing education programs
Link active programs (nature walks, litter cleanups) with passive ones (signs, museum
   displays)

Table 8.2. (*cont.*)

Use lifeguards and wardens as teachers where possible
Specify role of human actions in both degradation and restoration of natural environments
Explain the discrepancy between natural characteristics and the characteristics visitors see
Hold frequent meetings between municipal managers and stakeholders
Encourage stakeholders to educate themselves
Publish and distribute information newsletters
Establish environmental displays in local museums
Incorporate programs into the curricula of public schools
Encourage resident participation in restoration activities

Maintaining and evaluating restored environments
Develop formal but affordable plans for maintenance, monitoring, and adaptive
    management
Assess whether goals are achieved and are better than alternatives; require follow-up action
Document plant survival, need for replanting and provision of special functions
Evaluate erosion, changes in topography, need for renourishment and off-site changes
Provide evaluations rigorous enough to learn from past mistakes
Make inferences from key readily measured variables (e.g., species richness, percent cover)
Reexamine landforms and biota several years after restoration to have realistic expectations
Institute projects requiring little funding (e.g., removing litter, suspending incompatible
    uses)

and visitors accept the image of a naturally functioning beach. The beach could be separated into management zones that have a different method or frequency of cleaning (Kelly 2016). Storm wrack lines that are high on the backshore can have the greatest amount of litter (Bowman et al. 1998), and the backshore has the greatest concentration of beach plant communities and nesting sites for turtles and shorebirds (Kelly 2014, 2016). Accordingly, the seaward portion of the dry beach that is most intensively used by beach visitors could be cleaned while the upper wrack line and vegetation communities on the upper backshore could be left intact (Kelly 2016). The uppermost storm wrack line is critical because it has the greatest probability of surviving subsequent small storms and evolving into incipient dune habitat.

Unraked segments can be established as enclaves along the shore. The intensity of recreational beach use can decrease rapidly at greater distances from beach access ways, lifeguard stations, parking lots, and commercial or township centers, even within the most populated areas (Kelly 2016), lessening demands for raking in those areas. The no-rake segment at Avalon, identified in Section 8.4, was located where housing densities are minimal. The raked beaches on both sides of this segment are closer to businesses and subject to greater use by tourists. The unraked segment at Ocean City that resulted in creation of the dune field depicted in Figure 3.1 was established after the state endangered species program found piping plovers nesting on the nourished beach. This site is in an intensively developed segment of the shore that was heavily used by tourists. Its conversion to a naturally functioning habitat demonstrates the effectiveness of allowing nature to evolve simply by ceasing a common management action.

Cleaning operations could be restricted to summer rather than preventing them altogether (Nordstrom et al. 2000). At times, there may be an overabundance of wrack associated with algal blooms linked to increased nutrients from anthropogenic origin (Morand and Merceron 2005) or massive fish kills. Cleaning after fish kills may be one of the few times when mechanical cleaning is appropriate. Suspension of cleaning operations in winter, when aeolian transport is favored by strong winds, would allow incipient dunes to form. Cleaning in summer is compatible with many recreational uses, but it would be of little value to fauna and nature-based tourism. One solution would be to remove the cultural material manually and leave the natural material to provide the basis for new habitat. The strengths and weaknesses of mechanical and manual methods are compared in Zielinski et al. (2019), who suggest that manual cleaning should replace mechanical cleaning on low and moderately used beaches, with removal only in specific cases of large accumulations. Mechanical cleaning may be more efficient than manual cleaning, but manual cleaning is more feasible than often perceived. The general public views natural litter, such as driftwood and seaweed, as the least offensive kind of litter (Tudor and Williams 2003), and success using volunteer labor for removing cultural litter on a long-term basis has been noted (Breton et al. 2000; Nordstrom 2003). Mechanical cleaning may fail to clear small anthropogenic litter from the beach anyway, requiring manual cleaning to remove smaller items (Somerville et al. 2003). Some municipalities where tourists pay a fee for use of the beach have their beach wardens manually clean the beach when they are not monitoring users.

8.5.2 Grading

Grading (reshaping) by earth-moving equipment is one of the most ubiquitous ways landforms are altered by humans (Nordstrom 2000). Many municipalities have their own equipment and implement regularly scheduled and ad hoc programs for a variety of purposes. Grading to move sediment from accreting parts of the beach to eroding parts, to move sediment from the beach to create flood protection dunes, or to move overwash deposits back to the beach appears to have some justification for shore protection, although these actions interfere with natural evolution. Grading to remove dunes to provide views of the sea, to create wider beaches as recreation platforms, or to create higher backshores to support recreation structures appears less justifiable.

Grading eliminates naturally evolving surfaces and can prevent growth of vegetation, even if conducted at annual frequencies (Conaway and Wells 2005; Kelly 2014). Most of the new deposits depart in form, location, and function from natural landforms; they interfere with natural processes; and they convey an image of the shore as a cultural artifact. Steepening the beach by massive random scraping can destroy the dynamic equilibrium of the system and appears to increase the time required for natural recovery (Wells and McNinch 1991). If grading must be done, natural landforms should be mimicked to the extent possible. This may include using hummocky or undulating shapes rather than linear shapes and not creating oversized features out of spatial context. Fill can be placed adjacent to evolving environments rather than on them. Managers should avoid stockpiling sediment or creating beach features to accomplish temporary human utility functions that convey the impression that beach sand is a construction material. Guidelines for grading activities should restrict grading to actions required for safety (e.g., Wells and McNinch 1991), not some undefined or untested conception of convenience. Grading operations are highly experimental, so monitoring programs are required to determine whether the projects are working and how they can be designed better (Wells and McNinch 1991).

8.5.3 Vehicles on the Beach

There is little reason to allow private vehicles on beaches in developed municipalities when road networks exist landward. Undeveloped shoreline segments within municipalities are often so short and narrow that vehicle access to them from nearby paved roads is not necessary. Where driving is allowed, sign posts can be placed to prevent driving on the landward part of the backshore to protect the most sensitive upper beach areas, with little discernable detriment to beach use (Kelly 2016). The biggest problem in some areas may be use of government vehicles. Kelly (2014) found that government off-road vehicle use was ubiquitous at his sites (87 %), occurring on most municipal beaches for garbage collection, maintenance, law enforcement, or lifeguard or safety patrols. Municipal vehicles not only damage habitat; they create an unnatural landscape image that can undermine attempts to instill an appreciation of the beach as a natural environment. Instituting a "haul in / haul out" policy for garbage generated by tourists would encourage use of trash receptacles (Zielinski et al. 2019) or even eliminate the need for trash receptacles on the beach. Vehicles used for public safety could be confined to emergency operations and be subject to the same constraints as private vehicles, where they are authorized (Section 4.4).

8.5.4 Access Paths

Human trampling through dunes can have little immediate impact on large-scale geomorphic features but may locally change vegetation and increase the likelihood of sediment entrainment or slope failure. The visual effects of trampling on dunes can be obvious, but that does not mean that trampling is adverse to dunes. The creation of bare sand surfaces and small increases in dune mobility are not long-term threats to the natural value of dunes, but paths that are continually used will retain their human origin in size, shape, location, and function and encourage more trampling that could increase path size and number. Confining access to marked, well-placed paths will help limit the number of off-track crossings that evolve into unauthorized paths.

Access to the beach through dunes in developed municipalities can be seaward of every shorefront building or at greater distances, such as at the seaward end of shore perpendicular roads. Different sizes and configurations of pathways have a strong visual impact on beach users and influence willingness to use them. Paths can vary from narrow, winding trails that conform to the topographic contours of the dune to wide, straight conduits intentionally cut or trampled down to the elevation of the backshore. Many public and private pedestrian access ways are raised wooden walkover structures. These structures allow for a more continuous vegetation cover than occurs adjacent to sand paths and can better maintain the integrity of the dune crest beneath them (Purvis et al. 2015). Walkover structures make sense in intensively used access corridors, particularly in urban locations where tourists often do not think about environmental preservation. Paths used by shorefront homeowners who know the significance of minimizing bare ground can take up less space than walkover structures. Raised structures create a more obtrusive cultural feature than paths at ground surface, and they convey an image of the coast that is not natural, making raised structures less suitable for every shorefront lot. Optimum management of local access ways appears to be a function of the expectations of beach users and their willingness to properly use paths or walkovers.

8.5.5 Structures on the Beach and Dune

Reducing the physical and visual impact of structures is a desirable goal in obtaining a more naturally appearing and functioning beach and dune environment. Many regulations designed to reduce the hazard potential for coastal construction are compatible with restoration goals, such as requiring that buildings are located as far landward as possible, reducing the size or number of dwelling units (apartments/flats) of buildings, and elevating buildings on pilings. Small buildings that are set back from the beach/dune contact diminish the human imprint on the landscape and provide space for natural landforms and vegetation, while houses on pilings reinforce the image of the coast as a hazardous environment. The structures frequently permitted on a publicly managed beach are sand fences (Chapter 3) and moveable structures such as chairs, tables, and cabanas. Moveable structures may seem relatively benign because they are temporary, but they enhance the feeling that the beach is a recreation platform. Actions taken to employ and use moveable structures, such as grading surfaces flat and building walkways for access to them, can increase the level of physical and visual disturbance. Continued use of moveable structures can lead to calls to increase their numbers and make them permanent.

The principal structures just landward of a beach include shore parallel protection structures (seawalls, revetments and bulkheads), promenades and boardwalks, buildings, and roads. Seawalls and revetments that provide the only protection where beaches and dunes have been eliminated are not an issue in a restoration context, but bulkheads are intended as backup protection, and are more likely to exist where dunes could form. The vertical face of an exposed bulkhead (and seawall or revetment where a beach still exists) can be a trap for wind-blown sand and prevent transport of sand onto infrastructure. A vegetated municipal dune just seaward of or on top of the structure can prevent sand from blowing onto infrastructure and also add dune habitat. The expression "the only good bulkhead is a buried bulkhead" applies whether the rationale for its burial is shore protection, habitat creation, or aesthetic appeal (Mitteager et al. 2006). Shore parallel paths, troughs, and bulkheads (Figure 6.1) perpetuate the feeling that wild nature and human habitation are separable and even incompatible (Nordstrom and Mauriello 2001) and they disrupt the natural environmental gradient.

Elevated boardwalks are a visual and psychological barrier and can be a physical barrier to humans and natural processes, depending on their height. They can be built high enough to allow for exchanges of sediment and biota, but ground-level boardwalks are less obtrusive in the landscape, and sediment transport is not necessarily decoupled by a small structure near ground level (Figure 6.6).

Managers of public beaches often use fences for keeping visitors off the dune. Fences may be justified because of the vulnerability of the dune to trampling and the ambiguity of where the dune actually begins (Charbonneau et al. 2019). Sand-trapping fences now used for this purpose could be replaced by symbolic fences that are usually less expensive and allow for free movement of biota. Continued use of sand-trapping fences for crowd control may be based on the inertia of managers who already have these fences at their disposal or their impression that visitors will not respect symbolic fences that are easy to cross. Respect for symbolic fences may occur after adoption of better public awareness programs.

Municipalities that resist allowing dunes to form or increase in height often do so because buildings, promenades, and boardwalks are built near the elevation of the backshore (Nordstrom and Mauriello 2001). Dunes can be compatible with provision of views if viewing areas are built on upper floors of buildings and if multiple unit structures have dwelling units on two floors rather than as flats on each floor. This method of building should include homes that are rebuilt, and permits and zoning regulations should allow rebuilding to this design (Nordstrom and Mauriello 2001).

Residents on private lots landward of the beach should be convinced that placing cultural features on their lots is not actually an improvement in a coastal environment prized for its natural values (Mitteager et al. 2006). Impermeable barriers are often used to prevent inundation by wind-blown sand, obtain privacy, or demarcate property. Dune grasses can be used to prevent inundation, and the shrubs that can grow on the landward side of foredunes can be used for privacy or property demarcation. If fences are used to demarcate property lines, post and rail construction would be preferable to sand-trapping fences. Nearly any objects used to demarcate property lines, even natural vegetation, sea shells, and rocks occurring in the area create an unnatural appearance if placed in a line, and their use can encourage other artificial landscaping within the boundaries they create.

Private lots may have decks raised above the ground surface, but the decks are not always high enough to allow natural vegetation to form under them. Raised decks are preferable to the asphalt and pebble surfaces that are alternatives to them because raised decks are usually small. The psychological value of having decks above the surface (as a reminder of landform mobility or respect for the integrity of the natural surface) is another benefit. If a lot owner insists on a flat platform, a small raised structure that retains the natural topography is preferable to a large platform that replaces the natural surface (Mitteager et al. 2006).

8.5.6 Use of Vegetation for Landscaping

Sustainable green spaces in publicly managed areas should use native plants so that secondary plant succession can proceed (Cranz and Boland 2004). This option may not be easy to implement on private lots. Aesthetics plays an important role in the landscaping choices of shorefront property owners and managers of hotels and condominiums, whose preferences are often conditioned by experiences obtained in a noncoastal setting. Thus, lawn grass and trimmed hedges are often considered the landscaping ideal. Grasses are well adapted to sandy soil. The problem is not the appropriateness of grass species in a dune, but the way grass is managed as a lawn, which is an icon of communal order, a form of manufactured consent, and a cultural artifact that is taken for granted (Feagan and Ripmeester 2001). The common conception that a lawn must be flat and green argues against topographic diversity and results in overuse of water.

Color is a major aesthetic concern among homeowners, making flowering plants a popular decorative vegetation. Color of native shrubs can be used to wean people away from lawns. Plants that have the most attractive flowers or foliage in the summer will be seen by the greatest number of visitors to the coast. Native perennial plants bloom every year and require little effort, but they tend to flower in the spring or late summer, so they may not be at their optimum when the highest visitation rates occur. Many trees and high shrubs, such as pine, holly, and bayberry, do not flower, so their aesthetic appeal is in their height, texture, or berries. Several trees are adapted to the dune environment, including most cedars and junipers, and they can grow close enough to the beach to survive on a narrow backdune. Deciduous trees have too much surface area on their leaves to survive the desiccating winds and salt spray close to an ocean beach. Broadleaf evergreens, such as holly, have coriaceous leaves that protect them against desiccation. They can defoliate in a stressful year, which makes them less desirable to owners, but they can recover the next year (Mitteager et al. 2006).

Costs of plants that are not native can include their initial price, travel time to the nursery, losses in transplanting from nursery to the lot, ease of planting, maintenance after planting (watering, pruning, replacing unsuccessful plants), and preparing land for exotic species. Many plant species that owners ask landscapers to emplace have low survivability, but commercial landscapers may plant them anyway, especially if they are not required to guarantee survival. Natural dune vegetation is easy to maintain and thus is the cheapest alternative (Mitteager et al. 2006).

Owners often say that they want a specific plant, but when questioned further, they say they want a certain image (Mitteager et al. 2006). Working with the image offers landscapers greater opportunity to use native species. Factors considered most important in the selection of individual plants by owners are maintenance, visual interest (height, color, texture), and smell. Some plants, such as Rugosa Rose (*Rosa rugosa*), are clearly superior in many of these factors, although they may not be native. *Rosa rugosa* is a coarse shrub that grows in dense clumps, has attractive flowers with red, pink, or white petals, stands at a relatively high elevation, and has dark green, deeply furrowed leaflets (Duncan and Duncan 1987). Its rough appearance provides a contrast in texture from the surrounding grasses. It is relatively maintenance free, although pruning to remove the dead wood can make it more attractive. The shrub bayberry (*Myrica pennsylvanica*) provides contrast in height and texture, and it has a pleasing smell, making it a good plant to use in northeastern USA. Seashore elder (*Iva imbricata* Walt.) provides contrast in dunes in southern USA. Beach plum (*Prunus maritima*) has value for vertical relief and has attractive flowers and edible fruit, but it needs pruning (Mitteager et al. 2006). Not all species that have natural value are likely to be well received. *Toxicodendron radicans* produces berries eaten by a variety of birds, provides good nesting and hiding places for animals, and is a good stabilizer (Wootton et al. 2016), but it can create rashes on people. *Parthenocissus quinquefolia* occupies a similar niche and can be substituted for *T. radicans*.

The relatively slow growth of natural vegetation, its unkempt appearance, and lack of human tolerance for bare sand within vegetated areas may cause most property owners to reject the alternative of slow natural colonization. A slightly less natural alternative would be to plant native species directly and maintain them as aesthetic resources by removing dead material and debris and cutting back growth that would obscure the orderly appearance desired by many people. New vegetation could be planted and maintained to look natural by planting it in masses, revealing broad sweeps of color in irregular patterns (Mitteager et al. 2006).

Plants adapted to sandy soil require little attention. If planted, they may require watering in the first year but not after that. Property owners should be made aware that a change in color during drought conditions is not necessarily a problem. Greater appreciation of the change that natural vegetation undergoes throughout the year will reduce unnecessary maintenance costs.

Knowledge of the true characteristics of natural species will help gain acceptance for them. For example, residents in the northeastern USA think that the native seaside goldenrod (*Solidago sempervirons* L.) is ragweed (*Ambrosia artemisiifolia* L.), an allergenic plant. Seaside goldenrod grows well on dunes; it has attractive yellow flowers; and it has a height and texture different from the more common American beach grass. It would be a valuable addition to many municipal foredunes and dunes on private properties (Mitteager et al. 2006).

8.6 Developing and Implementing Public Education Programs

Beaches and dunes are critical landforms for increasing the social and ecological resilience of developed coasts. Public education programs are vital in the development of successful adaptation strategies and restoration programs that meet current and future challenges associated with changes in climate and sea level. Misconceptions and poor communication among stakeholders can restrict political and managerial progress (Charbonneau et al. 2019). For example, tourists will object to wrack on the beach unless they are informed of its value, and the reason trash receptacles are not placed on the beach may not be obvious unless the rationale is provided for visitors to take away their own trash. Projects to build dunes or increase their size are often opposed (Nordstrom and Mauriello 2001) as are alternative approaches to traditional methods of shore protection that augment natural features (Zelo et al. 2000). Even government employees in rural coastal communities can be relatively uninformed about environmental concerns, causing dunes to be perceived as wastelands (Moreno-Casasola 2004). Local stakeholders are often poorly informed about coastal processes and may misinterpret the rationale for management actions, for example not realizing that a beach and dune are part of the same sediment exchange system or thinking that the primary purpose of sand-trapping fences is for visitor control (Charboneau et al. 2019). Educational initiatives are needed to increase awareness about the natural resource values of beach vegetation and how natural beach features support endangered species conservation and reduce the potential for storm damage and erosion (Kelly 2014).

One of the best ways to incorporate stakeholders in management programs is through government supported "Beachcare" or "Dune Care" programs. Gesing (2019) provides examples of projects introduced in New Zealand, where the national Department of Conservation and territorial government agencies provide funding for acquisition of plants and education materials to local groups of different sizes and organizational structures that use volunteer labor to plant vegetation and install fences.

Education efforts for stakeholders who are not directly involved in management actions must be conducted via multiple means to reach different user groups, and they must be ongoing because the turnover in population can be rapid. Actions at the local level include instruction in public schools, tours at demonstration sites, displays in libraries, presentations at town meetings, mailings of information to property owners, and information signs at key field locations (Breton et al. 2000; Nordstrom et al. 2002; Nordstrom 2003; Hartley et al. 2015).

Use of demonstration sites for education may be active (bringing interested people to the sites for detailed information) or passive (posting signs for tourists passing by). Active use is most appropriate where specialists explain the feasibility of new management practices. Signs intended for less involved visitors are easy to implement, are low in cost, and require a medium level of maintenance (Carro et al. 2018). Signs must be presented at a level appropriate to casual learners with common terms, simple story lines, colorful word pictures, and bold graphics (Hose 1998). The chance to converse with an "expert" is much appreciated (Hose 1998), pointing to the value of linking demonstration sites to locations designated as critical habitat in endangered species programs and have wardens assigned to them. Wardens could provide information about restoration and comprehensive land management instead of simply telling people to walk around nesting sites. Lifeguards at public bathing areas could provide information as well, providing they are properly informed.

Museums can complement or supplement demonstration sites (Breton et al. 2000), and they have the most value when the role of humans is specified in the processes of both degradation and restoration. A discrepancy often exists between the natural characteristics of beaches displayed in environmentally oriented museums and the human-influenced characteristics of beaches that are normally seen by visitors in coastal resorts. The shells displayed in museums and the vegetation zonation commonly portrayed in pedagogical diagrams of beaches and dunes are not readily found on artificially cleaned and reshaped beaches. Museums can play a critical role in educating the public about the environmental losses in converting beaches and dunes to maximize recreational uses and the need for a return to a more dynamic and naturally functioning coast. A concentration on the way human-altered shores appeared prior to development, with development, and after natural values have been restored may be more enlightening than identifying how beautiful natural beaches may be in places that people will never visit. Museum displays can complement demonstration sites by providing more detailed information than can be communicated by signs on the beach. Directions to demonstration sites can be provided, so visitors can complete their experience by going there, and signs at the beach can direct tourists to the museums for more information (Nordstrom 2003).

The education program developed in the borough of Avalon, New Jersey, USA (Nordstrom et al. 2009), revealed the many ways that information can be conveyed to stakeholders, and it provided a model for transfer of results to other municipalities. Education of residents and landowners in Avalon was considered crucial to acceptance of dunes as an important resource. Meetings between representatives of the borough and landowners were held frequently because of the rapid turnover in resident population. Stakeholders were encouraged to be active in the community and to educate themselves about coastal protection and attend meetings of the Chamber of Commerce, Realtors Association, and Land and Homeowners Association every year. A borough newsletter (*The Avalon Reporter*) and flood hazard information were regularly mailed to property owners. Weekly guided tours of dunes in the dedicated undeveloped area described in Section 8.4 were offered by the nearby university research and education facility. A municipal historical museum traced the history of the dunes from the earliest photos from the 1890s to the present. Dune stewardship was incorporated into the fourth-grade curriculum, and students participated in planting dunes in the spring. The director of public works conducted periodic television interviews on the beach, discussing the significance of dunes and the value of actions to manage them (Nordstrom et al. 2009).

8.7 Maintaining and Evaluating Restored Environments

Assessing the success of restoration projects is critical to justify use of restoration and improve best practice (Wortley et al. 2013). Few locally based beach projects with a restoration component have guidelines for monitoring or are actually monitored, and there may be little information beyond qualitative observations of distinctive features, such as erosion scarps on nourished beaches (Zelo et al. 2000). Considerable ambiguity exists about the availability of funds for long-term monitoring and adaptive management for projects at any level (Shipman 2001). The way a beach is maintained and the size, location, and function of dunes that are intentionally built or are allowed to form can differ greatly, depending on the attitudes and actions of local managers. The development of formal management plans that codify programs and procedures for beach and dune resources increases the likelihood that maintenance and monitoring will continue after initial construction.

8.7.1 Monitoring and Adaptive Management

Monitoring is important in assessing (1) whether the goals of restoration projects have been achieved; (2) when follow-up actions are required; and (3) whether environmentally compatible but nontraditional shore protection approaches are working better than traditional (structural) alternatives. Unanticipated impacts of restoration will occur. The interactions between vegetation types and groundwater discharges are easily disturbed by even small changes in hydrological conditions, possibly causing decline of endangered species (Lammerts et al. 1995). Baeyens and Martínez (2004) provide an example of the complexity of biotic interactions and the ways these interactions can obscure the original intent and results of management decisions. Monitoring provides a way for regulatory agencies to evaluate effects within project areas and in adjacent environments and obtain critical information that can lead to better future projects (Zelo et al. 2000). If goals need to change, adaptive management provides a formalized framework for revising them. Gann et al. (2019) suggest that monitoring designs occur right at the planning stage of the project to ensure that goals and objectives and their indicators are measurable and that there are clear triggers for action if the objectives are not met.

Gann et al. (2019) suggest that each restoration objective should clearly articulate the following: (1) the indicators to be measured (e.g., percent cover of native plants); (2) desired outcomes (increase, decrease, or maintenance of habitats); (3) desired magnitude of effect (e.g., 40% increase); and time frame (e.g., 5 years). Biological monitoring is difficult because of the complexity of the types of observations and statistical procedures required to draw valid conclusions and provide guidance at a practical level for small projects (Zelo et al. 2000). Even evaluations of many large projects reveal fundamental flaws in sampling design, statistical analysis, and critical review (Peterson and Bishop 2005). Nevertheless, monitoring of some kind can be written into every project.

Guidelines for monitoring and quality control will depend on the specific goals of the project. Assessments of dune restorations can include measurement of (1) diversity of plants and insects; (2) vegetation structure, including abundance of ecosystem engineer species, plant biomass, and cover; and (3) belowground characteristics, such as soil nutrients and mycorrhizal abundance, with comparison of these attributes with reference dune communities serving as target conditions (Emery and Rudgers 2010). Funding constraints and lack of scientific expertise at municipal levels limit the rigor of assessments of success and failure, but many critical changes are readily observable, such as expansion of overwash or blowout areas, growth or dieback of key vegetation species, or encroachment of exotics. Documentation of small-scale projects should be rigorous enough to allow managers to learn from past mistakes (Zelo et al. 2000). Time series photography is a way of providing visual evidence to stakeholders and regulators that goals are being met (Gann et al. 2019). Advances in high-resolution dGPS, LiDAR, ARGUS cameras, terrestrial laser scanners, and drones make monitoring easier than in the past (Conlin et al. 2018; Guisado-Pintado et al. 2019; Garcin et al. 2020; Bossard and Nicolae Lerma 2020). Drones are especially useful for local use (Seymour et al. 2017).

Some key variables are readily measured without sophisticated field or lab equipment. Vegetation can be easily sampled for species richness, percent cover of each species, and percent thatch cover, while topographic profiles can provide data on overall morphology and heights and volumes of beaches and dunes. The influence of variables that are more difficult to measure, such as soil moisture and soil nutrient levels, can be inferred from site characteristics, such as changes in species richness and cover in relation to topography within a dune field. Data on simple physical and biological variables that are readily gathered in the field can account for nearly 50% of the variability in species richness in recently restored dunes (Nordstrom et al. 2007a).

Resilience to disturbance in dune restoration projects is rarely measured because most restoration projects have a short time frame (Lithgow et al. 2013). Evaluations of the characteristics of newly created landforms and biota should occur several years after dune building begins, so managers can appreciate the time required for a self-sustaining species-appropriate vegetation cover to become established and have realistic expectations of restoration outcomes. Even where the original target state is achieved, the landscape will continue to evolve. Monitoring after only a few years is likely to reveal few species other than pioneer vegetation and early dune colonizers (Freestone and Nordstrom 2001). Checklists for monitoring dunes can be developed, based on presence of stress factors, degrees of perturbation and vulnerability, the abiotic and biotic elements that facilitate restoration, the levels of management response that can be applied, and the ecosystem services that can be provided using variables that can be quantified using specific or relative values depending on the resources available (Davies et al. 1995; Lithgow et al. 2014; Garcia-Lozano et al. 2020). The variables useful for monitoring dune conditions presented in Table 8.3 represent only the most conspicuous aspects of dunes, and the significance of each variable is multifaceted. Evidence of sand movement within the dunes, for example, may be considered beneficial in terms of habitat rejuvenation and biodiversity but problematic in terms of the potential for inundation of human infrastructure or loss of an existing valued species that requires a stable surface. The sand-trapping and vegetation-planting programs initiated by managers to address sand movement can be evaluated in several ways and with different perspectives colored by the disciplinary backgrounds of the evaluators. Visitor information programs may be examined in terms of their usefulness in controlling visitor uses on site, enriching their experience while there, or providing messages about environmental stewardship that can shape their attitudes about use of resources in the future. The more comprehensive each set of evaluations is, the more useful the results will be for creating a multiple-use resource and documenting its value.

8.7.2 Creating a Stable Source of Funding

Projects for monitoring and adaptive management must be affordable. Stable funding is key to ensuring support but is often difficult to obtain, even for large-scale shore protection projects (Smith 1991; Aceti and Avendaño 1999; Woodruff and Schmidt 1999). Initiatives for obtaining stable funding for beach nourishment can often take place after damaging storms (Nordstrom and Mauriello 2001), and programs for restoration and adaptive management can be identified as part of these initiatives. Hotel and rental property taxes or real estate transfer taxes provide ongoing sources of funds that can be used to make a case for nourishment, although dedicating some of these funds for restoration initiatives may require extra effort to convince stakeholders of values beyond shore protection and active recreation. Dune building and maintenance often uses volunteer labor, and projects have been funded through governmental unemployment relief programs (Knutson 1978; van der Meulen et al. 2004). Regional-scale restoration programs that use volunteer labor and require little funding involve manually removing cultural litter from wrack lines while leaving natural litter, removing exotic species, and replacing sand-trapping fences used for crowd control with symbolic fences. Environmentally favorable actions, such as suspending raking, require no additional funding, but actions such as prohibiting off-road vehicle use may require additional personnel resources for enforcement. Clearly much can be

Table 8.3. *Variables useful in assessing condition of coastal dunes*

Variable	Significance/indicator of
Beach/dune characteristics	
Beach width and height; orientation to wind	Sand source and runup protection
Dune height, volume, surface area	Shore protection, habitat size, sheltering
Number of ridges	Habitat variety, accretion trend
Steepness of slope	Substrate stability, faunal movement
Area of slacks or other moist environments	Habitat variety, biodiversity
% of surface unvegetated	Variety and change of habitat
Evidence of sand movement within dune	Landform mobility, habitat change
Evidence of overwash	Hazard potential, system change
Beach sediment size	Transport potential to dune
Resistant (coarse) lag deposits on surface	Deflation potential, surface stability
Mineralogy affecting nutrients (lime rich or poor)	Growing conditions for vegetation
% of foredune cliffed by swash erosion	Erosional trend, system change
Scarp height and location relative to crest	Erosional trend, system change
Number and width of breaches in dune	Hazard potential, system change
Evidence of new dune growth in breaches	Recovery potential, habitat rejuvenation
Wrack on beach	Habitat variety, food source, new dune line
Vegetation colonization on backshore	Accretion/stability, dune growth
% new dunes forming seaward	Accretion trend
% and density of vegetation on seaward front	Potential new foredune location
Completeness of cross-shore gradients	Ecosystem health, restoration success
Connectivity of subenvironments	Ecosystem health, restoration success
Endemic species	Ecosystem health, restoration success
Variety of soil conditions	Habitat diversity
Presence and location of exotic vegetation	Disturbance, target for removal
Human pressure and use	
Interest of owner or manager	Reason for management approach
Number and size of access roads and parking lots	Potential human impact/degradation
Activities allowed (walking, driving, etc.)	Potential damage to surface cover
Path density and dimensions	Potential/actual damage to surface cover
Presence of houses	Potential for human use, exotic invasion
External influence and nature of boundaries	Sediment budgets, pollutants, exotics
Extraction activities	Surface or water table change, subsidence
Grazing; other uses by livestock, mammals, birds	Vegetation density/health, surface stability
Recent protection/management measures	
Protective legislation	Potential for development, habitat change
Surveillance and maintenance programs	Diagnosis/correction of system changes
% of area in restricted access	Conservation potential
Controls on parking, walking, driving, riding	Degradation/preservation potential
Controls on beach cleaning	Degradation/preservation potential

Table 8.3. (*cont.*)

Variable	Significance/indicator of
Managed paths	Degradation/preservation potential
Sand trapping and vegetation planting programs	Hazard control, dune growth or stabilization
Visitor information programs	Enrichment, crowd control, stewardship
Shore protection structures	Human dominance, desire for stability

(*Sources*: Williams et al. 1993; Davies et al. 1995; Heslenfeld et al. 2004; Lithgow et al. 2014; Garcia-Lozano et al. 2020).

accomplished with minimal financial resources, but policies may have to change, requiring adjustment of attitudes and expectations.

8.8 Developing Policies and Regulations

Community-wide adoption of restoration actions is enhanced by land use policies tailored to local physical and social conditions. Flexible land use policies may be necessary to conform to changes in perception of the value of natural features and changes in human and physical processes through time. Incremental change in policy and practice may be necessary in locations where past precedents favored a static shore and land use more typical of inland locations. Initial regulations may be successfully adopted to favor dune building, using shore protection as the primary rationale. State/provincial regulations based on shore protection alone may not ensure a naturally functioning dune if the regulations allow uses that are incompatible with restoration goals and discourage new environmentally friendly actions by municipal managers and private residents. Modification for habitat enhancement may not be allowed with the kind of general state permit that only allows installing sand fences to build up dunes or reshaping the beach after a storm. Once dunes are established, reshaping or relocating dunes by grading or excavating them or removing vegetation (even exotics) may be restricted by regulations, whether these actions are done for flood protection or for habitat creation and enhancement. Municipalities may implement their own special dune ordinances which may be more restrictive. Greater flexibility in accommodating environmentally friendly actions may be required to enhance environmental values, as indicated in Section 7.4.5.

In some cases, success may be obtained by combining municipal policies with policies at higher political levels to suit highly specific local constraints. The municipality of Bay Head, New Jersey, USA, provides an interesting example of the way the surface characteristics of a dune were maintained where the beach was backed by a seawall and was too narrow to provide a sufficient source of wind-blown sand. State and municipal regulations allowed shorefront property owners in Bay Head to mechanically remove beach sand to a depth of 0.3 m above mean high water and place it over the seawall, providing that the owner obtained a municipal permit. Owners were required to install sand fencing or vegetation plantings at their own expense if the dune became lower than the elevation approved by the municipal dune consultant, even if the lowering was by natural causes. These actions maintained the image and partial functions of a dune despite the narrow beach and presence of the seawall. A new Corps of Engineers beach nourishment project subsequently restored beach width and created a protective dune farther seaward than the seawall. The former precedent of maintaining a dune surface by municipal regulation could be applied to locations that lack a commitment to nourishment. The Bay Head precedent implies that burial of structures and maintenance of dunes by private residents is possible, given appropriate political will (Nordstrom 2019).

8.9 Planning for the Future

Many restoration opportunities could become available as components of actions to address climate change and sea level rise. It is likely that protection of human infrastructure will initially take precedence over restoration actions as coastal storms and sea level rise place more facilities at risk. Frequent alteration of beaches and dunes by physical processes and earth-moving equipment will occur until maintenance of facilities is considered untenable. At that point, incremental retreat from the coast may begin. Changes that can be put in place to speed acceptance of adaptation and restoration strategies that may be required in the future include (1) discouraging new development close to the shoreline; (2) allocating authority and resources between levels of governance according to their effectiveness at that level; (3) identifying and working with key individuals or groups within communities who can influence public opinion and initiate actions but are not constrained by government appointments; (4) expecting the unexpected and incorporating mechanisms for changing rules and regulations; (5) adding incentives to influence stakeholder decisions; and (6) making use of a major storm to call attention to issues, generate motivation for adopting new policies, or set new norms (Abel et al. 2011; Kousky 2014). Storms best work as catalysts if rules and incentives are formulated in anticipation of the storm because the instinctive response after a storm would be to rebuild and protect infrastructure by traditional means. The advantage of having local restoration programs and projects in place now is that stakeholders will have reference sites to evaluate the benefits of allowing coasts to evolve by natural processes.

9

Research Needs

9.1 Introduction

Restoration of developed coastal systems is challenging because it requires research collaboration among different experts (scientists, engineers, planners, managers, property owners) to develop outcomes that are sound and sustainable. The broader backdrop to research needs identified in this chapter includes the following considerations:

1. Climate change could lead to uncertainty of restoration outcomes
2. Restoration alternatives can be adapted to support sustainability goals
3. Restoration generates a coupled natural–human system
4. Restoring coastal landforms and biota is an ongoing, inclusive process

Viewing developed coasts as a coupled natural–human system is offering new insight to the types of restoration outcomes that can develop, but further research is warranted and a wider range of implementable designs will be needed in the future. Change that occurs at restoration sites will introduce new research questions that must be added to the ongoing research needs. Many suggestions in this book represent feasible restoration alternatives that can be evaluated for refinement and application elsewhere. Contemporary strategies such as the Sand Motor, hybrid structures, and managed retreat, in particular, have received much attention recently but have been undertaken with limited evaluation of applicability to locations where they have not been employed. Further studies are required to document the practicality of many of these alternatives so they can be implemented to counterbalance the many degradation actions.

Some of the many investigative questions that can be addressed (Table 9.1) are generic and broad scale; some are more site specific and applicable to individual projects; some address negative side effects; some are research oriented; and some are intended to guide managers. Responses to many of these questions have been

Table 9.1. *Research questions for issues related to restoring degraded coastal environments*

Evaluating and addressing adverse impacts of beach nourishment projects
Are evaluations of environmental losses supported by adequate evidence and analysis?
Are losses revealed in these studies adequately compensated?
Do evaluations extend beyond dredge and fill areas?
How can study of impacts be extended to time frames of decades?
Are smaller, more frequent projects or offshore nourishments better for recovery and
    sustainability of biota than larger upper beach projects?
Under what circumstances is mega nourishment a better alternative than traditional
    methods?
Can fauna in front of dredges be relocated?
How can nourished beaches be designed to accommodate wave action across the new
    backshore without compromising protection to human facilities landward?
How can dunes built and maintained for shore protection be modified to resemble natural
    dunes in form and function?
What are the advantages, opportunities, and constraints on using fill with grain size or
    mineral composition different from native sediment?
How can backpass/bypass operations be made more feasible than mining new borrow
    areas?
How can obstacles to sediment sharing be overcome to enable better distribution
    alongshore?
What measurements of physical variables are required to provide a process-based
    understanding of changes to borrow and fill areas?

Evaluating the scope of nourishment projects
What interdisciplinary studies are required to identify measures of success beyond those
    anticipated in design plans?
How can projects designed for shore protection or recreation be modified to restore
    landforms and ecosystems?
How do adjustments to nearshore topography to enhance surfing, dissipate wave energy, or
    supply sediment onshore increase or decrease success of nourishment programs?
What does each case study contribute to generic understanding of beach nourishment?
What are the criteria for assessing impacts and opportunities of nourishment on gravel
    beaches?
How do criteria for low-energy beaches differ from high-energy beaches?
How would different sediment characteristics improve or degrade nourished sites for
    habitat?
What outcomes of past projects provide insight for modifying future projects to enhance
    natural resources?
What is the future viability of nourishment given long-term sea level rise?
Are land use controls, construction setbacks, or retreat policies better solutions than
    nourishment?

Overcoming cost constraints of nourishment projects
How can restoration efforts be linked to nourishment operations to add value to taxpayers?
How can monitoring and adaptive management be built into projects and adequately
    funded?

Table 9.1. (*cont.*)

How are losses in the borrow area mitigated by environmental gains in the fill area?
Do mitigation projects have built-in criteria for performance, monitoring, or maintenance?
How is adaptive management used to ensure that project goals are achieved, compensation
  and mitigation are addressed, and unnecessary monitoring costs are avoided?

Building dunes
How can designs for artificially created dunes make them better resemble natural
  landforms?
How does the function of dunes built by different means compare?
What is the effect of scraping on backshore habitats?
What is the significance of vegetation in resisting erosion?
What are tradeoffs between creating a dune with ridges and slacks rather than a single
  ridge?
What is the effect of remnant fences in the dune, and how can adverse effects be overcome?
Where can symbolic fences be used to replace sand-trapping fences?
What are realistic planting options for property owners who want to plant backdune
  vegetation?
How can programs for removal of exotic species from private lots be conducted?

Restoring processes, structure, and functions
How and when should sand fences or vegetation plantings be used to seal gaps in
  foredunes?
Are seed banks and refuge areas available to ensure colonization of restored dunes?
How can a dune maintained as a dynamic system be more resistant to erosion?
Do instabilities in landforms and habitats represent temporary or permanent losses?
How does dune reactivation alter predator–prey relationships?
Is there sufficient bare surface, space, and dune mobility for reinitiating cycles of
  evolution?
How does reactivation of dunes affect height, width, volume, or type?
How much dynamism and complexity are critical to maintain long-term viability?
Does raking have a positive effect on initiating aeolian transport and increasing the value of
  the beach as a source for dune building?
Does depositing raked material at the landward edge of the beach create an exotic
  microhabitat?
Where is it possible to reduce or eliminate beach cleaning operations?
How can programs that protect endangered species best include non-endangered species?
How can critical niches that need time or space be maintained on an evolving landscape?

Altering or removing shore protection structures
How can modification of structures increase connectivity at local and regional scales?
How can the need for altering structures be presented in the most positive light?
How can shore protection structures be modified to accommodate restoration goals?
Where can managed retreat be implemented on exposed developed coasts?
What are the advantages of removing existing shore protection structures versus reducing
  their dimensions, allowing them to deteriorate or burying them in place?
Which nature-based alternatives are appropriate for high-energy coastal environments?

Table 9.1. (*cont.*)

Options in spatially restricted environments

How can planting vegetation and deploying fences help overcome temporal and spatial restrictions to vegetation growth?

What are the criteria for managing undeveloped dune enclaves in developed areas?

What are the meaningful differences between bulldozed dunes and dunes created by aeolian processes?

How can shore parallel and shore perpendicular management zones be demarcated or protected without interfering with flows of sediment and biota?

How can the height, width, and spacing of buildings be modified to have less effect on sediment transport and dune forms?

What are the criteria for planting and maintaining backdune species close to the water?

Addressing stakeholder concerns and needs

What are the alternative reference conditions, guidelines, and protocols for documenting the feasibility of return to more dynamic systems?

How do guidelines for nourished beaches differ from unnourished beaches?

How can stakeholder preferences be changed to favor restoring natural features?

How can communication of the advantages of beach nourishment and restoration be enhanced?

Are all stakeholder groups represented in final designs?

How can social science be used to get stakeholders to accept natural landforms and habitats?

Which interesting species or charismatic fauna are best used to encourage restoration efforts?

Which human-altered environments best retain species diversity and richness in the most naturally functioning habitats?

What is gained or lost in selecting a naturally appearing landscape over a naturally functioning one?

How should the concept of "natural" be defined in human-altered systems?

What are best educational initiatives for obtaining stakeholder involvement and commitment?

How can native species, natural litter, and dynamic landforms and habitats be made more acceptable to local stakeholders?

What are best regulatory and non-regulatory practices by national, state/provincial, and local governments that promote and enhance restoration practices and outcomes?

What are the best options for conditioning stakeholders to accept managed retreat?

Maintaining and evaluating restored environments

What are appropriate guidelines for restoration at the local level where scientific expertise and funds are scarce?

What are the limits to human actions to create or maintain naturally functioning environments?

How can unplanned beneficial side effects be incorporated into design of future projects?

What provisions are there for evaluating the benefits that evolve beyond the target state?

How do local restoration activities affect updrift or downdrift communities?

How can several jurisdictions coordinate efforts to obtain more successful restoration outcomes?

---

addressed in preceeding chapters, but site-specific conditions may require alternative solutions and result in new suggestions for application. The issues presented in this chapter represent only a few of the many that should be addressed in comprehensive programs for restoring degraded environments in ways acceptable to stakeholders.

9.2 Nourishing Beaches

9.2.1 Evaluating and Addressing Impacts

Studies evaluating environmental effects of beach nourishment projects reveal the ongoing need for high-quality site-specific investigations of dredge and fill activities, including pre- and post-operation differences and rates of recovery of sediment characteristics and biota. Increasing evidence that sediment from nourished beaches is transported offshore and alongshore beyond project limits indicates that biological sampling should occur in locations previously considered outside the area of interest. Few studies are funded for long periods, so cumulative effects are not documented. Long-term assessment (e.g., scales of decades) should be made whether a project is considered a one-time operation or one of a series of phases in a long-term modification of the environment.

Identifying the tradeoffs involved in using alternative dredge and fill technologies and methods, such as offshore deposition or backshore deposition, is an ongoing need. The benefits of dredging fewer large areas or multiple small areas, dredging in patches or strips, conducting smaller but more frequent projects, and other changes have been hypothesized (Cutter et al. 2000; Minerals Management Service 2001; Bishop et al. 2006), but the feasibility and probabilities of success of these options are not documented.

The amount and rate of application of beach fill can influence the size of the nourished area and the number of subenvironments in a cross-shore transect, as well as the rate of erosion and number of evolutionary stages represented within subenvironments. Large-scale fills are useful for creating new landforms and habitats, but preservation of existing landforms and habitats argues for small, frequent maintenance nourishment operations. More effort should be devoted to assessing how rates of fill determine both landform creation and longevity. The Sand Motor has garnered considerable attention, but the advantages and disadvantages of this kind of nourishment have not been sufficiently compared to traditional methods to determine applicability to other locations. Mega nourishments can intentionally or unintentionally enhance habitat for some species but may not be accurate representations of natural environments and may not be feasible as a way of supplying sediment to the downdrift shoreline where wave conditions are not favorable to longshore transport.

Ways must be found to overcome adverse changes in morphology or substrate conditions in fill areas, such as formation of scarps or impermeable layers, and to make constructed landforms more natural in appearance and function. Designs for nourishment projects can be made compatible with natural habitat requirements if beaches are nourished at the height of natural backshores and a dune is constructed to provide the height and volume for acceptable levels of protection against erosion and overwash. The linear bulldozed dikes that are built to perform the function of dunes and flood control can be redesigned to more closely resemble the hummocky shapes or ridge and swale topography of natural landforms, but designs for more topographically diverse dunes that still provide flood protection have not been evaluated.

Suitable borrow sediment is being depleted in some locations. Fine grain sediment (silts and clays) is usually no longer added to beaches in nourishment operations, but sand and gravel with grain size or mineral composition different from native sediment is likely to be increasingly used. More insight into the positive and negative aspects of using exotic fill is needed, including stakeholder concerns about aesthetics and heritage values.

Backpassing and bypassing operations can conserve sediment that is already in the longshore transport system and reduce the need to mine new borrow areas. Bypassing operations have been subject to considerable study, but evidence of the practicality of backpassing operations is required to make them a viable alternative to mining sources outside the beach environment. The significance of longshore losses from nourished areas is rarely addressed in terms of the potential for gains in landforms and habitats downdrift of the fill area or the degree to which returning this sediment to fill areas in backpassing operations is feasible (Jackson and Nordstrom 2020). Opposition can occur to sharing sediment resources alongshore (Kana and Kaczkowski 2019). Stakeholder demands can thwart recycling plans if the sediment is transferred locally rather than obtaining it from outside sources. Coastal engineers have long known that local interventions against coastal erosion have updrift and downdrift consequences, but the spatial and temporal scales of those effects may be large and transcend local jurisdictions, with each jurisdiction making decisions based on its own goals and objectives (Lazarus et al. 2016). Much work is left to be done in overcoming obstacles to delivery of sediment alongshore.

Most ecological assessments have been conducted as before and after impact studies of biota, without simultaneous measurements of the controlling physical variables (wave height, current velocity, sediment movement). Process-based understanding of changes to borrow and fill areas would help identify the causes of negative impacts to biota and ways of mitigating them (Peterson et al. 2000).

9.2.2 Expanding the Scope of Nourishment Projects

The success of beach nourishment programs is most often considered in terms of the lifespan of the fill and how actual performance conforms to predicted performance. Other measures of success should be defined, so nourishment can serve many objectives. Investigation of effects of dredge and fill activities is inherently interdisciplinary, and biological, geomorphological, sedimentological, engineering, economic, and regulatory expertise should be included. Input from these disciplines should be fully integrated, not produced as discrete, unrelated sections within technical reports, as commonly occurs.

The large volume of sediment that will be placed on beaches in the future in some locations represents an invaluable resource, but the full potential of this sediment will not be realized without a multi-objective management approach that adds habitat improvement and nature- and heritage-based tourism to the traditional goals of protection from erosion and flooding and provision of recreation space. The spatial scope of traditional assessments should be extended by synthesizing studies of discrete sites to make generic statements about the long-term viability and significance of beach nourishment as a general policy for states/provinces and nations.

Many nourishment projects now involve placement of fill in the nearshore to make operations more cost effective and allow sediment to move onshore by natural processes. Nourishment in the nearshore is also being suggested to intentionally create artificial bars, fill holes in bars, or build artificial surfing reefs. The operations designed to deliberately change offshore topography are more poorly studied than operations designed to supply sediment to the upper beach, and further research on creative alterations to nearshore topography seems warranted.

Interest in nourishing beaches with gravel is increasing, but few ecological evaluations of gravel fill projects exist, and technical information on the applied geomorphic and engineering aspects of gravel beaches and sand and gravel beaches is still limited. Experience is increasing in the United Kingdom, but little progress is being made elsewhere. Much of the gravel used in the Mediterranean Sea and Puget Sound is from upland sources (Pacini et al. 1997; Shipman 2001), and studies are required to evaluate the significance of sediment that is more angular and poorly sorted than sediment from beach and nearshore environments. Despite the increasing body of knowledge of the morphodynamics and habitat characteristics of gravel beaches and interest in the morphodynamics of mixed sand and gravel beaches, evaluation of the tradeoffs involved in exchanging gravel for sand is lacking.

Assessing the effects of beach nourishment on nesting processes and hatching success of many species that use the beach is difficult because the physical characteristics of optimal beach habitat are still largely unknown (Dickerson et al.

2007; Jackson et al. 2007). The significance of some of the characteristics of fill for turtle and shorebird nesting, egg viability, and hatchling success has been documented (Crain et al. 1995; Steinitz et al. 1998; Rumbold et al. 2001; Maslo et al. 2011). A similar level of effort is required for other target species and for the non-target species that will use the same beach, so beaches can be nourished as habitats for multiple purposes. Siting nourishment projects for ecological enhancement can have different criteria from siting projects to protect infrastructure. Issues of timing, location, grain size, and topography may take on added significance if creating habitat is important. Species move to take advantage of new opportunities, and there could be a mismatch between the optimum characteristics for species and for shore protection.

Studies of borrow and fill areas to minimize losses of existing habitat are more extensive than studies of ways to maximize the values and uses of new habitat. The major criterion for assessing ecological effects is often how the altered environments resemble preexisting conditions. Some positive by-products of beach nourishment are unanticipated, such as recolonization of the new beach by species that are not specific targets of the nourishment. Evaluations of the significance of unanticipated resources require study of the characteristics of beaches several years after fill operations occur, adopting the attitude that unanticipated differences between the previous or planned beach and the newly evolving beach are not necessarily bad, using adaptive management to maximize the value of the new resource.

9.2.3 Overcoming Cost Constraints

Altering traditional nourishment projects to enhance conditions for biota or to use a more natural template for constructing beaches may add to their cost, as will implementing better programs for long-term monitoring and adaptive management. The added cost should be included in initial estimates and set aside for use in ensuring project viability. Society may be willing to bear the cost of environmental improvements if access to the environments is provided and if sufficient information is presented on the reasons for the cost (van der Meulen et al. 2004). Linking restoration efforts to publicly funded beach nourishment operations in developed areas makes the cost of nourishment palatable to the many taxpayers who do not own shorefront properties or have business interests at the shore.

There are few studies of follow-up efforts to ensure that landforms and habitats created by beach nourishment operations continue to have natural value and that damage to habitats in the borrow area are compensated. For example, tilling the surface of a nourished beach may help overcome the adverse effects of sand compaction (Crain et al. 1995), but the practicality of tilling requires further assessment. Mitigation may be required to offset persistent or long-term adverse impacts. Loss of habitat in one area can never be exactly replaced in another area, so restoration as mitigation is rarely done on an in-kind basis. Loss of biota in a borrow area to create a beach for shore protection or recreation can be at least partially mitigated by allowing portions of the fill to provide habitat or evolve into later stages of natural habitat rather than managing the entire fill area as a recreation platform. Natural evolution of a nourished beach is not axiomatic, given the human preference for maintaining backshore environments as cultural landscapes. A potential problem of mitigation is that it can be performed as only one construction project without specifying future performance criteria or subsequent monitoring or maintenance. Mitigation criteria are required to identify how subsequent monitoring, maintenance, and adaptive management are to be accomplished and funded.

9.3 Building Dunes

Finding ways to increase the rate of growth of sand dunes and the diversity of habitat on them is important in human-occupied areas where time and space do not allow for long-term evolution. Ways must be found to increase the natural function of dunes built using sand-trapping fences and bulldozers. Guidelines for use of sand fences appear adequate for identifying dune heights, widths, and volumes, but alternative fence configurations can result in considerable variety in foredune topography and vegetation, varying from slacks and intervening ridges to higher continuous dunes. Guidelines for creating alternative dune types and managing dunes after they are built are lacking. Shaping dune dikes built by direct deposit of sediment to resemble natural dunes or deploying sand fences in a nonlinear configuration can contribute to variability of microhabitats and greater diversity of species, potentially resulting in greater future resilience. Studies are required of ways to construct dunes with variable topography and surface cover without sacrificing shore protection functions; in this case, allowing the feature to function as a dune rather than a dike.

Future increases in rates of sea level rise and erosion and decreases in amounts of sediment delivered to the coast by streams and bluff erosion are likely to increase the number and frequency of dune building projects using earth-moving equipment to transfer sand from beach to dune or to create storm berms. Many of the landforms built by beach scraping damage backshore habitat and are built as temporary features that provide little habitat value. Finding ways to overcome collateral damage and use the sediment to enhance dunes in more appropriate locations would be helpful.

Guidelines on use of sand fences rarely provide information on when to stop using them in an existing dune field. Placing multiple rows of sand-trapping fences seaward of an existing dune may make the dune wider and keep beach visitors out of sensitive areas, but this practice can be counterproductive in creating naturally functioning dune habitat. Symbolic fences may be a better option to control visitors while allowing for free passage of fauna. Partially buried fences that remain in stabilized dunes also should be evaluated to determine whether they should be removed or whether constructing biodegradable fences or fences with built-in corridors (Figure 3.2d$_3$) would be a better option.

Guidelines for planting vegetation are largely limited to use of primary foredune builders, such as *Ammophila* spp. The large protective foredunes that can be built and maintained by human efforts allow species adapted to locations well landward of the beach to survive close to the water and become viable landscaping elements on private properties. Species characteristic of backdunes and later stages in dune evolution must be evaluated for their suitability closer to the beach than they would naturally occur, and they must be commercially available at local nurseries. The need to eliminate exotic species is ongoing. Public agencies are involved in removing exotic species from their lands, but there are few incentives or guidelines for educating property owners about the problems of using exotic species and advantages of removing them from private shorefront lots.

9.4 Restoring Processes, Structure, and Functions

Attempts to immediately create a mature community structure in restored environments and attempts to categorically stabilize bare areas may not be necessary if beaches and dune fields are wide enough to accommodate dynamism or time is available to allow landforms to evolve. There is often no need to deploy sand fences in some of the undeveloped parks and conservation areas where they are now used. More careful evaluation of how and when to install sand fences and plant vegetation is required, both in building initial foredunes and repairing bare areas and gaps in existing dunes.

The height, width, volume, or type of dune that will occur following reactivation is not always easy to specify. Research on determining how much dynamism is required to maintain natural features and how much dynamism can be tolerated by stakeholders is required to allow for retention of human uses and survival of habitats where dunes are located close to buildings and infrastructure. Local managers will have to feel comfortable in allowing bare areas to form and portions of dunes to be mobile.

Studies of beach raking and evaluations of differences between the effects of natural litter and human litter are required to determine alternatives to removing all wrack by mechanical means and raking all portions of the beach. Evaluations of effects of raking on aeolian transport and creating exotic environments where raked material is dumped would contribute to arguments about the advisability of raking.

Initiatives for protecting target species by controlling active human uses reveal great potential for restoring ecological processes on beaches. Examination of ways these programs can be expanded to apply to species that are not presently endangered will greatly extend the proportion of coast that is allowed to evolve naturally. The rejuvenation projects for Dutch landscapes (Chapter 4) document only a few of the many human actions that can be taken to convert overly stabilized landscapes to more dynamic ones to accommodate target species. Attention should also be given to steps that should be taken to favor endangered species that occupy a different time-dependent or spatially dependent niche. Principles for managing beaches to remain in dynamic states are more poorly developed than principles for stabilizing them. Maintaining developed coastal landscapes in a state of controlled dynamism (e.g., providing analogs to washover formations for plovers) while ensuring that human infrastructure retains a sufficient level of protection is one of the greatest challenges to coastal scientists and engineers.

9.5 Altering or Removing Shore Protection Structures

Ways can be found to modify shore protection structures to allow landforms to be more dynamic, following the precedent of actions implemented or planned for some locations (Nordstrom et al. 2016; Pranzini et al. 2018b). The long-established precedent of using static structures for shore protection is difficult to overcome. The feasibility of converting standard armoring solutions to beach restoration projects or altering or removing structures in managed retreat projects is documented for low-energy shorelines (Section 5.3.1). Many of these projects involve using beach fill to mitigate erosion or enhance the value of the beaches, so they are not technically managed retreat. True managed retreat would allow the newly exposed upland to erode to supply the sediment for creation of landforms and habitats. Planning for retreat on exposed coasts is well advanced (Abel et al. 2011; Niven and Bardsley 2013; Kousky 2014). Many steps are required to successfully implement these programs, beginning with explaining the need for erosion to supply sediment and space for natural features to form, the value of goods and services supplied by these features, and incorporation of designs that work with natural processes. Programs for educating stakeholders and setting institutional arrangements are also necessary. Case studies of actual retreat on exposed coasts are needed to document feasibility.

Structures can be removed, allowed to deteriorate, reduced in dimensions to allow landforms and habitats to form landward of them, or modified to enhance colonization or use of vegetation and fauna on them. Decisions on which action to take will be site specific and will depend on stakeholder interests and legal or economic constraints. Greater evidence of the advantages of all of these alternatives will help speed adoption of the best solution for a given location. Burial of hard shore protection structures can occur as a by-product of dune building by aeolian processes following beach nourishment or by direct burial by earth-moving equipment. Studies of the significance of buried structures as habitat or enhancing recreational or aesthetic resources are lacking. There is still a lack of functional design guidance for hybrid structures (including, wave runup, overtopping, erosion, and overwash criteria), and research into their effectiveness in preventing storm damages is scarce (Almarshed et al. 2020).

Nature-based alternatives offer multiple social, economic, and environmental benefits that static structures alone do not provide. Nature-based options appear well suited to low-energy coasts. Opportunities for combining natural features and human structures are many and varied, but the suitability of these options has yet to be tested for application to high-energy coasts.

9.6 Options in Spatially Restricted Environments

More field investigations are required to determine how human actions can overcome temporal and spatial restrictions to habitats in developed areas. Small undeveloped dune enclaves at the scale of one or more individual lots can remain as isolated remnants between intensively developed lots. These sites may provide suitable natural coastal habitat or they may be occupied by upland species, exotics from adjacent landscaped properties, or predators of valued coastal species. Some landforms in these enclaves are deposited by earth-moving machinery or augmented using mulches and topsoil. Data on the characteristics and environmental significance of these enclaves are lacking.

More information is needed about the size and extent to which microhabitats in the urbanized matrix need to be connected by corridors of movement. Human-constructed shore parallel barriers, such as boardwalks, bulkheads, fences, rows of planted exotic vegetation, and buildings interfere with exchanges of sediment and biota, truncate the cross-shore environmental gradient, and convert the image of natural landscapes into cultural landscapes. Scientific criteria are required to document the rationale for reducing the physical impact of these features (by burying them, modifying their resistance to natural processes, or using alternatives to them) and creating a more continuous cross-shore gradient while making the resulting landscape acceptable to stakeholders.

The backdune environment of developed shores is often replaced by human facilities. If vegetation and fauna that occupy this niche on a natural shore are reintroduced to developed areas, they are likely to be confined to a zone that is closer to the water and narrower than under natural conditions. Human efforts will be required to retain a relatively stable surface that is protected from inundation by sand, water, and salt spray. Criteria are required to develop guidelines for planting and maintaining target species and determining whether the resulting compressed gradient with its aesthetically pleasing landscape and reassuring image of geomorphic stability is more valuable than the dynamic truncated gradient that would occupy this location on a natural shore.

9.7 Addressing Stakeholder Concerns and Needs

Societal issues (land use, private property rights, economic prosperity) and their connection and coupling to geomorphology and ecology are important to maintaining a resilient developed coastal system. Considerable effort will be required to get stakeholders to accept natural landforms and habitats as appropriate elements in a human-modified coastal landscape, especially where these landforms would interfere with views of the sea or ready access to the shoreline. The establishment of appropriate demonstration sites and development of realistic guidelines and protocols for restoration are important steps in convincing stakeholders that return to a more natural system is achievable and desirable. A spectrum of target states may be required to suit different levels of commitment and varying degrees of exposure to the beach and natural coastal processes. Guidelines for nourished beaches will differ from unnourished beaches because nourished beaches will provide a wider source of sand, more space for dunes, and more time for dunes to evolve.

Gaps in the process of communicating knowledge about the advantages and constraints to beach nourishment and other means of restoring natural habitat must be identified and addressed. Planning and execution of nourishment projects should be done with adequate input by all interest groups. Interviews with stakeholders normally outside the formal planning process can reveal concerns that are not incorporated in current designs and identify needs for changes in planning and practice.

Greater use of exotic sediment as beach fill can be expected in the future as suitable sources become unavailable or are not cost effective (Figures 2.8 and 2.9), or if stakeholders want more aesthetically pleasing sediment to change the image of their resorts. Restoration projects often change the image or function of the coast. User surveys that evaluate change may be needed to supplement surveys that focus on reasons people prefer certain environments. Addressing the negative perceptions that discourage acceptance of natural forms is important, especially when implementing strategies that will obscure views, or create environments that interfere with past uses or simply appear different. Assessments should also be conducted well after changes are made to determine whether stakeholders eventually accept the new situation. Issues of preference and acceptability are especially critical in debates about beach raking. Orderliness and cleanliness of landscape features are often viewed positively, but a better definition of how cleanliness applies to beach litter is required, especially whether a distinction can be made between human-generated and natural litter.

Obtaining support for restoration alternatives may be a multistage process. A human-modified nature (e.g., the compressed environmental gradient in Section 6.4) may be more acceptable to stakeholders than a purely natural one. Identification of the most charismatic species may help determine initial designs. What is gained or lost when stakeholders select a naturally appearing landscape rather than a naturally functioning landscape to gain initial favor for more comprehensive restoration actions later would provide insight to the feasibility of this staged approach. Some stakeholders may develop creative adaptation responses on their own, and promising new practices should be encouraged and favored by regulatory programs.

Landforms and habitats that evolve by natural processes will be low, hummocky, and poorly vegetated in initial stages of evolution, and the seaward dune forms will be subject to erosion. The value of these features as habitat, seed sources, and examples of the cycles of growth and destruction that occur on natural coasts must be identified and explained to local stakeholders to make dynamic landforms more acceptable. In turn, scientists must determine how the concept of "natural" should be redefined, how research frameworks must be adjusted to the temporal and spatial scales of small management units, and how restoration goals can be achieved under a scenario of no retreat. Eventually, climate change and sea level rise will make occupation of some portions of coast untenable. More can be done to prepare stakeholders for accepting the retreat option.

9.8 Maintaining and Evaluating Restored Environments

The lack of formal guidelines for constructing or monitoring small restoration projects or funds for adaptive management require the development of protocols that are rigorous but affordable and capable of being sustained. These protocols should reflect multidisciplinary perspectives and represent a variety of interests of local stakeholders. A critical issue is how much human action (e.g., bulldozing or use of structures) will be tolerated in achieving restoration goals. The appropriate levels of human input will be site specific and reflect both the physical constraints and the desires of competing stakeholders, requiring many compromises between interest groups. Local restoration activities will also cause changes in updrift and downdrift communities, indicating a need to coordinate efforts of several jurisdictions along a reach to maximize restoration outcomes.

Social and natural sciences can be considered of equal importance in coastal management. The need for collaboration between scientists with different expertise is evident, and many challenges must be addressed through interdisciplinary research. Incorporating cultural sites and physical landscape elements into green spaces and incorporating cultural heritage into coastal management programs (Ferreira et al. 2006; Gore 2007) can improve the value of restored sites, but it is not clear how cultural features should interact with natural physical or biological features. The attractive image of the marble beach at Marina di Pisa (Figure 2.10) and the importance of marble to the economy and cultural heritage of the region make that beach an evocative symbol with great potential value in a social and economic context (Nordstrom et al. 2008), but the beach may be undesirable from the standpoint of ecological restoration. Investigation of these natural–human hybrids that are obvious departures from natural settings will provide perspective on the limits to compromise solutions.

Some desirable effects occur unintentionally, such as the development of species-rich pioneer dune slacks landward of artificial sand dikes on large sand flats on the Dutch coast (Grootjans et al. 2004) or the rare slacks (Figure 4.9) that can develop landward of new dunes forming on overwash platforms (Nordstrom et al. 2002). Identification of these unplanned restoration successes not envisioned in original designs reveals options for incorporation into future projects. Locations restored to target states representing relatively stable conditions will be reworked by waves and winds and may become more dynamic than envisioned in restoration plans. Dynamism and landscape change may provide additional benefits that should be evaluated through new scientific eyes. In some cases, specialists with expertise that is different from project designers may be required to evaluate the alternative values that new dynamic resources provide.

Managers at the local level who wish to initiate restoration activities or use experimental designs are often restricted in the actions they can take by regulations at higher levels of authority. Institutional change in regulations, permit procedures, and oversight and monitoring may be necessary to provide greater flexibility.

9.9 Concluding Statement

Restoration is an ongoing process of promoting conditions for the development, longevity, and sustainability of landforms, biota, and the economy of developed coasts. The ability of humans to ensure diversity of landforms and biota will be measured by the ingenuity of individuals to effectuate change. Contributions can be made in small projects confined to one or two contiguous properties or large projects that extend beyond municipal borders. Case studies elucidate the benefits of projects that work with the natural processes at all scales. Building on these successes will ensure the refinement of existing methods and design of a new generation of methods that reflect future process regimes.

The restoration actions addressed in this book apply to locations that are developed to the point where a return to truly natural conditions is unlikely. The lack of emphasis on undeveloped natural landscapes in this book should not divert attention from the need to preserve large tracts of undeveloped coastal land, which should be the initial stage in a comprehensive nature protection strategy (Grumbine 1994). Return to fully natural conditions is not likely on developed coasts, so ongoing human actions of some kind will be required to maintain landforms and habitats. This does not mean that new development or intensification of existing human uses should be encouraged.

Successful restoration can be difficult to achieve on developed coasts, where space is at a premium and the time dimension is often shortened to achieve a restoration end point quickly. There is often no quick fix for restoring landforms to an optimum condition, even if that optimum could be identified. The restoration projects in this book represent prototypes of strategies that can become increasingly important to adapt to further human occupation of the coast under conditions of climate change and sea level rise. The number and scale of restoration efforts will have to be greatly increased if naturally functioning landforms and habitats are to be retained in the face of a coastal population unwilling to move away from the coast (Jackson et al. 2013).

References

Abel, N., Gorddard, R., Harman, B., Leitch, A., Langridge, J., Ryan, A., and Heyenga, S. (2011). Sea level rise, coastal development and planned retreat: analytical framework, governance principles and an Australian case study. *Environmental Science and Policy* **14**: 279–288.

Abraham, R. (2000). Die Renaturierung des Polders Friedrichshagen – zweites Deichrückbauprojekt in Ostvorpommern. *Naturschutzarbeit in Mecklenburg-Vorpommern* **43**: 70–73.

Aceti, S. and Avendaño, C. (1999). California's coastal communities organize to increase state funding for beaches. *Shore and Beach* **67**(4): 3–6.

Acosta, A., Carranza, M. L., and Izzi, C. F. (2009). Are there habitats that contribute best to plant species diversity in coastal dunes? *Biodiversity and Conservation* **18**: 1087–1098.

Adriani, M. J. and Terwindt, J. H. J. (1974). *Sand Stabilization and Dune Building*. Rijkswaterstaat Communications 19. The Hague: Rijkswaterstaat.

Alexander, D. E. (2013). Resilience and disaster risk reduction: an etymological journey. *Natural Hazards and Earth System Sciences* **13**: 2707–2716.

Alexandrakis, G., Ghionis, G., Poulos, S. E., and Kampanis, N. A. (2013). Greece. In Pranzini, E. and Williams, A. (Eds.), *Coastal Erosion and Protection in Europe*. London: Routledge, pp. 355–377.

Almarshed, B., Figlus, J., Miller, J., and Verhagen, J. (2020). Innovative coastal risk reduction through hybrid design: combining sand cover and structural defenses. *Journal of Coastal Research* **36**: 174–188.

Altomare, C. and Gentile, G. M. (2013). An innovative methodology for the re-naturalization process of a shingle beach. *Journal of Coastal Research* **SI65**: 1456–1460.

Alves, B., Ballester, R., Rigall-I-Torrent, R., Ferreira, Ó., and Benavente, J. (2017). How feasible is coastal management? A social benefit analysis of a coastal destination in SW Spain. *Tourism Management* **60**: 188–200.

Aminti, P., Cipriani, L. E., and Pranzini, E. (2003). Back to the beach: converting seawalls into gravel. In Goudas, C. L., Katsiaris, G., May, V., et al. (Eds.), *Soft Shore Protection*. Dordrecht: Kluwer Academic Publishers, pp. 261–274.

Anders, F. J. and Leatherman, S. P. (1987). Effects of off-road vehicles on coastal foredunes at Fire Island, New York, USA. *Environmental Management* **11**: 45–52.

Anderson, P. and Romeril, M. G. (1992). Mowing experiments to restore species-rich sward on sand dunes in Jersey, Channel Islands, GB. In Carter, R. W. G., Curtis, T.

G. F., and Sheehy-Skeffington, M. J. (Eds.), *Coastal Dunes: Geomorphology, Ecology and Management for Conservation*. Rotterdam: A.A. Balkema, pp. 219–234.

Andrade, C., Lira, F., Pereira, M. T., Ramos, R., Guerreiro, J., and Feirtas, M. C. (2006). Monitoring the nourishment of Santo Amaro estuarine beach (Portugal). *Journal of Coastal Research* **SI39**: 776–782.

Andrews, C. (2016). Local fiscal impacts of Hurricane Sandy. In O'Neil, K. M. and van Abs, D. J. (Eds.), *Taking Chances: The Coast after Hurricane Sandy*. New Brunswick, NJ: Rutgers University Press, pp. 190–207.

Anfuso, G. and Gracia, F.-J. (2005). Morphodynamic characteristics and short-term evolution of a coastal sector in SW Spain: implications for coastal erosion management. *Journal of Coastal Research* **21**: 1139–1153.

Anthony, E. J. and Cohen, O. (1995). Nourishment solutions to the problem of beach erosion in France: the case of the French Riviera. In Healy, M. G. and Doody, J. P. (Eds.), *Directions in European Coastal Management*. Cardigan: Samara Publishing, pp. 199–212.

Anthony, E. J., Vanhee, S., and Ruz, M.-H. (2007). An assessment of the impacts of experimental brushwood fences on foredune sand accumulation based on digital elevation models. *Ecological Engineering* **31**: 41–46.

Anthony, E. J., Cohen, O., and Sabatier, F. (2011). Chronic offshore loss of nourishment on Nice beach, French Riviera: a case of over-nourishment of a steep beach. *Coastal Engineering* **58**: 374–383.

Anthony, E. J. and Sabatier, F. (2013). France. In Pranzini, E. and Williams, A. (Eds.), *Coastal Erosion and Protection in Europe*. London: Routledge, pp. 227–253.

Anthony, E. J., Marriner, N., and Morhange, C. (2014). Human influence and the changing geomorphology of Mediterranean deltas and coasts over the last 6000 years: from progradation to destruction phase? *Earth-Science Reviews* **139**: 336–361.

Antunes do Carmo, J., Schreck Reis, C., and Freitas, H. (2010). Working with nature by protecting sand dunes: lessons learned. *Journal of Coastal Research* **26**: 1068–1078.

Arba, P., Arisci, A., de Waele, A., et al. (2002). Environmental impact of artificial nourishment of the beaches of Cala Gonone (Central – East Sardinia). *Littoral* **2002**: 465–468.

Archetti, R. (2009). Quantifying the evolution of a beach protected by low crested structures using video monitoring. *Journal of Coastal Research* **25**: 884–899.

Ardeshiri, A., Swait, J., Heagney, E. C., and Kovac, M. (2019). Willingness-to-pay for coastline protection in New South Wales: beach preservation management and decision making. *Ocean & Coastal Management* **178**: 104805.

Arens, S. M. and Wiersma, J. (1994). The Dutch foredunes: inventory and classification. *Journal of Coastal Research* **10**: 189–202.

Arens, S. M., Jungerius, P. D., and van der Meulen, F. (2001). Coastal dunes. In Warren, A. and French, J. R. (Eds.), *Habitat Conservation: Managing the Physical Environment*. London: John Wiley & Sons, pp. 229–272.

Arens, S. M., Slings, Q., and de Vries, C. N. (2004). Mobility of a remobilised parabolic dune in Kennemerland, The Netherlands. *Geomorphology* **59**: 175–188.

Arens, S. M., Geelen, L., Slings, R., and Wondergem, H. (2005). Restoration of dune mobility in the Netherlands. In Herrier, J.-L., Mees, J., Salman, A., et al. (Eds.), *Dunes and Estuaries 2005 – International Conference on Nature Restoration Practices in European Coastal Habitats*. Koksijde, Belgium: VLIZ Special Publication, pp. 129–138.

Arens, S. M., Mulder, J. P. M., Slings, Q. L., Geelen, L. H. W. T., and Damsma, P. (2013a). Dynamic dune management, integration objectives of nature development and coastal safety: examples from the Netherlands. *Geomorphology* **199**: 205–213.

Arens, S. M., Slings, Q. L., Geelen, L. H. W. T., and van der Hagen, H. G. J. M. (2013b). Restoration of dune mobility in The Netherlands. In Martínez, M. L., Gallego-Fernández, J. B., and Hesp, P. A. (Eds.), *Restoration of Coastal Dunes*. New York: Springer, pp. 107–124.

Arens, S. M., de Vries, S., Geelen, L. H. W. T., Ruessink, G., van der Hagen, H. G. J. M., and Groenendijk, D. (2020). Comment on Is "remobilisation" nature restoration or nature destruction? A commentary by I. Delgado-Fernandez, R. G. D. Davidson-Arnott & P. A. Hesp. *Journal of Coastal Conservation* **24**: 17.

Ariza, E. (2011). An analysis of beach management framework in Spain. Study case: the Catalonian coast. *Journal of Coastal Conservation* **15**: 445–455.

Ariza, E., Sarda, R., Jimenez, J. A., Mora, J., and Avila, C. (2008). Beyond performance assessment measurements for beach management: application to Spanish Mediterranean beaches. *Coastal Management* **36**: 47–66.

Ariza, E., Lindeman, K. C., Mozumder, P., and Suman, D. O. (2014). Beach management in Florida: assessing stakeholder perceptions on governance. *Ocean & Coastal Management* **96**: 82–93.

Arler, F. (2000). Aspects of landscape or nature quality. *Landscape Ecology* **15**: 291–302.

Armstrong, J. W., Staude, C. P., Thom, R. M., and Chew, K. K. (1976). Habitats and relative abundances of intertidal macrofauna at five Puget Sound beaches in the Seattle area. *Syesis* **9**: 277–290.

Armstrong, S. B., Lazarus, E. D., Limber, P. W., Goldstein, E. B., Thorp, C., and Ballinger, R. C. (2016). Indications of a positive feedback between coastal development and beach erosion. *Earths Future* **4**: 626–635.

Aronson, J., Floret, C., Le Floc'h, E., Ovalle, C., and Pontanier, R. (1993). Restoration and rehabilitation of degraded ecosystems in arid and semi-arid lands. I. A view from the south. *Restoration Ecology* **1**: 8–17.

Aronson, J., Dhillion, S., and Le Floc'h, E. (1995). On the need to select an ecosystem of reference, however imperfect: a reply to Pickett and Parker. *Restoration Ecology* **3**: 1–3.

Arthurton, R. (1998). Resource, evaluation and net benefit. In Hooke, J. (Ed.), *Coastal Defense and Earth Science Conservation*. Bath: The Geological Society, pp. 151–161.

Austin, M. J. and Masselink, G. (2006). Swash-groundwater interaction on a steep gravel beach. *Continental Shelf Research* **26**: 2503–2519.

Autorita' di Bacino Del Fiume Arno. (1994). *L'evoluzione e la dinamica del litorale prospiciente i bacini dell'Arno e del Serchio e i problemi di erosione della costa*, 3. Autorità di Bacino dell'Arno e del Serchio.

Avis, A. M. (1995). An evaluation of the vegetation developed after artificially stabilizing South African coastal dunes with indigenous species. *Journal of Coastal Conservation* **1**: 41–50.

Baeyens, G. and Martínez, M. L. (2004). Animal life on coastal dunes: from exploitation and prosecution to protection and monitoring. In Martínez, M. L. and Psuty, N. P. (Eds.), *Coastal Dunes, Ecology and Conservation*. Berlin: Springer-Verlag, pp. 279–296.

Balestri, E., Vallerini, F., and Lardicci, C. (2006). A qualitative and quantitative assessment of the reproductive litter from *Posidonia oceanica* accumulated on a sand beach following a storm. *Estuarine, Coastal and Shelf Science* **66**: 30–34.

Balestri, E., Vallerini, F., Seggiani, M., et al. (2019). Use of bio-containers from seagrass wrack with nursery planting to improve the eco-sustainability of coastal habitat restoration. *Journal of Environmental Management* **251**: 109604.

Balletto, J. H., Heimbuch, M. V., and Mahoney, H. K. (2005). Delaware Bay salt marsh restoration: mitigation for a power plant cooling water system in New Jersey, USA. *Ecological Engineering* **25**: 204–213.

Barbier, E. B., Hacker, S. D., Kennedy, C., Koch, E. W., Stier, A. C., and Silliman, B. R. (2011). The value of estuarine and coastal ecosystem services. *Ecological Monographs* **81**: 169–193.

Barbour, E. and Kueppers, L. M. (2012). Conservation and management of ecological systems in a changing California. *Climatic Change* **111**: 135–163.

Barbour, M. G. (1990). The coastal beach plant syndrome. In Davidson-Arnott, R. G. D. (Ed.), *Proceedings of the Symposium on Coastal Sand Dunes*. Ottawa: National Research Council Canada, pp. 197–214.

Barnard, P. L., Erikson, L. H., Foxgrover, A. C., et al. (2019). Dynamic flood modeling essential to assess the coastal impacts of climate change. *Scientific Reports* **9**.

Barragán Muñoz, J. M. (2003). Coastal zone management in Spain (1975–2000). *Journal of Coastal Research* **19**: 314–325.

Barrett, C. B. and Grizzle, R. (1999). A holistic approach to sustainability based on pluralism stewardship. *Environmental Ethics* **21**: 23–42.

Barton, M. E. (1998). Geotechnical problems with the maintenance of geological exposures in clay cliffs subject to reduced erosion rates. In Hooke, J. (Ed.), *Coastal Defense and Earth Science Conservation*. Bath: The Geological Society, pp. 32–45.

Basco, D. R. (1998). The economic analysis of "soft" versus "hard" solutions for shore protection: an example. In *Coastal Engineering: Proceedings of the Twenty-Sixth Coastal Engineering Conference*. New York: American Society of Civil Engineers, pp. 1449–1460.

Basco, D. R. and Pope, J. (2004). Groin functional design guidance from the Coastal Engineering Manual. *Journal of Coastal Research* **SI33**: 121–130.

Baye, P. (1990). Ecological history of an artificial foredune ridge on a northeastern barrier spit. In Davidson-Arnott, R. G. D. (Ed.), *Proceedings of the Symposium on Coastal Sand Dunes*. Ottawa: National Research Council Canada, pp. 389–403.

Beachler, K. E. and Mann, D. W. (1996). Long range positive effects of the Delray beach nourishment program. In *Coastal Engineering 1996: Proceedings of the Twenty-Fifth International Conference*. New York: American Society of Civil Engineers, pp. 4613–4620.

Beatley, T. (1991). Protecting biodiversity in coastal environments: introduction and overview. *Coastal Management* **19**: 1–19.

Beck, T. M. and Wang, P. (2019). Morphodynamics of barrier-inlet systems in the context of regional sediment management, with case studies from west-central Florida, USA. *Ocean & Coastal Management* **177**: 31–51.

Belcher, C. R. (1977). Effect of sand cover on the survival and vigor of *Rosa rugosa* Thunb. *International Journal of Biometeorology* **21**: 276–280.

Bell, J., Sauders, M. I., Leon, J. X., et al. (2014). Maps, laws and planning policy: working with biophysical and spatial uncertainty in the case of sea level rise. *Environmental Science and Policy* **44**: 247–257.

Bell, M. C. and Fish, J. D. (1996). Fecundity and seasonal changes in reproductive output of females of the gravel beach amphipod *Pectenogammarus planicrurus*. *Journal of the Marine Biological Association of the United Kingdom* **76**: 37–55.

Benavente, J., Anfuso, G., Del Rio, L., Gracia, F. J., and Reyes, J. L. (2006). Evolutive trends of nourished beaches in SW Spain. *Journal of Coastal Research* **SI39**: 765–769.

Benedet, L., Finkl, C. W., Campbell, T., and Klein, A. (2004). Predicting the effect of beach nourishment and cross-shore sediment variation on beach morphodynamic assessment. *Coastal Engineering* **51**: 839–861.

Benedet, L., Finkl, C. W., and Dobrochinski, J. P. H. (2013). Optimization of nearshore dredge pit design to reduce impacts on adjacent beaches. *Journal of Coastal Research* **29**: 519–525.

Bennett, A. F. (1991). Roads, roadsides and wildlife conservation: a review. In Saunders, D. A. and Hobbs, R. J. (Eds.), *Nature Conservation 2: The Role of Corridors.* Chipping Norton, NSW: Surrey Beatty and Sons, pp. 71–84.

Bergillos, R. J., Ortega-Sánchez, M., Masselink, G., and Losada, M. A. (2016). Morpho-sedimentary dynamics of a micro-tidal mixed sand and gravel beach, Playa Granada, southern Spain. *Marine Geology* **379**: 28–38.

Berkowitz, J. F., van Zomeren, C. M., and Priestas, A. M. (2018). Potential color change dynamics of beneficial use sediments. *Journal of Coastal Research* **34**: 1149–1156.

Berlanga-Robles, C. A. and Ruiz-Luna, A. (2002). Land use mapping and change detection in the coastal zone of northwest Mexico using remote sensing techniques. *Journal of Coastal Research* **18**: 514–522.

Berry, A., Fahey, S., and Meyers, N. (2013). Changing of the guard: adaptation options that maintain ecologically resilient sandy beach ecosystems. *Journal of Coastal Research* **29**: 899–908.

Bertness, M. D. and Callaway, R. (1994). Positive interactions in communities. *Trends in Ecology and Evolution* **9**: 191–193.

Bertoni, D. and Sarti, G. (2011). On the profile evolution of three artificial pebble beaches at Marina di Pisa, Italy. *Geomorphology* **130**: 244–254.

Bessette, S. R., Hicks, D. W., and Fierro-Cabo, A. (2018). Biological assessment of dune restoration in south Texas. *Ocean & Coastal Management* **163**: 466–477.

Best, P. N. (2003). Shoreline management areas: a tool for shoreline ecosystem management. *Puget Sound Notes* **47**: 8–11.

Biel, R. G., Hacker, S. D., Ruggiero, P., Cohn, N., and Seabloom, E. W. (2017). Coastal protection and conservation on sandy beaches and dunes: context-dependent trade-offs in ecosystem service supply. *Ecosphere* **8**: e01791.

Bilhorn, T. W., Woodard, D. W., Otteni, L. C., Dahl, B. E., and Baker, R. L. (1971). *The Use of Grasses for Dune Stabilization along the Gulf Coast with Initial Emphasis on the Texas Coast.* Report GURC-114. Galveston, TX: Gulf Universities Research Consortium.

Bilkovic, D. M., Mitchell, M., Mason, P., and Duhring, K. (2016). The role of living shorelines as estuarine habitat conservation strategies. *Coastal Management* **44**: 161–174.

Bilodeau, A. L. and Bourgeois, R. P. (2004). Impact of beach restoration on the deep-burrowing ghost shrimp *Callichirus islagrande. Journal of Coastal Research* **20**: 931–936.

Binder, S. B., Baker, C. K., and Barile, J. P. (2015). Rebuild or relocate? Resilience and postdisaster decision-making after Hurricane Sandy. *American Journal of Community Psychology* **56**: 180–196.

Bishop, M. J., Peterson, C. H., Summerson, H. C., Lenihan, H. S., and Grabowski, J. H. (2006). Deposition and long-shore transport of dredge spoils to nourish beaches:

impacts on benthic infauna of an ebb-tidal delta. *Journal of Coastal Research* **22**: 530–546.

Bissett, S. N., Zinnert, J. C., and Young, D. R. (2014). Linking habitat with associations of woody vegetation and vines on two Mid-Atlantic barrier islands. *Journal of Coastal Research* **30**: 843–850.

Bitan, M. and Zviely, D. (2020). Sand beach nourishment: experience from the Mediterranean Coast of Israel. *Journal of Marine Science and Engineering* **8**: 273.

Blackstock, T. (1985). Nature conservation within a conifer plantation on a coastal dune system, Newborough Warren, Anglesey. In Doody, P. (Ed.), *Sand Dunes and Their Management*. Peterborough: Nature Conservancy Council, pp. 145–149.

Blanco, B., Whitehouse, R., Holmes, P., and Clarke, S. (2003). Mixed beaches (sand/gravel): process understanding and implications for management. In *Proceedings of the 38th DEFRA Flood and Coastal Management Conference*. London: Department for Environment, Food and Rural Affairs, pp. 3.1–3.12.

Blindow, I., Gauger, D., and Ahlhaus, M. (2017). Management regimes in a coastal heathland – effects on vegetation, nutrient balance, biodiversity and gain of bio-energy. *Journal of Coastal Conservation* **21**: 273–288.

Blott, S. J. and Pye, K. (2004). Morphological and sedimentological changes on an artificially nourished beach, Lincolnshire, UK. *Journal of Coastal Research* **20**: 214–233.

Bobbink, R., Hornung, M., and Roelofs, J. G. M. (1998). The effects of air-borne nitrogen pollutants on species diversity in natural and semi-natural European vegetation. *Journal of Ecology* **86**: 717–738.

Bocamazo, L. M., Grosskopf, W. G., and Buonuiato, F. S. (2011). Beach nourishment, shoreline change, and dune growth at Westhampton Beach, New York, 1996–2009. *Journal of Coastal Research* **SI59**: 181–191.

Bonte, D., Maelfait, J.-P., and Hoffmann, M. (2000). The impact of grazing on spider communities in a mesophytic calcareous dune grassland. *Journal of Coastal Conservation* **6**: 135–144.

Boon, J. D. (2012). Evidence of sea level acceleration at U.S. and Canadian tide stations, Atlantic Coast, North America. *Journal of Coastal Research* **28**: 1437–1445.

Boorman, L. A. (1989). The grazing of British sand dune vegetation. *Proceedings of the Royal Society of Edinburgh* **96B**: 75–88.

Borja, A. and Elliott, M. (2019). So when will we have enough papers on microplastics and ocean litter? *Marine Pollution Bulletin* **146**: 312–316.

Borsje, B. W., van Wesenbeeck, B. K., Dekker, F., et al. (2011). How ecological engineering can serve in coastal protection. *Ecological Engineering* **37**: 113–122.

Bossard, V. and Nicolae Lerma, A. (2020). Geomorphologic characteristics and evolution of managed dunes on the South West Coast of France. *Geomorphology* **367**: 107312.

Bossuyt, B., Cosyns, E., and Hoffmann, M. (2007). The role of seed banks in the restoration of dry acidic dune grassland after burning of *Ulex europaeus* scrub. *Applied Vegetation Science* **10**: 131–138.

Bourman, R. P. (1990). Artificial beach progradation by quarry waste disposal at Rapid Bay, South Australia. *Journal of Coastal Research* **SI6**: 69–76.

Bowman, D., Manor-Samsonov, N., and Golik, A. (1998). Dynamics of litter pollution on Israeli Mediterranean beaches: a budgetary, litter flux approach. *Journal of Coastal Research* **14**: 418–432.

Brampton, A. H. (1998). Cliff conservation and protection: methods and practices to resolve conflicts. In Hooke, J. (Ed.), *Coastal Defense and Earth Science Conservation*. Bath: The Geological Society, pp. 21–31.

Bray, M. and Hooke, J. (1998). Spatial perspectives in coastal defence and conservation strategies. In Hooke, J. (Ed.), *Coastal Defence and Earth Science Conservation*. Bath: The Geological Society, pp. 115–132.

Brennan, J. (2011). *Powel Shoreline Restoration Design Project: Final Report*. Bainbridge Island, WA: Bainbridge Island Land Trust.

Breton, F. and Esteban, P. (1995). The management and recuperation of beaches in Catalonia. In Healy, M. G. and Doody, J. P. (Eds.), *Directions in European Coastal Management*. Cardigan: Samara Publishing Ltd., pp. 511–517.

Breton, F., Clapés, J., Marqués, A., and Priestly, G. K. (1996). The recreational use of beaches and consequences for the development of new trends in management: the case of the beaches in the Metropolitan Region of Barcelona. *Ocean & Coastal Management* **32**: 153–180.

Breton, F., Esteban, P., and Miralles, E. (2000). Rehabilitation of metropolitan beaches by local administrations in Catalonia: new trends in sustainable coastal management. *Journal of Coastal Conservation* **6**: 97–106.

Bridges, T. S., Wagner, P. W., Burks-Copes, K. A., et al. (2015). *Use of Natural and Nature-Based Features (NNBF) for Coastal Resilience*. ERDC SR-15-1. Vicksburg, MS: U.S. Army Engineer Research and Development Center.

Brock, K. A., Reece, J. S., and Ehrhart, L. M. (2009). The effects of artificial beach nourishment on marine turtles: differences between Loggerhead and green turtles. *Restoration Ecology* **17**: 297–307.

Brodie, K., Conery, I., Cohn, N., Spore, N., and Palmsten, M. (2019). Spatial variability of coastal foredune evolution, Part A: timescales of months to years. *Journal of Marine Science and Engineering* **7**(124): 1–28.

Brooks, S. M. and Spencer, T. (2010). Temporal and spatial variations in recession rates and sediment release from soft rock cliffs, Suffolk, coast, UK. *Geomorphology* **124**: 26–41.

Broome, S. W., Seneca, E. D., Woodhouse, W. W., and Griffin, C. (1982). *Building and Stabilizing Coastal Dunes with Vegetation*. UNC-SG-82-05. Raleigh: North Carolina University Sea Grant.

Brown, A. C. (1996). Behavioural plasticity as a key factor in the survival and evolution of the macrofauna on exposed sandy beaches. *Revista Chilena de Historia Natural* **69**: 469–474.

Brown, A. C. and McLachlan, A. (2002). Sandy shores ecosystems and the threats facing them: some predictions for the year 2025. *Environmental Conservation* **29**: 62–77.

Brown, A. C., Nordstrom, K. F., McLachlan, A., Jackson, N. L., and Sherman, D. J. (2008). The future of sandy shores. In Polunin, N. (Ed.), *The Waters, Our Future. Prospects for the Integrity of Aquatic Ecosystems*. Cambridge: Cambridge University Press, pp. 263–280.

Brown, J. M., Phelps, J. J. C., Barkwith, A., Hurst, M. D., Ellis, M. A., and Plater, A. J. (2016). The effectiveness of beach mega-nourishment, assessed over three management epochs. *Journal of Environmental Management* **184**: 400–408.

Browne, M. A. and Chapman, M. G. (2011). Ecologically informed engineering reduces loss of intertidal biodiversity on artificial shores. *Environmental Science and Technology* **45**: 8204–8207.

Brunbjerg, A. K., Jørgensen, G. P., Nielsen, K. M., Pedersen, M. L., Svenning, J.-C., and Ejrnaes, R. (2015). Disturbance in dry coastal dunes in Denmark promotes diversity of plants and arthropods. *Biological Conservation* **182**: 243–253.

Brutsché, K. E., Wang, P., Rosati, J. D., and Beck, T. M. (2015). Evolution of a swash zone berm nourishment and influence of berm elevation on the performance of beach-nearshore nourishments along Perdido Key, Florida, USA. *Journal of Coastal Research* **31**: 964–977.

Buchanan, J. K. (1995). Managing heritage coasts. In Salman, A. H. P. M., Berends, H., and Bonazountas, M. (Eds.), *Coastal Management and Habitat Conservation*. Leiden: EUCC, pp. 153–159.

Buckley, E. C. B., Hilton, M. J., Konlechner, T. M., and Lord, J. M. (2016). Downwind sedimentation and habitat development following *Ammophila arenaria* removal and dune erosion, Mason Bay, New Zealand. *Journal of Coastal Research* **SI75**: 268–272.

Burger, J., Howe, M. A., Hahn, D. C., and Chase, J. (1977). Effects of tide cycles on habitat selection and habitat partitioning by migrating shorebirds. *Auk* **4**: 743–758.

Burger, J., O'Neill, K. M., Handel, S. N., Hensold, B., and Ford, G. (2017). The shore is wider than the beach: ecological planning solutions to sea level rise for the Jersey shore USA. *Landscape and Urban Planning* **157**: 512–522.

Burgess, K., Hunt, P., Cali, M., Jay, H., and Campbell, D. (2016). Planning for change on the Lincolnshire coast. In Baptiste, A. (Ed.), *Coastal Management: Changing Coast, Changing Climate, Changing Minds*. International Coastal Management Conference. London: ICE Publishing, pp. 425–436.

Burke, S. M. and Mitchell, N. (2007). People as ecological participants in ecological restoration. *Restoration Ecology* **15**: 348–350.

Burkitt, J. and Wootton, L. (2011). Effects of disturbance and age of invasion on the impact of invasive sand sedge, Carex kobomugi, on native dune plant populations in New Jersey's coastal dunes. *Journal of Coastal Research* **27**: 182–193.

Caetano, C. H. S., Cardoso, R. S., Veloso, V. G., and Silva, E. S. (2006). Population biology and secondary production of *Excirolana braziliensis* (Isopoda: cirolanidae) in two sandy beaches of southeastern Brazil. *Journal of Coastal Research* **22**: 825–835.

Cahoon, L. B., Carey, E. S., and Blum, J. E. (2012). Benthic microalgal biomass on ocean beaches: effects of sediment grain size and beach nourishment. *Journal of Coastal Research* **28**: 853–859.

Caldwell, J. M. (1966). Coastal processes and beach erosion. *Journal of the Society of Civil Engineers* **53**: 142–157.

Callaway, R. M. (1995). Positive interactions among plants. *Botanical Review* **61**: 306–349.

Cammelli, C., Jackson, N. L., Nordstrom, K. F., and Pranzini, E. (2006). Assessment of a gravel-nourishment project fronting a seawall at Marina di Pisa, Italy. *Journal of Coastal Research* **SI39**: 770–775.

Campbell, T. and Benedet, L. (2007). A dedicated issue on the "storm protective value of coastal restoration." *Shore and Beach* **75**(1): 2–3.

Capobianco, M., Hanson, H., Larson, M., et al. (2002). Nourishment design and evaluation: applicability of model concepts. *Coastal Engineering* **47**: 113–135.

Caputo, C., Chiocci, F. L., Ferrante, A., La Monica, G. B., Landini, B., and Pugliese, F. (1993). La ricostituzione dei litorali in erosione mediante ripascimento artificiale e il problema del reperimento degli inerti. In *La difesa dei litorali in Italia*. Roma: Edizioni delle Autonomie, pp. 121–151.

Carro, I., Seijo, L., Nagy, G. J., Lagos, X., and Gutiérez, O. (2018). Building capacity on ecosystem-based adaptation strategy to cope with extreme events and sea-level rise on the Uruguayan coast. *International Journal of Climate Change Strategies and Management* **10**: 504–522.

Carter, R. W. G. and Orford, J. D. (1984). Coarse clastic barrier beaches: a discussion of the distinctive dynamic and morphosedimentary characteristics. *Marine Geology* **60**: 377–389.

Castelle, B., Turner, I. L., Bertin, X., and Tomlinson, R. (2009). Beach nourishments at Coolangatta Bay over the period 1987–2005. *Coastal Engineering* **56**: 940–950.

Castelle, B., Laporte-Fauret, Q., Marieu, V., et al. (2019). Nature-based solution along high-energy eroding sandy coasts: preliminary tests on the reinstatement of natural dynamics in reprofiled coastal dunes. *Water* **11**: 2518.

Castillo, S. A. and Moreno-Casasola, P. (1996). Coastal sand dune vegetation: an extreme case of species invasion. *Journal of Coastal Conservation* **2**: 13–22.

Castley, J. G., Bruton, J.-S., Kerley, G. I. H., and McLachlan, A. (2001). The importance of seed dispersal in the Alexandria coastal dune field, South Africa. *Journal of Coastal Conservation* **7**: 57–70.

Catma, S. (2020). Non-market valuation of beach quality: using spatial hedonic price modeling in Hilton Head Island, SC. *Marine Policy* **115**: 103866.

Cazenave, A. and Le Cozannet, G. (2013). Sea level rise and its coastal impacts. *Earth's Future* **2**: 15–34.

Cerra, J. F. (2017). Emerging strategies for voluntary urban ecological stewardship on private property. *Landscape and Urban Planning* **157**: 586–597.

Chandramohan, P., Kumar, S. J., Kumar, V. S., and Ilangovan, D. (1998). Fine particle deposition at Vainguinim tourist beach, Goa, India. *Journal of Coastal Research* **14**: 1074–1081.

Chapman, D. M. (1989). *Coastal Dunes of New South Wales: Status and Management.* Sydney: University of Sydney Coastal Studies Unit Technical Report 89/3.

Chapman, M. G. and Underwood, A. J. (2011). Evaluation of ecological engineering of "armored" shorelines to improve their value as habitat. *Journal of Experimental Marine Biology and Ecology* **400**: 302–313.

Charbonneau, B. R., Wnek, J. P., Langley, J. A., Lee, G., and Balsamo, R. A. (2016). Above vs. belowground plant biomass along a barrier island: implications for dune stabilization. *Journal of Environmental Management* **182**: 126–133.

Charbonneau, B. R., Wootton, L. S., Wnek, J. P., Langley, J. A., and Posner, M. A. (2017). A species effect on storm erosion: invasive sedge stabilized dunes more than native grass during Hurricane Sandy. *Journal of Applied Ecology* **54**: 1385–1394.

Charbonneau, B. R., Cochran, C., and Avenarius, C. (2019). What we know and what we think we know: revealing misconceptions about coastal management for sandy beaches along the U.S. Atlantic Seaboard. *Journal of Environmental Management* **245**: 131–142.

Cheney, D., Oestman, R., Volkhardt, G., and Getz, J. (1994). Creation of rocky intertidal and shallow subtidal habitats to mitigate for the construction of a large marina in Puget Sound, Washington. *Bulletin of Marine Science* **55**: 772–782.

Chittora, A., Joshi, V. B., and Kumar, M. P. (2017). Assessment of beach nourishment through analysis of beach profiles. *International Journal of Civil Engineering and Technology* **8**: 153–159.

Chiva, L., Pagán, J. I., López, I., Tenza-Abril, A. J., Aragonés, L., and Sánchez, I. (2018). The effects of sediment used in beach nourishment: study case El Portet de Moraira beach. *Science of the Total Environment* **628–629**: 64–73.

Choi, T.-J., Choi, J.-Y., Park, J.-Y., and Yang, Y.-J. (2020). Long-term temporal and spatial morphological variability of a nourished beach using the EOF analysis. *Journal of Coastal Research* **SI95**: 428–432.

Choi, Y. D. (2007). Restoration ecology to the future: a call for new paradigm. *Restoration Ecology* **15**: 351–353.

Choi, Y. D. and Pavlovic, N. B. (1998). Experimental restoration of native vegetation in Indiana Dunes National Lakeshore. *Restoration Ecology* **6**: 118–129.

Christensen, S. N. and Johnsen, I. (2001). The lichen-rich coastal heath vegetation on the isle of Anholt, Denmark – conservation and management. *Journal of Coastal Conservation* **7**: 13–22.

Cialone, M. A. and Stauble, D. K. (1998). Historical findings on ebb shoal mining. *Journal of Coastal Research* **14**: 537–563.

Cipriani, L. E., Dreoni, A. M., and Pranzini, E. (1992). Nearshore morphological and sedimentological evolution induced by beach restoration: a case study. *Bolletino di Oceanologia Teorica ed Applicata* **10**: 279–295.

Clarke, M. L. and Rendell, H. M. (2015). "This restless enemy of all fertility": exploring paradigms of coastal dune management in Western Europe over the last 700 years. *Transactions of the Institute of British Geographers* **40**: 414–429.

Clausner, J. E., Gebert, J. A., Rambo, A. T., and Watson, K. D. (1991). Sand bypassing at Indian River Inlet, Delaware. In *Coastal Sediments 91*. New York: American Society of Civil Engineers, pp. 1177–1191.

Clewell, A. and Rieger, J. P. (1997). What practitioners need from restoration ecologists. *Restoration Ecology* **5**: 350–354.

Coastal Engineering Research Center (CERC). (1984). *Shore Protection Manual*. Ft. Belvoir, VA: U.S. Army Corps of Engineers.

Colantoni, P., Menucci, D., and Nesci, O. (2004). Coastal processes and cliff recession between Gabicce and Pesaro (northern Adriatic Sea): a case history. *Geomorphology* **62**: 257–268.

Colenbrander, D. and Bavinck, M. (2017). Exploring the role of bureaucracy in the production of coastal risks, City of Capetown, South Africa. *Ocean & Coastal Management* **150**: 35–50.

Colombini, I. and Chelazzi, L. (2003). Influence of marine allochthonous input on sandy beach communities. *Oceanography and Marine Biology: An Annual Review* **41**: 115–159.

Coltorti, M. (1997). Human impact in the Holocene fluvial and coastal evolution of the Marche region, Central Italy. *Catena* **30**: 311–335.

Conaway, C. A. and Wells, J. T. (2005). Aeolian dynamics along scraped shorelines, Bogue Banks, North Carolina. *Journal of Coastal Research* **21**: 242–254.

Conger, T. and Chang, S. E. (2019). Developing indicators to identify coastal green infrastructure potential: the case of the Salish Sea region. *Ocean & Coastal Management* **175**: 53–69.

Conlin, M., Cohn, N., and Ruggiero, P. (2018). A quantitative comparison of low-cost structure from motion (SfM) data collection platforms on beaches and dunes. *Journal of Coastal Research* **34**: 1341–1357.

Connors, P. G., Myers, J. P., Connors, C. S. W., and Pitelka, F. A. (1981). Interhabitat movements by sanderlings in relation to foraging profitability and the tidal cycle. *Auk* **98**: 49–64.

Conway, T. M. and Nordstrom, K. F. (2003). Characteristics of topography and vegetation at boundaries between the beach and dune on residential shorefront lots in two municipalities in New Jersey, USA. *Ocean & Coastal Management* **46**: 635–648.

Cooke, B. C., Jones, A. R., Goodwin, I. D., and Bishop, M. J. (2012). Nourishment practices on Australian sandy beaches: a review. *Journal of Environmental Management* **113**: 319–327.

Cooper, J. A. G. and Lemckert, C. (2012). Extreme sea-level rise and adaptation options for coastal resort cities: a qualitative assessment from the Gold Coast, Australia. *Ocean & Coastal Management* **64**: 1–14.

Cooper, J. A. G. and McKenna, J. (2008). Working with natural processes: the challenge for coastal protection strategies. *The Geographical Journal* **174**: 315–331.

Cooper, J. A. G. and Pile, J. (2014). The adaptation-resistance spectrum: a classification of contemporary adaptation approaches to climate-related coastal change. *Ocean & Coastal Management* **94**: 90–98.

Cooper, J. A. G. and Pilkey, O. H. (2004). Alternatives to mathematical modeling of beaches. *Journal of Coastal Research* **20**: 641–644.

Cooper, N. J. and Pethick, J. S. (2005). Sediment budget approach to addressing coastal erosion problems in St. Ouen's Bay, Jersey, Channel Islands. *Journal of Coastal Research* **21**: 112–122.

Cooper, N., Benson, N., McNeill, A., and Siddle, R. (2017). Changing coastlines in NE England: a legacy of colliery spoil tipping and the effects of its cessation. *Proceedings of the Yorkshire Geological Society* **61**: 217–229.

Cooper, W. S. (1958). The coastal sand dunes of Oregon and Washington. *Geological Society of America Memoir* **72**.

Corbella, S. and Stretch, D. D. (2012). Geotextile sand filled containers as coastal defence: South African experience. *Geotextiles and Geomembranes* **35**: 120–130.

Corbin, J. D. and D'Antonio, C. M. (2012). Gone but not forgotten: invasive plant's legacies on community and ecosystem properties. *Invasive Plant Science and Management* **5**: 117–124.

Cordshagen, H. (1964). *Der Küstenschutz in Mecklenburg*. Schwerin: Petermänken-Verlag.

Corlett, R. T. (2016). Restoration, reintroduction, and rewilding in a changing world. *Trends in Ecology and Evolution* **31**: 453–462.

Correll, D. L. (1991). Human impact on the functioning of landscape boundaries. In Holland, M. M., Naiman, R. J., and Risser, P. G. (Eds.), *Ecotones: The Role of Landscape Boundaries in the Management and Restoration of Changing Environments*. New York: Chapman and Hall, pp. 90–109.

Costas, S., Ferreira, O., and Martinez, G. (2015). Why we decide to live with risk at the coast. *Ocean & Coastal Management* **118**: 1–11.

Council of Europe. (1999). *European Code of Conduct for Coastal Zones*. CO-DBP(99)11. Strasbourg: Council of Europe.

Cowell, P. J., Thom, B. G., Jones, R. A., Everts, C. H., and Simanovic, D. (2006). Management of uncertainty in predicting climate-change impacts on beaches. *Journal of Coastal Research* **22**: 232–245.

Cox, D. (1997). On the value of natural relations. *Environmental Ethics* **19**: 173–183.

Crain, A. D., Bolten, A. B., and Bjorndal, K. A. (1995). Effects of beach nourishment on sea turtles: review and research initiatives. *Restoration Ecology* **3**: 95–104.

Cranz, G. and Boland, M. (2004). Defining the sustainable park: a fifth model for urban parks. *Landscape Journal* **23**: 102–120.

Crawford, K. M., Busch, M. H., Locke, H., and Luecke, N. C. (2020). Native soil microbial amendments generate trade-offs in plant productivity, diversity, and soil stability in coastal dune restorations. *Restoration Ecology* **28**: 328–336.

Creer, J., Litt, E., Ratcliffe, J., Rees, S., Thomas, N., and Smith, P. (2020). A comment on some of the conclusions made by Delgado-Fernandez et al. (2019). "Is 're-mobilisation' nature conservation or nature destruction: a commentary". *Journal of Coastal Conservation* **24**: 29.

Cristiano, S. C., Portz, L., Nasser, P. C., Pinto, A. C., da Silva, P. R., and Barboza, E. G. (2018). Strategies for the management of the marine shoreline in the Orla Araranguá project (Santa Catarina, Brazil). In Botero, C. M., Cervantes, O., and Finkl, C. W.

(Eds.), *Beach Management Tools – Concepts, Methodologies and Case Studies*. Cham, Switzerland: Springer International Publishing, pp. 735–754.

Crowe, S. E., Bergquist, D. C., Sanger, D. M., and van Dolah, R. F. (2016). Physical and biological alterations following dredging in two beach nourishment borrow areas in South Carolina's coastal zone. *Journal of Coastal Research* **32**: 875–889.

Cruz, H. da. (1996). Tourism and environment in the Mediterranean. In Salman, A. H. P. M., Langeveld, M. J., and Bonazountas, M. (Eds.), *Coastal Management and Habitat Conservation*, vol. II. Leiden: European Union for Coastal Conservation, pp. 113–116.

Cullen, P. and Bird, E. C. F. (1980). *The Management of Coastal Sand Dunes in South Australia*. Black Rock, VIC: Geostudies.

Cunniff, S. E. (1985). Impacts of severe storms on beach vegetation. In *Coastal Zone 85*. New York: American Society of Civil Engineers, pp. 1022–1037.

Cunningham, D. J. and Wilson, S. P. (2003). Marine debris on beaches of the greater Sydney region. *Journal of Coastal Research* **19**: 421–430.

Currin, C. A., Chappell, W. S., and Deaton, A. (2010). Developing alternative shoreline armoring strategies: the living shoreline approach in North Carolina. In: Shipman, H., Dethier, M. N., Gelfenbaum, G., Fresh, K. L., and Dinicola, R. S. (Eds.), Puget Sound Shorelines and the Impacts of Armoring: Proceedings of a State of the Science Workshop, May 2009. U.S. Geological Survey Scientific Investigations Report 2010-5254, pp. 91–102.

Cutter, G. R., Diaz, R. J., Musick, J. A., et al. (2000). *Environmental Survey of Potential Sand Resource Sites Offshore Delaware and Maryland*. U.S. Department of the Interior, Minerals Management Service, OCS Study 2000-055.

Cutter, S., Nordstrom, K. F., and Kucma, G. (1979). Social and environmental factors influencing beach site selection. In *Resource Allocation Issues in the Coastal Environment*. Arlington, VA: The Coastal Society.

d'Angremond, K., van den Berg, E. J. F., and de Jager, J. H. (1992). Use and behavior of gabions in coastal protection. In *Proceedings of the Twenty-Third Coastal Engineering Conference*. New York: American Society of Civil Engineers, pp. 1748–1757.

Dahl, B. E. and Woodard, D. W. (1977). Construction of Texas coastal foredunes with sea oats (*Uniolo paniculata*) and bitter panicum (*Panicum amarum*). *International Journal of Biometeorology* **21**: 267–275.

Dally, W. R. and Osiecki, D. A. (2018). Evaluating the impact of beach nourishment on surfing: Surf City, Long Beach Island, New Jersey, USA. *Journal of Coastal Research* **34**: 793–805.

Daly, H. (2020). A note in defense of the concept of natural capital. *Ecosystem Services* **41**: 101051.

Damgaard, C., Thomsen, M. P., Borchsenius, F., Nielsen, K. E., and Strandberg, M. (2013). The effect of grazing on biodiversity in coastal dune heathlands. *Journal of Coastal Conservation* **17**: 663–670.

Darke, I. B., Eamer, J. B. R., Beaugrand, H. E. R., and Walker, I. J. (2013). Monitoring considerations for a dynamic dune restoration project: Pacific Rim National Park Reserve, British Columbia, Canada. *Earth Surface Processes and Landforms* **38**: 983–993.

Darke, I. B., Walker, I. J., and Hesp, P. A. (2016). Beach-dune sediment budgets and dune morphodynamics following coastal dune restoration, Wickkaninnish Dunes, Canada. *Earth Surface Processes and Landforms* **41**: 1370–1385.

Davenport, J. and Davenport, J. L. (2006). The impact of tourism and personal leisure transport on coastal environments: a review. *Estuarine, Coastal and Shelf Science* **67**: 280–292.

Davidson, A. T., Nicholls, J., and Leatherman, S. P. (1992). Beach nourishment as a coastal management tool: an annotated bibliography on developments associated with the artificial nourishment of beaches. *Journal of Coastal Research* **8**: 984–1022.

Davies, P., Curr, R., Williams, A. T., Hallégouët, B., Bodéré, J. C., and Koh, A. (1995). Dune management strategies: a semi-quantitative assessment of the interrelationships between coastal dune vulnerability and protection measures. In Salman, A. H. P. M., Berends, H., and Bonazountas, M. (Eds.), *Coastal Management and Habitat Conservation*. Leiden: EUCC, pp. 313–331.

Davis, J. H. (1975). *Stabilization of Beaches and Dunes by Vegetation in Florida*. Florida Sea Grant-7. Gainesville: Florida Sea Grant College Program.

Davis, J. L., Currin, C., O'Brien, C., Raffenburg, C., and Davis, A. (2015). Living shorelines: coastal resilience with a blue carbon benefit. *PLoS ONE* **10**(11): 1–18.

Davis, R. A. (1991). Performance of a beach nourishment project based on detailed multi-year monitoring: Redington Beach, FL. In *Coastal Sediments 91*. New York: American Society of Civil Engineers, pp. 2101–2115.

Davis, R. A. Jr., FitzGerald, M. V., and Terry, J. (1999). Turtle nesting on adjacent nourished beaches with different construction styles: Pinellas County, Florida. *Journal of Coastal Research* **15**: 111–120.

Dawson, R. J., Dickson, M. E., Nicholls, R. J., et al. (2009). Integrating analysis of risks of coastal flooding and cliff erosion under scenarios of long term change. *Climatic Change* **95**: 249–288.

De Bonte, A. J., Boosten, A., van der Hagen, H. G. J. M., and Sýkora, K. V. (1999). Vegetation development influenced by grazing in the coastal dunes near The Hague, The Netherlands. *Journal of Coastal Conservation* **5**: 59–68.

De Jong, B., Keijsers, J. G. S., Riksen, M. J. P. M., Krol, J., and Slim, P. A. (2014). Soft engineering vs. A dynamic approach in coastal dune management: a case study on the North Sea barrier island of Ameland, The Netherlands. *Journal of Coastal Research* **30**: 670–684.

de la Vega-Leinert, C., Stoll-Kleemann, S., and Wegener, E. (2018). Managed realignment (MR) along the eastern German Baltic Sea: a catalyst for conflict or for a coastal zone management consensus. *Journal of Coastal Research* **34**: 586–601.

De Lillis, M., Costanzo, L., Bianco, P. M., and Tinelli, A. (2004). Sustainability of sand dune restoration along the coast of the Tyrrhenian Sea. *Journal of Coastal Conservation* **10**: 93–100.

De Raeve, F. (1989). Sand dune vegetation and management dynamics. In van der Meulen, F., Jungerius, P. D., and Visser, J. H. (Eds.), *Perspectives in Coastal Dune Management*. The Hague: SPB Academic Publishing, pp. 99–109.

de Ruig, J. H. M. (1996). Seaward coastal defence: limitations and possibilities. In Salman, A. H. P. M., Langeveld, M. J., and Bonazountas, M. (Eds.), *Coastal Management and Habitat Conservation*. Leiden: EUCC, pp. 453–464.

de Ruig, J. H. M. and Hillen, R. (1997). Developments in Dutch coastline management: conclusions from the second governmental coastal report. *Journal of Coastal Conservation* **3**: 203–210.

De Ruyck, A. M. C., Ampe, C., and Langohr, R. (2001). Management of the Belgian coast: opinions and solutions. *Journal of Coastal Conservation* **7**: 129–144.

de Schipper, M. A., de Vries, S., Ruessink, G., et al. (2016). Initial spreading of a mega feeder nourishment: observations of the Sand Engine pilot project. *Coastal Engineering* **111**: 23–38.

De Vincenzo, A., Covelli, C., Molino, A. J., Pannone, M., Ciccaglione, M., and Molino, B. (2019). Long-term management policies of reservoirs: possible re-use of dredged sediments for coastal nourishment. *Water* **11**: 15.

Dean, R. G. (1997). Models for barrier island restoration. *Journal of Coastal Research* **13**: 694–703.

Dean, R. G. (2002). *Beach Nourishment: Theory and Practice.* World Scientific Publishing Company.

Dech, J. P. and Maun, M. A. (2005). Zonation of vegetation along a burial gradient on the leeward slopes of Lake Huron. *Canadian Journal of Botany* **83**: 227–236.

Defeo, O. and McLachlan, A. (2005). Patterns, processes and regulatory mechanisms in sandy beach macrofauna: a multi-scale analysis. *Marine Ecology Progress Series* **295**: 1–20.

Defeo, O., McLachlan, A., Schoeman, D. S., et al. (2009). Threats to sandy beach ecosystems: a review. *Estuarine, Coastal and Shelf Science* **81**: 1–12.

Deguchi, I., Ono, M., Araki, S., and Sawaragi, T. (1998). Motions of pebbles on pebble beach. In *Coastal Engineering.* Reston, VA: American Society of Civil Engineers, pp. 2654–2667.

Del Vecchio, S., Marba, N., Acosta, A., Vignolo, C., and Traveset, A. (2013). Effects of *Posidonia oceanica* beach-cast on germination, growth and nutrient uptake of coastal dune plants. *PLoS ONE* **8**: e70607.

Del Vecchio, S., Jucker, T., Carboni, M., and Acosta, A. T. R. (2017). Linking plant communities on land and at sea: the effects of *Posidonia oceanica* wrack on the structure of dune vegetation. *Estuarine, Coastal and Shelf Science* **184**: 30–36.

Delgado-Fernandez, I., Davidson-Arnott, R. G. D., and Hesp, P. A. (2019). Is "re-mobilisation" nature restoration or nature destruction? A commentary. *Journal of Coastal Conservation* **23**: 1093–1103.

Demirayak, F. and Ulas, E. (1996). Mass tourism in Turkey and its impact on the Mediterranean coast. In Salman, A. H. P. M., Langeveld, M. J., and Bonazountas, M. (Eds.), *Coastal Management and Habitat Conservation*, vol. II. Leiden: European Union for Coastal Conservation, pp. 117–123.

Denevan, W. M. (1992). The pristine myth: the landscape of the Americas in 1492. *Annals of the Association of American Geographers* **82**: 369–385.

Des Roches, C. T. (2020). The preservation paradox and natural capital. *Ecosystem Services* **41**: 101058.

Dette, H.-H., Führböter, A., and Raudkivi, A. J. (1994). Interdependence of beach fill volumes and repetition intervals. *Journal of Waterway, Port, Coastal, and Ocean Engineering* **120**: 580–593.

Diaz, H. (1980). The mole crab *Emerita talpoida* (say); a case study of changing life history pattern. *Ecological Monographs* **50**: 437–456.

Diaz, R. J., Cutter, G. R., Jr., and Hobbs, C. H. (2004). Potential impacts of sand mining offshore of Maryland and Delaware: Part 2 – biological considerations. *Journal of Coastal Research* **20**: 61–69.

Dickerson, D. D., Smith, J., Wolters, M., Theriot, C., Reine, K. J., and Dolan, J. (2007). A review of beach nourishment impacts on marine turtles. *Shore and Beach* **75**(1): 49–56.

Dlamini, L. Z. D. and Xulu, S. (2019). Monitoring mining disturbance and restoration over RBM site in South Africa using LandTrendr algorithm and landsat data. *Sustainability* **11**: 6916.

Dodkin, M. and McDonald, T. (2019). Witnessing four decades of change in coastal park management: interview with Mike Dodkin. *Ecological Management and Restoration* **20**: 192–201.

Doing, H. (1985). Coastal foredune zonation and succession in various parts of the world. *Vegetatio* **61**: 65–75.

Donnelly, C., Kraus, N., and Larson, M. (2006). State of knowledge on measurement and modeling of coastal overwash. *Journal of Coastal Research* **22**: 965–991.

Donohue, K. A., Bocamazo, L. M., and Dvorak, D. (2004). Experience with groin notching along the northern New Jersey coast. *Journal of Coastal Research* **SI33**: 198–214.

Doody, J. P. (1989). Management for nature conservation. *Proceedings of the Royal Society of Edinburgh* **96B**: 247–265.

Doody, J. P. (1995). Infrastructure development and other human influences on the coastline of Europe. In Salman, A. H. P. M., Berends, H., and Bonazountas, M. (Eds.), *Coastal Management and Habitat Conservation*. Leiden: EUCC, pp. 133–151.

Doody, J. P. (2001). *Coastal Conservation and Management: An Ecological Perspective*. Dordrecht: Kluwer Academic Publishers.

Dornbusch, U. (2017). Design requirements for mixed sand and gravel beach defences under scenarios of sea level rise. *Coastal Engineering* **124**: 12–24.

Dornbusch, U., Williams, R. B. G., Moses, C., and Robinson, D. A. (2002). Life expectancy of shingle beaches: measuring in situ abrasion. *Journal of Coastal Research* **SI36**: 249–255.

Drius, M., Malavasi, M., Acosta, A. T. R., Ricotta, C., and Carranza, M. L. (2013). Boundary-based analysis for the assessment of coastal dune landscape integrity over time. *Applied Geography* **45**: 41–48.

Drucker, B. S., Waskes, W., and Byrnes, M. R. (2004). The US Minerals Management Service Outer Continental Shelf Sand and Gravel Program: environmental studies to assess the potential effects of offshore dredging operations in federal waters. *Journal of Coastal Research* **20**: 1–5.

Dugan, J. E. and Hubbard, D. M. (2010). Loss of coastal strand habitat in southern California: the role of beach grooming. *Estuaries and Coasts* **33**: 67–77.

Dugan, J. E., Hubbard, D. M., McCrary, M. D., and Pierson, M. O. (2003). The response of macrofauna communities and shorebirds to macrophyte wrack subsidies on exposed sandy beaches of southern California. *Estuarine, Coastal and Shelf Science* **58S**: 25–40.

Dugan, J. E., Hubbard, D. M., Rodil, I. F., Revell, D. L., and Schroeter, S. (2008). Ecological effects of coastal armoring on sandy beaches. *Marine Ecology* **29** (Suppl. 1): 160–170.

Dugan, J. E., Airoldi, L., Chapman, M. G., Walker, S. J., and Schlacher, T. (2011). Estuarine and coastal structures: environmental effects, a focus on shore and near-shore structures. *Treatise on Estuarine and Coastal Science* **8**: 17–41.

Dugan, J. E., Emery, K. A., Alber, M., et al. (2018). Generalizing ecological effects of shoreline armoring across soft sediment environments. *Estuaries and Coasts* **41**: S180–S196.

Duncan, W. H. and Duncan, M. B. (1987). *Seaside Plants of the Gulf and Atlantic Coasts*. Washington, DC: Smithsonian Institution Press.

Dzhaoshvili, Sh. V. and Papashvili, I. G. (1993). Development and modern dynamics of alluvial-accumulative coasts of the eastern Black Sea. In Kos'yan, R. (Ed.), *Coastlines of the Black Sea*. New York: American Society of Civil Engineers, pp. 224–233.

Eastwood, D. A. and Carter, R. W. G. (1981). The Irish dune consumer. *Journal of Leisure Research* **13**: 273–281.

Edge, B. L., Dowd, M., Dean, R. G., and Johnson, P. (1994). The reconstruction of Folly Beach. In *Coastal Engineering: Proceedings of the Twenty-Fourth Coastal Engineering Conference*. New York: American Society of Civil Engineers, pp. 3491–3506.

Ehrenfeld, J. G. (1990). Dynamics and processes of barrier island vegetation. *Aquatic Science* **2**: 437–480.

Ehrenfeld, J. G. (2000). Evaluating wetlands within an urban context. *Ecological Engineering* **15**: 253–265.

Eitner, V. (1996). The effect of sedimentary texture on beach fill longevity. *Journal of Coastal Research* **12**: 447–461.

Ellis, J. T. and Román-Rivera, M. A. (2019). Assessing natural and mechanical dune performance in a post-hurricane environment. *Journal of Marine Science and Engineering* **7**(126): 1–15.

Ellison, J. C. (2018). Pacific island beaches: values, threats and rehabilitation. In Botero, C. M., Cervantes, O., and Finkl, C. W. (Eds.), *Beach Management Tools – Concepts, Methodologies and Case Studies*. Cham, Switzerland: Springer International Publishing, pp. 679–700.

Emery, S. M. and Rudgers, J. A. (2010). Ecological assessment of dune restorations in the Great Lakes region. *Restoration Ecology* **18**(S1): 184–194.

Environmental Protection Agency. (1995). *Rehabilitation and Revegetation: Best Practice. Environmental Management in Mining*. Canberra: Australian Federal Environment Department.

Erwin, R. M., Truitt, B. R., and Jiménez, J. E. (2001). Ground-nesting waterbirds and mammalian carnivores in the Virginia barrier island region: running out of options. *Journal of Coastal Research* **17**: 292–296.

Escofet, A. and Espejel, I. (1999). Conservation and management-oriented ecological research in the coastal zone of Baja California, Mexico. *Journal of Coastal Conservation* **21**: 43–50.

Escudero, M., Silva, R., and Mendoza, E. (2014). Beach erosion driven by natural and human activity at Isla del Carmen barrier island, Mexico. *Journal of Coastal Research* **SI71**: 62–74.

Espejel, I. (1993). Conservation and management of dry coastal vegetation. In Fermán-Almada, J. L., Gómez-Morin, L., and Fischer, D. W. (Eds.), *Coastal Zone Management in Mexico: The Baja California Experience*. New York: American Society of Civil Engineers, pp. 119–136.

Espejel, J., Ahumada, B., Cruz, Y., and Heredia, A. (2004). Coastal vegetation as indicators for conservation. In Martínez, M. L. and Psuty, N. P. (Eds.), *Coastal Dunes, Ecology and Conservation*. Berlin: Springer-Verlag, pp. 297–318.

Esteves, L. S. and Thomas, K. (2014). Managed realignment in practice in the UK: results from two independent surveys. . *Journal of Coastal Research* **SI70**: 407–413.

European Commission. (2004). *Living with Coastal Erosion in Europe – Sediment and Space for Sustainability*. Luxemburg: Office for Official Publications of the European Commission.

Evans, A. J., Firth, L. B., Hawkins, S. J., Morris, E. S., Goudge, H., and Moore, P. J. (2016). Drill-cored rock pools: an effective method of ecological enhancement on artificial structures. *Marine and Freshwater Research* **67**: 123–130.

Everard, M., Jones, L., and Watts, B. (2010). Have we neglected the societal importance of sand dunes? An ecosystem services perspective. *Aquatic Conservation: Marine and Freshwater Ecosystems* **20**: 476–487.

Everts, C. H., Eldon, C. D., and Moore, J. (2002). Performance of cobble berms in southern California. *Shore and Beach* **70**(4): 5–14.

Ewel, J. J. (1990). Restoration is the ultimate test of ecological theory. In Jordan, W. R. (Ed.), *Restoration Ecology – A Synthetic Approach to Ecological Research.* Cambridge: Cambridge University Press, p. 31.

Fairweather, P. G. and Henry, R. J. (2003). To clean or not to clean? Ecologically sensitive management of wrack deposits on sandy beaches. *Ecological Management and Restoration* **4**: 227–228.

Falk, D. A. (1990). Discovering the future, creating the past: some reflections on restoration. *Restoration and Management Notes* **8**: 71.

Fang, J., Lincke, D., Brown, S., et al. (2020). Coastal flood risks in China through the 21st century – an application of DIVA. *Science of the Total Environment* **704**: 135311.

Fanini, L., Marchetti, G. M., Scapini, F., and Defeo, O. (2007). Abundance and orientation responses of the sandhopper *Talitrus saltator* to beach nourishment and groynes building at San Rossore Regional Park, Tuscany, Italy. *Marine Biology* **152**: 1169–1179.

Feagan, R. and Ripmeester, M. (2001). Reading private green space: competing geographical identities at the level of the lawn. *Philosophy and Geography* **4**: 79–95.

Feagin, R. A. (2005). Artificial dunes created to protect property on Galveston Island, Texas: the lessons learned. *Ecological Restoration* **23**: 89–94.

Feagin, R. (2013). Foredune restoration before and after hurricanes: inevitable destruction, certain reconstruction. In: Martínez, M. L., Gallego-Fernández, J. B., and Hesp, P. A. (Eds.), *Restoration of Coastal Dunes.* New York: Springer, pp. 93–103.

Feagin, R., Figlus, J., Zinnert, J. C., et al. (2015). Going with the flow or against the grain? The promise of vegetation for protecting beaches, dunes, and barrier islands from erosion. *Frontiers in Ecology and Environment* **13**: 203–210.

Feagin, R. A., Furman, M., Salgado, K., et al. (2019). The role of beach and sand dune vegetation in mediating wave run up erosion. *Estuarine, Coastal and Shelf Science* **219**: 97–106.

Ferreira, J. C., Silva, C., Tenedório, J. A., Pontes, S., Encarnação, S., and Marques, L. (2006). Coastal greenways: interdisciplinarity and integration challenges for the management of developed coastal areas. *Journal of Coastal Research* **SI39**: 1833–1837.

Finkl, C. W. (2002). Long-term analysis of trends in shore protection based on papers appearing in the *Journal of Coastal Research*, 1984–2000. *Journal of Coastal Research* **18**: 211–224.

Firth, L. B., Thompson, R. C., Bohn, K., et al. (2014). Between a rock and a hard place: environmental and engineering considerations when designing coastal defence structures. *Coastal Engineering* **87**: 122–135.

Fischer, D. L. (1989). Response to coastal storm hazard: short-term recovery versus long-term planning. *Ocean and Shoreline Management* **12**: 295–308.

FitzGerald, D. M., van Heteren, S., and Montello, T. M. (1994). Shoreline processes and damage resulting from the Halloween Eve storm of 1991 along the north and south shores of Massachusetts Bay, USA. *Journal of Coastal Research* **10**: 113–132.

FitzGerald, D. M., Fenster, M. S., Argow, B. A., and Buynevich, I. V. (2008). Coastal impacts due to sea-level rise. *Annual Review of Earth and Planetary Sciences* **36**: 601–647.

FitzGerald, D. M., Georgiou, I., and Kulp, M. (2016). Restoration of the Chandeleur Barrier arc, Louisiana. *Journal of Coastal Research* **SI75**: 1282–1286.

Forman, R. T. T. (1995). *Land Mosaics: The Ecology of Landscapes and Regions.* Cambridge: Cambridge University Press.

Foster-Smith, J., Birchenough, A. C., Evans, S. M., and Prince, J. (2007). Human impacts on Cable Beach, Broome (Western Australia). *Coastal Management* **35**: 181–194.

Fouqueray, T., Trommetter, M., and Frascaria-Lacoste, N. (2018). Managed retreat of settlements and infrastructures: ecological restoration as an opportunity to overcome maladaptive coastal development in France. *Restoration Ecology* **26**: 806–812,

Freestone, A. L. and Nordstrom, K. F. (2001). Early evolution of restored dune plant microhabitats on a nourished beach at Ocean City, New Jersey. *Journal of Coastal Conservation* **7**: 105–116.

French, K. (2012). Competition strength of two significant invasive species in coastal dunes. *Plant Ecology* **213**: 1667–1673.

French, P. W. (2006). Managed realignment – the developing story of a comparatively new approach to soft engineering. *Estuarine, Coastal and Shelf Science* **67**: 409–423.

Frihy, O. E., Deabes, E. A., and Helmy, E.-E. D. F. (2016). Compatibility analysis of dredged sediments from routine pathways and maintenance of harbor's channels for reuse in nearshore nourishment in the Nile Delta, Egypt. *Journal of Coastal Research* **32**: 555–566.

Fuller, R. M. (1987). Vegetation establishment on shingle beaches. *Journal of Ecology* **75**: 1077–1089.

Furmanczyk, K. (2013). Poland. In Pranzini, E. and Williams, A. (Eds.), *Coastal Erosion and Protection in Europe.* London: Routledge, pp. 81–95.

Gallego-Fernández, J. B., Sánchez, I. A., and Ley, C. (2011). Restoration of isolated and small coastal sand dunes on the rocky coast of northern Spain. *Ecological Engineering* **37**: 1822–1832.

Gallego-Fernández, J. B., Morales-Sánchez, J. A., Martínez, M. L., García-Franco, J. G., and Zunzunegui, M. (2020). Recovery of beach-foredune vegetation after disturbance by storms. *Journal of Coastal Research* **SI95**: 34–38.

Gallien, T. W., O'Reilly, W. C., Flick, R. E., and Guza, R. T. (2015). Geometric properties of anthropogenic flood control berms on southern California beaches. *Ocean & Coastal Management* **105**: 35–47.

Gangaiya, P., Beardsmore, A., and Miskiewicz, T. (2017). Morphological changes following vegetation removal and foredune re-profiling at Woonona Beach, New South Wales, Australia. *Ocean & Coastal Management* **146**: 15–25.

Gann, G. D., McDonald, T., Walder, B., et al. (2019). International principles and standards for the practice of ecological restoration. Second edition. *Restoration Ecology* **27**: S1–S46.

Gao, J., Kennedy, D. M., and Konlechner, T. M. (2020). Coastal dune mobility over the past century: A global review. *Progress in Physical Geography* **44**: 814–836.

Garbutt, R. A., Reading, C. J., Wolters, M., Gray, A. J., and Rothery, P. (2006). Monitoring the development of intertidal habitats on former agricultural land after the managed realignment of coastal defenses at Tollesbury, Essex, UK. *Marine Pollution Bulletin* **53**: 155–164.

Garcia-Lozano, C., Pintó, J., and Daunis-i-Estadella, P. (2018). Changes in coastal dune systems on the Catalan shoreline (Spain, NW Mediterranean Sea). Comparing dune landscapes between 1890 and 1960 with their current status. *Estuarine, Coastal and Shelf Science* **208**: 235–247.

Garcia-Lozano, C., Pintó, J., and Roig-Munar, F. X. (2020). Set of indices to assess dune development and dune restoration potential in beach-dune systems on Mediterranean developed coasts. *Journal of Environmental Management* **259**: 109754.

García-Mora, M. R., Gallego-Fernández, J. B., and García-Novo, F. (2000). Plant diversity as a suitable tool for coastal dune vulnerability assessment. *Journal of Coastal Research* **16**: 990–995.

García Novo, F., Díaz Barradas, M. C., Zunzunegui, M., García Mora, R., and Gallego Fernández, J. B. (2004). Plant functional types in coastal dune habitats. In Martínez, M. L. and Psuty, N. P. (Eds.), *Coastal Dunes, Ecology and Conservation*. Berlin: Springer-Verlag, pp. 155–169.

García-Romero, L., Delgado-Fernández, I., Hesp, P. A., Hernández-Calvento, L., Hernández-Cordero, A. I., and Viera-Pérez, M. (2019). Biogeomorphological processes in an arid transgressive dunefild as indicators of human impact by urbanization. *Science of the Total Environment* **650**: 73–86.

Garcin, M., Desmazes, F., Lerma, A. N., Gouguet, L., and Météreau, V. (2020). Contributions of lightweight revolving laser scanner, HiRes UAV LiDARS and photogrammetry for characterization of coastal aeolian morphologies. *Journal of Coastal Research* **SI95**: 1094–1100.

Gardner, E. and Burningham, H. (2013). Ecology and conservation of the rare annual *Petrorhagia nanteuilii* (Childing Pink) on the vegetated shingle spits of Pagham Harbour, West Sussex. *Journal of Coastal Conservation* **17**: 589–600.

Gares, P. A. (1989). Geographers and public policy making: lessons learned from the failure of the New Jersey Dune Management Plan. *Professional Geographer* **41**: 20–29.

Gares, P. A. and Nordstrom, K. F. (1995). A cyclic model of foredune blowout evolution for a leeward coast, Island Beach, New Jersey. *Annals of the Association of American Geographers* **85**: 1–20.

Gauci, M. J., Deidun, A., and Schembri, P. J. (2005). Faunistic diversity of Maltese pocket sandy and shingle beaches: are these of conservation value? *Oceanologia* **47**: 219–241.

Gedan, K. B., Kirwan, M. L., Wolanski, E., Barbier, E. B., and Silliman, B. R. (2011). The present and future role of coastal wetland vegetation in protecting shorelines: answering recent challenges to the paradigm. *Climatic Change* **106**: 7–29.

Geelen, L. H. W. T., Kamps, P. T. W. J., and Olsthoorn, T. N. (2017). From exploitation to sustainable use, an overview of 160 years of water extraction in the Amsterdam dunes, the Netherlands. *Journal of Coastal Conservation* **21**: 657–668.

Gemma, J. N. and Koske, R. E. (1997). Arbuscular mycorrhizae in sand dune plants of the North Atlantic coast of the U.S.: field and greenhouse studies. *Journal of Environmental Management* **50**: 251–264.

Gerhardt, P. (1900). *Handbuch des Deutschen Dünenbaues*. Berlin: Paul Parey.

Gerlach, A. (1992). Dune cliffs: a buffered system. In Carter, R. W. G., Curtis, T. G. F., and Sheehy-Skeffington, M. J. (Eds.), *Coastal Dunes: Geomorphology, Ecology and Management for Conservation*. Rotterdam: A.A. Balkema, pp. 51–55.

Gesing, F. (2019). The politics of artificial dunes: sustainable coastal protection measures and contested socio-natural objects. *Die Erde* **150**: 145–157.

Ghate, S. D., Sridhar, K. R., and Karum, N. C. (2014). Macrofungi on the coastal sand dunes of south-western India. *Mycosphere* **5**: 144–151.

Gibeaut, J. C., Hepner, T. L., Waldinger, R., Andrews, J. R., Smyth, R. C., and Gutierrez, R. (2003). Geotubes for temporary erosion control and storm surge protection along the Gulf of Mexico shoreline of Texas. *Proceedings of the 13th Biennial Coastal Zone Conference*.

Gibson, D. J. and Looney, P. B. (1994). Vegetation colonization of dredge spoil on Perdido Key, Florida. *Journal of Coastal Research* **10**: 133–143.

Gibson, D. J., Ely, J. S., and Looney, P. B. (1997). A Markovian approach to modeling succession on a coastal barrier island following beach nourishment. *Journal of Coastal Research* **13**: 831–841.

Gilburn, A. S. (2012). Mechanical grooming and beach award status are associated with low strandline biodiversity in Scotland. *Estuarine, Coastal and Shelf Science* **107**: 81–88.

Gobster, P. H., Nassauer, J. I., Daniel, T. C., and Fry, T. (2007). The shared landscape: what does aesthetics have to do with ecology. *Landscape Ecology* **22**: 959–972.

Godfrey, P. J. (1977). Climate, plant response, and development of dunes on barrier beaches along the U.S. east coast. *International Journal of Biometeorology* **21**: 203–215.

Godfrey, P. J. and Godfrey, M. M. (1981). Ecological effects of off-road vehicles on Cape Cod. *Oceanus* **23**: 56–67.

Godfrey, P. J., Leatherman, S. P., and Zaremba, R. (1979). A geobotanical approach to classification of barrier beach systems. In Leatherman, S. P. (Ed.), *Barrier Islands from the Gulf of St. Lawrence to the Gulf of Mexico*. New York: Academic Press, pp. 99–126.

Goeldner-Gianella, L. (2007). Perceptions and attitudes toward de-polderisation in Europe: a comparison of five opinion surveys in France and the UK. *Journal of Coastal Research* **23**: 1218–1230.

Goeldner-Gianella, L., Bertrand, F., Oiry, A., and Grancher, D. (2015). Depolderization policy against coastal flooding on the French Atlantic coast: the case of Arachon Bay. *Ocean & Coastal Management* **116**: 98–107.

Goldin, M. R. and Regosin, J. V. (1998). Chick behavior, habitat use, and reproductive success of piping plovers at Goosewing Beach, Rhode Island. *Journal of Field Ornithology* **69**: 228–234.

Golfi, P. (1996). The future of tourism in the Mediterranean. In Salman, A. H. P. M., Langeveld, M. J., and Bonazountas, M. (Eds.), *Coastal Management and Habitat Conservation*, vol. II. Leiden: European Union for Coastal Conservation, pp. 133–140.

Gómez-Pina, G. (2004). The importance of aesthetic aspects in the design of coastal groins. *Journal of Coastal Research* **SI33**: 83–98.

Gómez-Pina, G., Muñoz-Pérez, J. J., Ramírez, J. L., and Ley, C. (2002). Sand dune management problems and techniques, Spain, *Journal of Coastal Research* **SI36**: 325–332.

Gómez-Pina, G., Muñoz-Pérez, J. J., Fages, L., Ramírez, J. L., Enriques, J., and de Sobrino, J. (2004). A critical review of urban beach nourishment projects in Cadiz City after twelve years. In *Coastal Engineering 2004: Proceedings of the 29th International Conference*. New York: American Society of Civil Engineers, pp. 3454–3466.

Gonçalves, D. S., Pinheiro, L. M., Silva, P. A., et al. (2014). Morphodynamic evolution of a sand extraction excavation offshore Vale do Lobo, Algarve, Portugal. *Coastal Engineering* **88**: 75–87.

Gopalakrishnan, S., Landry, C. E., Smith, M. D., and Whitehead, J. C. (2016). Economics of coastal erosion and adaptation to sea level rise. *Annual Review of Resource Economics* **8**: 119–139.

Gore, S. (2007). Framework development for beach management in the British Virgin Islands. *Ocean & Coastal Management* **50**: 732–753.

Gorzelany, J. F. and Nelson, W. G. (1987). The effects of beach nourishment on the benthos of a subtropical Florida beach. *Marine Environmental Research* **21**: 75–94.

Gosz, J. R. (1991). Fundamental ecological characteristics if landscape boundaries. In Holland, M. M., Naiman, R. J., and Risser, P. G. (Eds.), *Ecotones: The Role of Landscape Boundaries in the Management and Restoration of Changing Environments*. New York: Chapman and Hall, pp. 8–30.

Graetz, K. E. (1973). *Seacoast Plants of the Carolinas for Conservation and Beautification*. UNC-SG-73-06. Raleigh: North Carolina University Sea Grant.

Grafals-Soto, R. (2012). Effects of sand fences on coastal dune vegetation distribution. *Geomorphology* **145–146**: 45–55.

Grafals-Soto, R. and Nordstrom, K. F. (2009). Sand fences in the coastal zone: intended and unintended effects. *Environmental Management* **44**: 420–429.

Granja, H. M. and Carvalho, G. S. (1995). Is the coastline "protection" of Portugal by hard engineering structures effective? *Journal of Coastal Research* **11**: 1229–1241.

Greene, K. (2002). *Beach Nourishment: A Review of the Biological and Physical Impacts*. Washington, DC: Atlantic States Marine Fisheries Commission Habitat Management Series No. 7.

Gribbin, T. (1990). Sand dune rehabilitation and management in Prince Edward Island National Park. In Davidson-Arnott, R. G. D. (Ed.), *Proceedings of the Symposium on Coastal Sand Dunes*. Ottawa: National Research Council Canada, pp. 433–446.

Griggs, G. and Patsch, K. (2019). The protection/hardening of California's coast: times are changing. *Journal of Coastal Research* **35**: 1051–1061.

Grime, J. P. (1979). *Plant Strategies and Vegetation Processes*. London: John Wiley & Sons.

Grootjans, A. P., Geelan, H. W. T., Jansen, A. J. M., and Lammerts, E. J. (2002). Restoration of coastal dune slacks in the Netherlands. *Hydrobiologia* **478**: 181–203.

Grootjans, A. P., Adema, E. B., Bekker, R. M., and Lammerts, E. J. (2004). Why young coastal dune slacks sustain a high biodiversity. In Martínez, M. L. and Psuty, N. P. (Eds.), *Coastal Dunes, Ecology and Conservation*. Berlin: Springer-Verlag, pp. 85–101.

Grootjans, A. P., Dullo, B. S., Kooijman, A. M., Bekker, R. M., and Aggenbach, C. (2013). Restoration of dune vegetation in the Netherlands. In Martínez, M.L., Gallego-Fernández, J. B., and Hesp, P. A. (Eds.), *Restoration of Coastal Dunes*. Berlin: Springer, pp. 235–253.

Grottoli, E., Bertoni, D., and Ciavola, P. (2017). Short- and medium-term response to storms on three Mediterranean coarse-grained beaches. *Geomorphology* **295**: 738–748.

Grumbine, R. E. (1994). Wildness, wise use, and sustainable development. *Environmental Ethics* **16**: 241–249.

Gueben-Venière, S. (2016). How do civil engineers see the coast they manage? A comparative approach between The Netherlands, England and France. In Baptiste, A. (Ed.), *Coastal Management: Changing Coast, Changing Climate, Changing Minds*. International Coastal Management Conference. London: ICE Publishing, pp. 489–500.

Guisado-Pintado, E., Jackson, D. W. T., and Rogers, D. (2019). 3D mapping efficacy of a drone and terrestrial laser scanner over a temperate beach-dune zone. *Geomorphology* **328**: 157–172.

Halle, S. (2007). Present state on future perspectives of restoration ecology – introduction. *Restoration Ecology* **15**: 304–306.

Hamer, D., Belcher, C., and Miller, C. (1992). *Restoration of Sand Dunes along the Mid-Atlantic Coast*. Somerset, NJ: U.S. Department of Agriculture Natural Resources Conservation Service.

Hamm, L., Capobianco, M., Dette, H. H., Lechuga, A., Spanhoff, R., and Stive, M. J. F. (2002). A summary of European experience with shore nourishment. *Coastal Engineering* **47**: 237–264.

Hanley, M. E., Hoggart, S. P. G., Simmonds, D. J., et al. (2014). Shifting sands? Coastal protection by sand banks, beaches and dunes. *Coastal Engineering* **87**: 136–146.

Hanley, N. and Roberts, M. (2019). The economic benefits of invasive species management. *People and Nature* **1**: 124–137.

Hanson, H., Brampton, A., Capobianco, M., et al. (2002). Beach nourishment projects, practices, and objectives – a European overview. *Coastal Engineering* **47**: 81–111.

Harley, M. D. and Ciavola, P. (2013). Managing local coastal inundation risk using real-time forcasts and artificial dune placements. *Coastal Engineering* **77**: 77–90.

Harman, B. P., Heyenga, S., Taylor, B. M., and Fletcher, C. S. (2015). Global lessons for adapting coastal communities to protect against storm surge inundation. *Journal of Coastal Research* **31**: 790–801.

Harris, L. D. and Scheck, J. (1991). From implications to applications: the dispersal corridor principal applied to the conservation of biological diversity. In Saunders, D. A. and Hobbs, R. J. (Eds.), *Nature Conservation 2: The Role of Corridors*. Chipping Norton, NSW: Surrey Beatty and Sons, pp. 189–220.

Harris, L., Nel, R., Smale, M., and Schoeman, D. (2011). Swashed away? Storm impacts on sandy beach macrofaunal communities. *Estuarine, Coastal and Shelf Science* **94**: 210–221.

Harris, M. E. and Ellis, J. T. (2020). A holistic approach to evaluating dune cores. *Journal of Coastal Conservation* **24**: 42.

Harris, T. B., Rajakaruna, N., Nelson, S. J., and Vaux, P. D. (2012). Stressors and threats to the flora of Acadia National Park, Maine: current knowledge, information gaps, and future directions. *The Journal of the Torrey Botanical Society* **139**: 323–344.

Hartley, B. L., Thompson, R. C., and Pahl, S. (2015). Marine litter education boosts children's understanding and self-reported actions. *Marine Pollution Bulletin* **90**: 209–217.

Hatzikyriakou, A., Lin, N., Gong, J., Xian, S., Hu, X, and Kennedy, A. (2016). Component-based vulnerability analysis for residential structures subjected to storm surge impact from Hurricane Sandy. *Natural Hazards Review* **17**: 05015005-1-15.

Hemmingsen, M. A., Eikaas, H. S., and Marsden, D. C. J. (2019). A GIS approach to sediment displacement in mixed sand and gravel beach environments. *Journal of Environmental Management* **249**: 109083.

Henderson, S. P. B., Perkins, N. H., and Nelischer, M. (1998). Residential lawn alternatives: a study of their distribution, form and structure. *Landscape and Urban Planning* **42**: 135–145.

Hernández-Cordero, A. I., Hernández-Calvento, L., and Pérez-Chacon, E. (2017). Vegetation changes as an indicator of impact from tourist development in an arid transgressive coastal dune field. *Land Use Policy* **64**: 479–491.

Hertling, U. M. and Lubke, R. A. (1999). Use of *Ammophila arenaria* for dune stabilization in South Africa and its current distribution – perceptions and problems. *Environmental Management* **24**: 467–482.

Heslenfeld, P., Jungerius, P. D., and Klijn, J. A. (2004). European coastal dunes: ecological values, threats, opportunities and policy development. In Martínez, M. L. and Psuty, N. P. (Eds.), *Coastal Dunes, Ecology and Conservation*. Berlin: Springer-Verlag, pp. 335–351.

Hesp, P. A. (1989). A review of biological and geomorphological processes involved in the initiation and development of incipient foredunes. *Proceedings of the Royal Society of Edinburgh* **96B**: 181–201.

Hesp, P. A. (1991). Ecological processes and plant adaptations on coastal dunes. *Journal of Arid Environments* **21**: 165–191.

Hesp, P. A. and Hilton, M. J. (2013). Restoration of foredunes and transgressive dunefields: case studies from New Zealand. In: Martínez, M. L., Gallego-Fernández, J. B., and Hesp, P. A. (Eds.), *Restoration of Coastal Dunes*. New York: Springer, pp. 67–92.

Hickman, T. and Cocklin, C. (1992). Attitudes toward recreation and tourism development in the coastal zone: a New Zealand case study. *Coastal Management* **20**: 269–289.

Higgs, E. S. (1997). What is good ecological restoration? *Conservation Biology* **11**: 338–348.

Higgs, E. S. (2003). *Nature by Design: People, Natural Process, and Ecological Restoration*. Cambridge, MA: The MIT Press.

Higgs, E. S. (2006). Restoration goes wild: a reply to Throop and Purdom. *Restoration Ecology* **14**: 500–503.

Higgs, E., Harris, J., Murphy, S., et al. (2018). On principles and standards in ecological restoration. *Restoration Ecology* **26**: 399–403.

Hill-Spanik, K. M., Smith, A. S., and Plante, C. J. (2019). Recovery of benthic microalgal biomass and community structure following beach renourishment at Folly Beach, South Carolina. *Estuaries and Coasts* **42**: 157–172.

Hilton, M. J. (2006). The loss of New Zealand's active dunes and the spread of marram grass (*Ammophila arenaria*). *New Zealand Geographer* **62**: 105–120.

Hilton, M., Duncan, M., and Jul, A. (2005). Processes of *Ammophila arenaria* (Marram grass) invasion and indigenous species displacement, Stewart Island, New Zealand. *Journal of Coastal Research* **21**: 175–185.

Hilton, M., Harvey, N., Hart, A., James, K., and Arbuckle, C. (2006). The impact of exotic dune grass species on foredune development in Australia and New Zealand: a case study of *Ammophila arenaria* and *Thinopyrum junceiforme*. *Australian Geographer* **37**: 313–334.

Hilton, M., Woodley, D., Sweeney, C., and Konlechner, T. (2009). The development of a prograded foredune barrier following *Ammophila arenaria* eradication, Doughboy Bay, Stewart Island. *Journal of Coastal Research* **SI56**: 317–321.

Hilton, M., Konlechner, T., McLachlan, K., Lim, D., and Lord, J. (2019). Long-lived seed banks of *Ammophila arenaria* prolong dune restoration programs. *Journal of Coastal Conservation* **23**: 461–471.

Hirsch, S. E., Kedzuf, S., and Perrault, J. R. (2019). Impacts of a geotextile container dune core on marine turtle nesting in Juno Beach, Florida, United States. *Restoration Ecology* **27**: 431–439.

Hobbs, C. H. III (2002). An investigation of potential consequences of marine mining in shallow water: an example from the mid-Atlantic coast of the United States. *Journal of Coastal Research* **18**: 94–101.

Hobbs, R. J. and Norton, D. A. (1996). Towards a conceptual framework for restoration ecology. *Restoration Ecology* **4**: 93–110.

Hofstede, J. (2019). Küstenschutz in Schleswig-Holstein: ein Überblick über Strategien und Massnahmen. *Die Küste* **87**.

Holcomb, B. (2016). A sure thing: tourism recovery in New York and New Jersey after Hurricane Sandy. In O'Neil, K. M. and van Abs, D. J. (Eds.), *Taking Chances: The Coast after Hurricane Sandy*. New Brunswick, NJ: Rutgers University Press, pp. 177–189.

Holz, R., Hermann, C., and Müller-Motzfeld, G. (1996.) Vom Polder zum Ausdeichungsgebiet: Das Projekt Karrendorfer Wiesen und die Zukunft der Küstenüberflutungsgebiete in Mecklenburg-Vorpommern. *Natur und Naturschutz in Mecklenburg-Vorpommern* **32**: 3–27.

Hoogeboom, K. R. (1989). Restoration and development guidelines for ocean beach recreation areas. In *Coastal Zone 89*. New York: American Society of Civil Engineers, pp. 3120–3134.

Hooke, J. M. and Bray, M. J. (1995). Coastal groups, littoral cells, policies and plans in the UK. *Area* **27**: 358–368.

Hoonhout, B. and de Vries, S. (2017). Aeolian sediment supply at a mega nourishment. *Coastal Engineering* **123**: 11–20.

Hoonhout, B. and de Vries, S. (2019). Simulating spatiotemporal aeolian sediment supply at a mega nourishment. *Coastal Engineering* **145**: 21–35.

Hooton, N., Miller, D. L., Thetford, M., and Claypool, B. S. (2014). Survival and growth of planted *Uniola paniculata* and dune building using surrogate wrack on Perdido Key Florida, U.S.A. *Restoration Ecology* **22**: 701–707.

Horn, D. P. and Walton, S. M. (2007). Spatial and temporal variations of sediment size on a mixed sand and gravel beach. *Sedimentary Geology* **202**: 509–528.

Hornsey, W. P., Carley, J. T., Coghlan, I. R., and Cox, R. J. (2011). Geotextile sand container shoreline protection systems: design and application. *Geotextiles and Geomembranes* **29**: 425–439.

Hose, T. A. (1998). Selling coastal geology to visitors. In Hooke, J. (Ed.), *Coastal Defence and Earth Science Conservation*. Bath: The Geological Society, pp. 178–195.

Hotta, S., Kraus, N. C., and Horikawa, K. (1987). Function of sand fences in controlling wind-blown sand. In *Coastal Sediments 87*. New York: American Society of Civil Engineers, pp. 772–787.

Hotta, S., Kraus, N. C., and Horikawa, K. (1991). Functioning of multi-row sand fences in forming foredunes. In *Coastal Sediments 91*. New York: American Society of Civil Engineers, pp. 261–275.

Houser, C., Wernette, P., Rentschlar, E., Jones, H., Hammond, B., and Trimble, S. (2015). Post-storm beach and dune recovery: implications for barrier island resilience. *Geomorphology* **234**: 54–63.

Houston, J. R. (1996). Engineering practice for beach-fill designs. *Shore and Beach* **64**(3): 27–35.

Houston, J. R. (2013). The economic value of beaches – a 2013 update. *Shore and Beach* **81**(1): 3–11.

Houston, J. R. (2017). Shoreline change in response to sea-level rise on Florida's west coast. *Journal of Coastal Research* **336**: 1243–1260.

Howe, M. A. (2015). Coastal soft cliff invertebrates are reliant upon dynamic coastal processes. *Journal of Coastal Conservation* **19**: 809–820.

Howe, M. A., Knight, G. T., and Clee, C. (2010). The importance of coastal sand dunes for terrestrial invertebrates in Wales and the UK, with particular reference to aculeate Hymenoptera (bees, wasps & ants). *Journal of Coastal Conservation* **14**: 91–102.

Hu, X., Liu, B., Wu, Z. Y., and Gong, J. (2016). Analysis of dominant factors associated with hurricane damages to residential structures using the rough set theory. *Natural Hazards Review* **17**: 04016005-1-10.

Hummel, P., Thomas, S., Dillon, J., Johannessen, J., Schlenger, P., and Laprade, W. T. (2005). Seahurst Park: restoring nearshore habitat and reconnecting natural sediment supply processes. In *Puget Sound Georgia Basin Research Conference 2005*. Puget Sound Action Team, F7, pp. 1–5.

Huxel, G. R. and Hastings, A. (1999). Habitat loss, fragmentation, and restoration. *Restoration Ecology* **7**: 309–315.

Ibrahim, J. C., Holmes, P., and Blanco, B. (2006). Response of a gravel beach to swash zone hydrodynamics. *Journal of Coastal Research* **SI39**: 1685–1690.

Innocenti, R. A., Feagin, R. A., and Huff, T. P. (2018). The role of Sargassum macroalgal wrack in reducing coastal erosion. *Estuarine, Coastal and Shelf Science* **214**: 82–88.

Irish, J., Lynett, P. J., Weiss, R., Smallegan, S. M., and Cheng, W. (2013). Buried relic seawall mitigates Hurricane Sandy's impacts. *Coastal Engineering* **80**: 79–82.

Isermann, M. and Krisch, H. (1995). Dunes in contradiction with different interests. An example: the camping-ground prerow (Darss/Baltic Sea). In Salman, A. H. P. M., Berends, H., and Bonazountas, M. (Eds.), *Coastal Management and Habitat Conservation*. Leiden: EUCC, pp. 439–449.

Itzkin, M., Moore, L. J., Ruggiero, P., and Hacker, S. D. (2020). The effect of sand fencing on the morphology of natural dune systems. *Geomorphology* **352**: 106995.

Jackson, D. W. T., Costas, S., González-Villanueva, R., and Cooper, A. (2019). A global "greening" of coastal dunes: an integrated consequence of climate change? *Global and Planetary Change* **182**: 103026.

Jackson, L. L., Lopukhine, N., and Hillyard, D. (1995). Ecological restoration: a definition and comments. *Restoration Ecology* **3**: 71–76.

Jackson, N. L. and Nordstrom, K. F. (2011). Aeolian transport and landforms in managed coastal systems: a review. *Aeolian Research* **3**: 181–196.

Jackson, N. L. and Nordstrom, K. F. (2020). Trends in research on beaches and dunes on sandy shores, 1969–2019. *Geomorphology* **366**: 106737.

Jackson, N. L., Nordstrom, K. F., and Smith, D. R. (2002). Geomorphic-biotic interactions on beach foreshores in estuaries. *Journal of Coastal Research* **SI36**: 414–424.

Jackson, N. L., Smith, D. R., Tiyarattanachi, R., and Nordstrom, K. F. (2007). Use of a small beach nourishment project to enhance habitat suitability for horseshoe crabs. *Geomorphology* **89**: 172–185.

Jackson, N. L., Smith, D. R., and Nordstrom, K. F. (2008). Physical and chemical changes in the foreshore of an estuarine beach: implications for viability and development of horseshoe crab (*Limulus polyphemus*) eggs. *Marine Ecology Progress Series* **355**: 209–218.

Jackson, N. L., Nordstrom, K. F., Saini, S., and Smith, D. R. (2010). Effects of nourishment on the form and function of an estuarine beach. *Ecological Engineering* **36**: 1709–1718.

Jackson, N. L., Nordstrom, K. F., Feagin, R. A., and Smith, W. K. (2013). Coastal geomorphology and restoration. *Geomorphology* **199**: 1–7.

Jackson, N. L., Saini, S., Smith, D. R., and Nordstrom, K. F. (2020). Egg exhumation and transport on a foreshore under wave and swash processes. *Estuaries and Coasts* **43**: 286–297.

Janssen, M. P. (1995). Coastal management: restoration of natural processes in foredunes. In Healy, M. G. and Doody, J. P. (Eds.), *Directions in European Coastal Management*. Cardigan: Samara Publishing Ltd., pp. 195–198.

Janssen, M. P. J. M. and Salman, A. H. P. M. (1995). A national strategy for dune conservation in The Netherlands. In Salman, A. H. P. M., Berends, H., and Bonazountas, M. (Eds.), *Coastal Management and Habitat Conservation*. Leiden: EUCC, pp. 153–159.

Jantarasami, L. C., Lawler, J. J., and Thomas, C. W. (2010). Institutional barriers to climate change adaptation in U.S. national parks and forests. *Ecology and Society* **15**: 33.

Jaramillo, E., Contreras, H., and Bollinger, A. (2002). Beach and faunal response to the construction of a seawall in a sandy beach of south central Chile. *Journal of Coastal Research* **18**: 523–529.

Jayappa, K. S. and Deepika, B. (2018). Impacts of coastal erosion, anthropogenic activities and their management on tourism and coastal ecosystems: a study with reference to

Karnataka coast, India. In Botero, C. M., Cervantes, O., and Finkl, C. W. (Eds.), *Beach Management Tools – Concepts, Methodologies and Case Studies*. Cham, Switzerland: Springer International Publishing, pp. 421–440.

Jellinek, S., Wilson, K. A., Hagger, V., et al. (2019). Integrating diverse social and ecological motivations to achieve landscape restoration. *Journal of Applied Ecology* **56**: 246–252.

Jennings, R. and Shulmeister, J. (2002). A field based classification scheme for gravel beaches. *Marine Geology* **186**: 211–228.

Jentsch, A. (2007). The challenge to restore process in the face of nonlinear dynamics – on the crucial role of disturbance regimes. *Restoration Ecology* **15**: 334–339.

Jeschke, L. (1983). Landeskulturelle Probleme des Salzgraslandes an der Küste. *Naturschutzarbeit in Mecklenburg* **26**: 5–12.

Johnson, D. E. and Dagg, S. (2003). Achieving public participation in coastal zone environmental impact assessment. *Journal of Coastal Conservation* **9**: 13–18.

Johnson, L. and Bauer, W. (1987). Beach stabilization design. In *Coastal Zone 87*. New York: American Society of Civil Engineers, pp. 1432–1445.

Johnstone, C. A., Pastor, J., and Pinay, G. (1992). Quantitative methods for studying landscape boundaries. In Hansen, A. J. and di Castri, F. (Eds.), *Landscape Boundaries: Consequences for Biotic Diversity and Ecological Flows*. New York: Springer-Verlag, pp. 107–128.

Jolicoeur, S. and O'Carroll, S. (2007). Sandy barriers, climate change and long-term planning of strategic coastal infrastructures, Îles-de-la-Madeleine, Gulf of St. Lawrence (Québec, Canada). *Landscape and Urban Planning* **81**: 287–298.

Jones, M. L. M., Wallace, H. L., Norris, D., et al. (2004). Changes in vegetation and soil characteristics in coastal sand dunes along a gradient of atmospheric nitrogen deposition. *Plant Biology* **6**: 598–605.

Jones, M. L. M., Norman, K., and Rhind, P. M. (2010). Topsoil inversion as a restoration measure in sand dunes, early results from a UK field-trial. *Journal of Coastal Conservation* **14**: 139–151.

Jones, S. R. and Magnun, W. R. (2001). Beach nourishment and public policy after Hurricane Floyd: where do we go from here? *Ocean & Coastal Management* **44**: 207–220.

Judd, F. W., Lonard, R. I., Everitt, J. H., and Villarreal, R. (1989). Effects of vehicular traffic in the secondary dunes and vegetated flats of South Padre Island, Texas. In *Coastal Zone 89*. New York: American Society of Civil Engineers, pp. 4634–4645.

Jungerius, P. D., Koehler, H., Kooijman, A. M., Mücher, H. J., and Graefe, U. (1995). Response of vegetation and soil ecosystem to mowing and sod removal in the coastal dunes "Zwanenwater" The Netherlands. *Journal of Coastal Conservation* **1**: 3–16.

Kaczkowski, H. L., Kana, T. W., Traynum, S. B., and Visser, R. (2018). Beach-fill equilibrium and dune growth at two large-scale nourishment sites. *Ocean Dynamics* **68**: 1191–1206.

Kana, T. W. and Kaczkowski, H. L. (2019). Myrtle Beach: a history of shore protection and beach restoration. *Shore and Beach* **87**(3): 13–34.

Kana, T. W., Traynum, S. B., Gaudiano, D., Kaczkowski, H. L., and Hair, T. (2013). The physical condition of South Carolina beaches 1980–2010. *Journal of Coastal Research* **SI69**: 61–82.

Kar, D., Rhode, R., Snider, N. P., and Robichaux, E. (2020). Measuring success through outcome indicators for restoration efforts in Louisiana. *Journal of Coastal Research* **SI95**: 1128–1133.

Katz, E. (1999). A pragmatic re-consideration of anthropocentrism. *Environmental Ethics* **21**: 377–390.

Keddy, P. A. (1981). Experimental demography of a dune annual: *Cakile edentula* growing along an environmental gradient in Nova Scotia. *Journal of Ecology* **69**: 615–630.

Kelly, J. F. (2014). Effects of human activities (raking, scraping, off-road vehicles) and natural resource protections on the spatial distribution of beach vegetation and related shoreline features in New Jersey. *Journal of Coastal Conservation* **18**: 383–398.

Kelly, J. F. (2016). Assessing the spatial compatibility of recreational activities with beach vegetation and wrack in New Jersey: prospects for compromise management. *Ocean & Coastal Management* **123**: 9–17.

Kennedy, A., Rogers, S., Sallenger, A., et al. (2011). Building destruction from waves and surge on the Bolivar Peninsula during Hurricane Ike. *Journal of Waterway, Port, Coastal, and Ocean Engineering* **137**: 132–141.

Kenny, A. J. and Rees, H. L. (1994). The effects of marine gravel extraction on the macrobenthos: early post-dredging recolonization. *Marine Pollution Bulletin* **28**: 442–447.

Kenny, A. J. and Rees, H. L. (1996). The effects of marine gravel extraction on the macrobenthos: results 2 years post-dredging. *Marine Pollution Bulletin* **32**: 615–622.

Kessler, R. (2008). Sand dune stabilization at Pineda Ocean Club. *Land and Water* **52**(3): 13–22.

Ketner-Oostra, R. and Sýkora, K. V. (2000). Vegetation succession and lichen diversity on dry coastal calcium-poor dunes and the impact of management experiments. *Journal of Coastal Conservation* **6**: 191–206.

Kilibarda, Z., Graves, N., Dorton, M., and Dorton, R. (2014). Changes in beach gravel lithology caused by anthropogenic activities along the southern coast of Lake Michigan. *Environmental Earth Science* **71**: 1249–1266.

Kirk, R. M. (1980). Mixed sand and gravel beaches: morphology, processes and sediments: *Progress in Physical Geography* **4**: 189–210.

Klein, A. H. F., Aroujo, R. S., Polette, M., et al. (2009). Ameliorative strategies at Balneário Piçarras Beach. In Williams, A. and Micallef, A. (Eds.), *Beach Management: Principles & Practice*. London: Earthscan, pp. 247–261.

Klein, R. J. T., Smit, M. J., Goosen, H., and Hulsbergen, C. H. (1998). Resilience and vulnerability – coastal dynamics or Dutch dikes? *The Geographical Journal* **164**: 259–268.

Klein, Y. L., Osleeb, J. P., and Viola, M. R. (2004). Tourism-generated earnings in the coastal zone: a regional analysis. *Journal of Coastal Research* **20**: 1080–1088.

Klijn, J. A. (1990). The younger dunes in The Netherlands: chronology and causation. *Catena* Suppl **18**: 89–100.

Knevel, I. C., Venema, H. G., and Lubke, R. A. (2002). The search for indigenous dune stabilizers: germination requirements of selected South African species. *Journal of Coastal Conservation* **8**: 169–178.

Knight, R. L. (1999). Private lands: the neglected geography. *Conservation Biology* **13**: 223–224.

Knutson, P. L. (1978). Planting guidelines for dune creation and stabilization. In *Coastal Zone 78*. New York: American Society of Civil Engineers, pp. 762–779.

Kochnower, D., Reddy, S. M. W., and Flick, R. E. (2015). Factors influencing local decisions to use habitats to protect coastal communities from hazards. *Ocean & Coastal Management* **116**: 277–290.

Koehler, H., Munderloh, E., and Hofmann, S. (1995). Soil microarthropods (Acari and Collembola) from beach and dune: characteristics and ecosystematic context. In

Salman, A. H. P. M., Berends, H., and Bonazountas, M. (Eds.), *Coastal Management and Habitat Conservation*. Leiden: EUCC, pp. 371–383.

Komar, P. D., Allen, J. C., and Winz, R. (2003). Cobble beaches – the "design with nature" approach for shore protection. In *Coastal Sediments 03*. New York: American Society of Civil Engineers, pp. 1–13.

Kombiadou, K., Costas, S., Carrasco, A. R., Plomaritis, T. A., Ferreira, Ó., and Matias, A. (2019). Bridging the gap between resilience and geomorphology of complex coastal systems. *Earth-Science Reviews* **198**: 102934.

Konlechner, T. M. and Hilton, M. J. (2009). The potential for marine dispersal of *Ammophila arenaria* (marram grass) rhizome in New Zealand. *Journal of Coastal Research* **SI56**: 434–437.

Konlechner, T. M., Hilton, M. J., and Arens, S. M. (2014). Transgressive dune development following deliberate de-vegetation for dune restoration in the Netherlands and New Zealand. *Dynamic Environments* **33**: 141–154.

Konlechner, T. M., Ryu, W., Hilton, M. J., and Sherman, D. J. (2015a). Evolution of foredune texture following dynamic restoration, Doughboy Bay, Stewart Island, New Zealand. *Aeolian Research* **19**: 203–214.

Konlechner, T. M., Hilton, M. J., and Lord, J. M. (2015b). Plant community response following the removal of the invasive *Lupinus arboreus* in a coastal dune system. *Restoration Ecology* **23**: 607–614.

Kooijman, A. M. (2004). Environmental problems and restoration measures in coastal dunes in The Netherlands. In Martínez, M. L. and Psuty, N. P. (Eds.), *Coastal Dunes, Ecology and Conservation*. Berlin: Springer-Verlag, pp. 243–258.

Kooijman, A. M. and de Haan, M. W. A. (1995). Grazing as a measure against grass encroachment in Dutch dry dune grassland: effects on vegetation and soil. *Journal of Coastal Conservation* **1**: 127–134.

Kooijman, A. M. and Smit, A. (2001). Grazing as a measure to reduce nutrient availability and plant productivity in acid dune grasslands and pine forests in The Netherlands. *Ecological Engineering* **17**: 63–77.

Kooijman, A. M., van Til, M., Noordijk, E., Remke, E., and Kalbitz, K. (2017). Nitrogen deposition and grass encroachment in calcareous and acidic grey dunes (H2130) in NW-Europe. *Biological Conservation* **212**: 406–415.

Kopp, R. E., Gilmore, E. A., Little, C. M., Lorenzo-Trueba, J., Ramenzoni, V. C., and Sweet, W. V. (2019). Usable science for managing the risks of sea-level rise. *Earth's Future* **7**: 1235–1269.

Koske, R. E., Gemma, J. N., Corkidi, L., Sigüenza, C., and Rincón, E. (2004). Arbuscular mycorrhizas in coastal dunes. In Martínez, M. L. and Psuty, N. P. (Eds.), *Coastal Dunes, Ecology and Conservation*. Berlin: Springer-Verlag, pp. 173–187.

Koster, M. J. and Hillen, R. (1995). Combat erosion by law: coastal defense policy for The Netherlands. *Journal of Coastal Research* **11**: 1221–1228.

Kousky, C. (2014). Managing shoreline retreat: a US perspective. *Climatic Change* **124**: 9–20.

Kowalewski, M., Domènech, R., and Martinell, J. (2014). Vanishing clams on an Iberian beach: local consequences and global implications of accelerating loss of shells to tourism. *PLoS ONE* **9**: e83615.

Kratzmann, M. G. and Hapke, C. J. (2012). Quantifying anthropogenically driven morphologic changes on a barrier island: Fire Island National Seashore. *Journal of Coastal Research* **28**: 76–88.

Kraus, N. C. and Rankin, K. L. (2004). Functioning and design of coastal groins: the interaction of groins and the beach – process and planning. *Journal of Coastal Research* **SI33**.

Krelling, A. P., Williams, A. T., and Turra, A. (2017). Differences in perception and reaction of tourist groups to beach marine debris that can influence loss of tourism revenue in coastal areas. *Marine Policy* **85**: 87–99.

Kriesel, W., Keeler, A., and Landry, C. (2004). Financing beach improvements: comparing two approaches on the Georgia coast. *Coastal Management* **32**: 433–447.

Krogh, M. G. and Schweitzer, S. H. (1999). Least terns nesting on natural and artificial habitats in Georgia, USA. *Waterbirds* **22**: 290–296.

Kuang, C., Mao, X., Gu, J., et al. (2019). Morphological processes of two artificial submerged shore-parallel sandbars for beach nourishment in a nearshore zone. *Ocean & Coastal Management* **179**: 104870.

Kumar, A., Seralathan, P., and Jayappa, K. S. (2009). Distribution of coastal cliffs in Kerala, India: their mechanisms of failure and related human engineering response. *Environmental Geology* **58**: 815–832.

Kuriyama, Y., Mochizuki, N., and Nakashima, T. (2005). Influence of vegetation on Aeolian sand transport rate from a backshore to a foredune at Hasaki, Japan. *Sedimentology* **52**: 1123–1132.

Kutiel, P. B. (2013). Restoration of coastal sand dunes for conservation of biodiversity: the Israeli experience. In: Martínez, M. L., Gallego-Fernández, J. B., and Hesp, P. A. (Eds.), *Restoration of Coastal Dunes*. New York: Springer, pp. 173–185.

Kutiel, P., Peled, Y., and Geffen, E. (2000). The effect of removing shrub cover on annual plants and small mammals in a coastal sand dune ecosystem. *Biological Conservation* **94**: 235–242.

Lamb, F. H. (1898). Sand-dune reclamation on the Pacific Coast. *The Forester* **4**: 141–142.

Lamberti, A. and Mancinelli, A. (1996). Italian experience on submerged barriers as beach defence structures. In Edge, B. L. (Ed.), *Coastal Engineering 1996*. New York: American Society of Civil Engineers, pp. 2352–2365.

Lamberti, A., Archetti, R., Kramer, M., Paphilitis, D., Mosso, C., and Di Risio, M. (2005). European experience of low crested structures for coastal management. *Coastal Engineering* **52**: 841–866.

Lammerts, E. J., Grootjans, A., Stuyfzand, P., and Sival, F. (1995). Endangered dune slack plants: gastronovers in need of mineral water. In Salman, A. H. P. M., Berends, H., and Bonazountas, M. (Eds.), *Coastal Management and Habitat Conservation*. Leiden: EUCC, pp. 355–369.

Lapointe, M., Gurney, G. G., and Cumming, G. S. (2020). Urbanization alters ecosystem service preferences in a small island developing state. *Ecosystem Services* **43**: 101109.

Larson, M. and Kraus, N. C. (2000). Representation of non-erodible (hard) bottoms in beach profile change modeling. *Journal of Coastal Research* **16**: 1–14.

Latsoudis, P. K. (1996). The natural and artificial dunes of Cape Epanomi. In Salman, A. H. P. M., Langeveld, M. J., and Bonazountas, M. (Eds.), *Coastal Management and Habitat Conservation*. Leiden: EUCC, pp. 55–57.

Lawrenz-Miller, S. (1991). Grunion spawning versus beach nourishment: nursery or burial ground. In *Coastal Zone 91*. New York: American Society of Civil Engineers, pp. 2197–2208.

Lazarus, E. D. and Goldstein, E. B. (2019). Is there a bulldozer in your model? *Journal of Geophysical Research: Earth Surface* **124**: 696–699.

Lazarus, E. D., McNamara, D. E., Smith, M. D., Gopalakrishnan, S., and Murray, A. B. (2011). Emergent behavior in a coupled economic and coastline model for beach nourishment. *Nonlinear Processes in Geophysics* **18**: 989–999.

Lazarus, E. D., Ellis, M. A., Murray, A. B., and Hall, D. M. (2016). An evolving research agenda for human-coastal systems. *Geomorphology* **256**: 81–90.

Leafe, R., Pethick, J., and Townend, I. (1998). Realizing the benefits of shoreline management. *The Geographical Journal* **164**: 282–290.

Ledoux, L., Cornell, S., O'Riordan, T., Harvey, R., and Banyard, L. (2005). Towards sustainable flood and coastal management: identifying drivers of, and obstacles to, managed realignment. *Land Use Policy* **22**(2): 129–144.

Lee, E. M. (1993). The political ecology of coastal planning and management in England and Wales: policy responses to the implications of sea-level-rise. *The Geographical Journal* **159**: 169–178.

Lee, E. M. (1998). Problems associated with the prediction of cliff recession rates for coastal defence and conservation. In Hooke, J. (Ed.), *Coastal Defence and Earth Science Conservation*. Bath: The Geological Society, pp. 46–57.

Leege, L. M. and Kilgore, J. S. (2014). Recovery of foredune and blowout habitats in a freshwater dune following removal of invasive Austrian pine (*Pinus nigra*). *Restoration Ecology* **22**: 641–648.

Leege, L. M. and Murphy, P. G. (2000). Growth of the non-native *Pinus nigra* in four habitats on the sand dunes of Lake Michigan. *Forest Ecology and Management* **126**: 191–200.

Leewis, L., van Bodegom, P. M., Rozema, J., and Janssen, G. M. (2012). Does beach nourishment have long-term effects on intertidal macroinvertebrate species abundance? *Estuarine, Coastal and Shelf Science* **113**: 172–181.

Lehrer, D., Becker, N., and Kutiel, P. B. (2013). The value of coastal sand dunes as a measure to plan an optimal policy for invasive plant species: the case of the *Acacia saligna* at the Nizzanim LTER Coastal Sand Dune Nature Reserve, Israel. In: Martínez, M. L., Gallego-Fernández, J. B., and Hesp, P. A. (Eds.), *Restoration of Coastal Dunes*. New York: Springer, pp. 273–288.

Lemauviel, S. and Roze, F. (2000). Ecological study of pine forest clearings along the French Atlantic sand dunes: perspectives of restoration. *Acta Oecologica* **21**: 179–192.

Lemauviel, S., Gallet, S., and Roze, F. (2003). Sustainable management of fixed dunes: example of a pilot site in Brittany, France. *Comptes Rendus Biologies* **326**: S183–S191.

Lentz, E. E., Thieler, E. R., Plant, N. G., Stippa, S. R., Horton, R. M., and Gesch, D. B. (2016). Evaluation of dynamic coastal response to sea-level rise modifies inundation likelihood. *Nature Climate Change* **6**: 696–701.

Li, B. and Sherman, D. J. (2015). Aerodynamics and morphodynamics of sand fences: a review. *Aeolian Research* **17**: 33–48.

Li, D., van de Werfhorst, L. C., Dunne, T., Devarajan, N., Gomez Ayala, T., and Holden, P. A. (2020). Surf zone microbiological water quality following emergency beach nourishment using sediments from a catastrophic debris flow. *Water Research* **176**: 115733.

Light, A. and Higgs, E. S. (1996). The politics of ecological restoration. *Environmental Ethics* **18**: 227–247.

Lin, P. C-P., Hansen, I., and Sasso, R. H. (1996). Combined sand bypassing and navigation improvements at Hillsboro Inlet, Broward County, Florida: the importance of a regional approach. In Tait, L. S. (Ed.), *The Future of Beach Nourishment*. Tallahassee: Florida Shore and Beach Preservation Association, pp. 43–59.

Lindeman, K. C. and Snyder, D. B. (1999). Nearshore hardbottom fishes of southeast Florida and effects of habitat burial caused by dredging. *Fishery Bulletin* **97**: 508–525.

Lithgow, D., Martínez, M. L., Gallego-Fernández, J. B., et al. (2013). Linking restoration ecology with coastal dune restoration. *Geomorphology* **199**: 214–224.

Lithgow, D., Martínez, M. L., and Gallego-Fernández, J. B. (2014). The "ReDune" index (Restoration of coastal Dunes Index) to assess the need and viability of coastal dune restoration. *Ecological Indicators* **49**: 178–187.

Liu, G., Feng, C., Hongshuai, Q., et al. (2019). A method to nourished beach stability assessment: the case of China. *Ocean & Coastal Management* **177**: 166–178.

Long, Z. T., Fegley, S. R., and Peterson, C. H. (2013). Fertilization and plant diversity accelerate primary succession and restoration of dune communities. *Plant Ecology* **214**: 1419–1429.

Looney, P. B. and Gibson, D. J. (1993). Vegetation monitoring of beach nourishment. In Stauble, D. K. and Kraus, N. C. (Eds.), *Beach Nourishment: Engineering and Management Considerations*. New York: American Society of Civil Engineers, pp. 226–241.

Lorang, M. S. (2002). Predicting the crest height of a gravel beach. *Geomorphology* **48**: 87–101.

Lorenzoni, I. and Hulme, M. (2009). Believing is seeing: laypeople's views of future socio-economic and climate change in England and in Italy. *Public Understanding of Science* **18**: 383–400.

Lortie, C. J. and Cushman, J. H. (2007). Effects of a directional abiotic gradient on plant community dynamics and invasion in a coastal dune system. *Journal of Ecology* **95**: 468–481.

Lubke, R. A. (2004). Vegetation dynamics and succession on sand dunes of the eastern coasts of Africa. In Martínez, M. L. and Psuty, N. P. (Eds.), *Coastal Dunes, Ecology and Conservation*. Berlin: Springer-Verlag, pp. 67–84.

Lubke, R. A. (2013). Restoration of dune ecosystems following mining in Madagascar and Namibia: contrasting restoration approaches adopted in regions of high and low human population density. In Martínez, M. L., Gallego-Fernández, J. B., and Hesp, P. A. (Eds.), *Restoration of Coastal Dunes*. New York: Springer, pp. 199–215.

Lubke, R. A. and Avis, A. M. (1998). A review of the concepts and application of rehabilitation following heavy mineral dune mining. *Marine Pollution Bulletin* **37**: 546–557.

Lubke, R. A. and Hertling, U. M. (2001). The role of European marram grass in dune stabilization and succession near Cape Agulhas, South Africa. *Journal of Coastal Conservation* **7**: 171–182.

Lubke, R. A., Avis, A. M., and Hellstrom, G. B. (1995). Current status of coastal zone management in the Eastern Cape region, South Africa. In Salman, A. H. P. M., Berends, H., and Bonazountas, M. (Eds.), *Coastal Management and Habitat Conservation*. Leiden: EUCC, pp. 239–260.

Lubke, R. A., Avis, A. M., and Moll, J. B. (1996). Post-mining rehabilitation of coastal sand dunes in Zululand, South Africa. *Landscape and Urban Planning* **34**: 335–345.

Lück-Vogel, M., and Mbolambi, C. (2018). Assessment of coastal Strandveld integrity using WorldView-2 imagery in False Bay, South Africa. *South African Journal of Botany* **116**: 150–157.

Lucrezi, S., and van der Walt, M. F. (2016). Beachgoers' perceptions of sandy beach conditions: demographic and attitudinal influences, and the implications for beach ecosystem management. *Journal of Coastal Conservation* **20**: 81–96.

Ludka, B. C., Guza, R. T., and O'Reilly, W. C. (2018). Nourishment evolution and impacts at four southern California beaches: a sand volume analysis. *Coastal Engineering* **136**: 96–105.

Luisetti, T., Turner, R. K., Bateman, I. J., Morse-Jones, S., Adams, C., and Fonseca, L. (2011). Coastal and marine ecosystem services evaluation for policy and management: managed realignment case studies in England. *Ocean & Coastal Management* **54**: 212–224.

Lundberg, A., Kapfer, J., and Maren, I. E. (2017). Reintroduced mowing can counteract biodiversity loss in abandoned meadows. *Erdkunde* **71**: 127–142.

Luo, S., Liu, Y., Jin, R., Ahang, J., and Wei, W. (2016). A guide to coastal management: benefits and lessons learned of beach nourishment practices in China over the past two decades. *Ocean & Coastal Management* **134**: 207–215.

Lutz, K. (1996). Studie zum Generalplan Küstenschutz und zur Rekultivierung von Salzgrasländern. Unpublished report on behalf of the WWF Germany, Ostseeschutz project office Stralsund.

Maes, D., Ghesquiere, A., Logie, M., and Bonte, D. (2006). Habitat and mobility of two threatened coastal dune insects: implications for conservation. *Journal of Insect Conservation* **10**: 105–115.

Malvárez-García, G., Pollard, J., and Hughes, R. (2002). Coastal Zone Management on the Costa del Sol: a small business perspective. *Journal of Coastal Research* **SI36**: 470–482.

Manning, L. M., Peterson, C. H., and Fegley, S. R. (2013). Degradation of surf-fish foraging habitat driven by persistent sedimentological modifications caused by beach nourishment. *Bulletin of Marine Science* **89**: 83–106.

Manning, L. M., Peterson, C. H., and Bishop, M. J. (2014). Dominant macrobenthic populations experience sustained impacts from annual disposal of fine sediments on sandy beaches. *Marine Ecology Progress Series* **508**: 1–15.

Manno, G., Anfuso, G., Messina, E., Williams, A. T., Suffo, M., and Liguori, V. (2016). Decadal evolution of coastline armouring along the Mediterranean Andalusia littoral (South of Spain). *Ocean & Coastal Management* **124**: 84–99.

Marcomini, S. C. and López, R. A. (2006). Evolution of a beach nourishment project at Mar del Plata. *Journal of Coastal Research* **SI39**: 834–837.

Marinho, B., Coelho, C., Larson, M., and Hanson, H. (2018). Short- and long-term responses of nourishments: Barra-Vagueira coastal stretch, Portugal. *Journal of Coastal Conservation* **22**: 475–489.

Marqués, M. A., Psuty, N. P., and Rodriguez, R. (2001). Neglected effects of eolian dynamics on artificial beach nourishment: the case of Riells, Spain. *Journal of Coastal Research* **17**: 694–704.

Marsh, G. P. (1885). *Earth as Modified by Human Action*. New York: Charles Scribner.

Martin, K. L. and Adams, L. C. (2020). Effects of repeated sand replenishment projects on runs of a beach-spawning fish, the California grunion. *Journal of Marine Science and Engineering* **8**: 178.

Martinez, G., Armaroli, C., Costas, S., Harley, M. D., and Paolisso, M. (2018). Experiences and results from interdisciplinary collaboration: utilizing qualitative information to formulate disaster risk reduction measures for coastal regions. *Coastal Engineering* **134**: 62–72.

Martínez, M. L. and García-Franco, J. G. (2004). Plant-plant interactions in coastal dunes. In Martínez, M. L. and Psuty, N. P. (Eds.), *Coastal Dunes, Ecology and Conservation*. Berlin: Springer-Verlag, pp. 205–220.

Martínez, M. L., Maun, M. A., and Psuty, N. P. (2004). The fragility and conservation of the world's coastal dunes: geomorphological, ecological, and socioeconomic perspectives. In Martínez, M. L. and Psuty, N. P. (Eds.), *Coastal Dunes, Ecology and Conservation*. Berlin: Springer-Verlag, pp. 355–369.

Martínez, M. L., Chávez, V., Lithgow, D., and Silva, R. (2019). Integrating biophysical components in coastal engineering practices. *Journal of Coastal Research* **SI92**: 1–5.

Maslo, B., Handel, S. N., and Pover, T. (2011). Restoring beaches for Atlantic Coast piping plovers (*Charadrius melodus*): a classification and regression tree analysis of nest-site selection. *Restoration Ecology* **19**: 194–203.

Maslo, B., Leu., K., Pover, T., Weston, M. A., Gilby, B. L., and Schlacher, T. A. (2019). Optimizing conservation benefits for threatened beach fauna following severe natural disturbances. *Science of the Total Environment* **649**: 661–671.

Mason, T. and Coates, T. T. (2001). Sediment transport processes on mixed beaches: a review for shoreline management. *Journal of Coastal Research* **17**: 645–657.

Mason, T. J. and French, K. (2007). Management regimes for a plant invader differentially impact resident communities. *Biological Conservation* **136**: 246–259.

Masselink, G. and Lazarus, E. D. (2019). Defining coastal resilience. *Water* **11**: 2587.

Masselink, G. and Short, A. D. (1993). The effect of tide range on beach morphodynamics and morphology: a conceptual beach model. *Journal of Coastal Research* **9**: 785–800.

Masucci, G. D. and Reimer, J. D. (2019). Expanding walls and shrinking beaches: loss of natural coastline in Okinawa Island, Japan. *PeerJ* **7**: e7520.

Mather, A. S. and Ritchie, W. (1977). *The Beaches of the Highlands and Islands of Scotland*. Perth: Countryside Commission for Scotland.

Mathew, M. J., Sautter, B., Ariffin, E. H., et al. (2020). Total vulnerability of the littoral zone to climate change-driven natural hazards in north Brittany, France. *Science of the Total Environment* **706**: 135963.

Matias, A., Ferreira, Ó., Mendes, I., Dias, J. A., and Vila-Consejo, A. (2005). Artificial construction of dunes in the south of Portugal. *Journal of Coastal Research* **21**: 472–481.

Matsumoto, H., Young, A. P., and Guza, R. T. (2020). Observations of surface cobbles at two southern California beaches *Marine Geology* **419**: 106049.

Maun, M. A. (1998). Adaptation of plants to burial in coastal sand dunes. *Canadian Journal of Botany* **76**: 713–738.

Maun, M. A. (2004). Burial of plants as a selective force in sand dunes. In Martínez, M. L. and Psuty, N. P. (Eds.), *Coastal Dunes, Ecology and Conservation*. Berlin: Springer-Verlag, pp. 119–135.

Mauriello, M. N. (1989). Dune maintenance and enhancement: a New Jersey example. In *Coastal Zone 89*. New York: American Society of Civil Engineers, pp. 1023–1037.

Mauriello, M. N. (1991). Beach nourishment and dredging: New Jersey's policies. *Shore and Beach* **59**(3): 25–28.

Mauriello, M. N. and Halsey, S. D. (1987). Dune building on a developed coast. In *Coastal Zone 87*. New York: American Society of Civil Engineers, pp. 1313–1327.

May, V. (2015). Coastal cliff conservation and management: the Dorset and east Devon coast world heritage site. *Journal of Coastal Conservation* **19**: 821–829.

McArdle, S. B. and McLachlan, A. (1992). Sand beach ecology: swash features relevant to the macrofauna. *Journal of Coastal Research* **8**: 398–407.

McIntyre, A. F. and Heath, J. A. (2011). Evaluating the effects of foraging habitat restoration on shorebird reproduction: the importance of performance criteria and comparative designs. *Journal of Coastal Conservation* **15**: 151–157.

McLachlan, A. (1985). The biomass of macro- and interstitial fauna on clean and wrack-covered beaches in Western Australia. *Estuarine, Coastal and Shelf Science* **21**: 587–599.

McLachlan, A. (1990). The exchange of materials between dune and beach systems. In Nordstrom, K. F., Psuty, N. P., and Carter, R. W. G. (Eds.), *Coastal Dunes: Form and Process*. Chichester: John Wiley & Sons, pp. 201–215.

McLachlan, A. (1996). Physical factors in benthic ecology: effects of changing sand particle size on beach fauna. *Marine Ecology Progress Series* **131**: 205–217.

McLachlan, A. and Burns, M. (1992). Headland bypass dunes on the South African coast: 100 years of (mis)management. In Carter, R. W. G., Curtis, T. G. F., and Sheehy-Skeffington, M. J. (Eds.), *Coastal Dunes: Geomorphology, Ecology and Management for Conservation*. Rotterdam: A.A. Balkema, pp. 71–79.

McLean, R. F. and Kirk, R. M. (1969). Relationship between grain size, size-sorting and fore-shore slope on mixed sand-shingle beaches. *New Zealand Journal of Geology and Geophysics* **12**: 138–155.

McLean, R. and Shen, J.-S. (2006). From foreshore to foredune: foredune development over the last 30 years at Moruya Beach, New South Wales, Australia. *Journal of Coastal Research* **22**: 28–36.

McLouth, M. E., Lapolla, J., and Bodge, K. (1994). Port Authority's role in inlet management and beneficial use of dredged material. In *Dredging '94*. New York: American Society of Civil Engineers, pp. 971–980.

McNamara, D. E., Gopalakrishnan, S., Smith, M. D., and Murray, A. B. (2015). Climate adaptation and policy-induced inflation of coastal property value. *PLoS ONE* **10**: e0121278.

McNinch, J. E. and Wells, J. T. (1992). Effectiveness of beach scraping as a method of erosion control. *Shore and Beach* **60**(1): 13–20.

Melvin, S. M., Griffin, C. R., and MacIvor, L. H. (1991). Recovery strategies for piping plovers in managed coastal landscapes. *Coastal Management* **19**: 21–34.

Mendelssohn, I. A., Hester, M. W., Monteferrante, F. J., and Talbot, F. (1991). Experimental dune building and vegetative stabilization in a sand-deficient barrier island setting on the Louisiana coast, USA. *Journal of Coastal Research* **7**: 137–149.

Meyer-Arendt, K. J. (1990). Recreational business districts in Gulf of Mexico seaside resorts. *Journal of Cultural Geography* **11**: 39–55.

Meyer-Arendt, K. J. (1993). Geomorphic impacts of resort evolution along the Gulf of Mexico coast; applicability of resort cycle models. In Wong, P. P. (Ed.), *Tourism vs Environment: The Case for Coastal Areas*. Dordrecht: Kluwer Academic Publishers, pp. 125–138.

Mielck, F., Hass, H. C., Michaelis, R., Sander, L., Papenmeier, S., and Wiltshire, K. H. (2019). Morphological changes due to marine aggregate extraction for beach nourishment in the German Bight (SE North Sea). *Geo-Marine Letters* **39**: 47–58.

Miller, D. L., Thetford, M., and Yager, L. (2001). Evaluating sand fence and vegetation for dune building following overwash by Hurricane Opal on Santa Rosa Island, Florida. *Journal of Coastal Research* **17**: 936–948.

Miller, D. L., Thetford, M., Dupree, J., and Atwood, L. (2014). Influence of seasonal changes and shifting substrate on survival of restoration plantings of *Schizachyrium maritimum* (Gulf Bluestem) on Santa Rosa Island, Florida. *Journal of Coastal Research* **30**: 237–247.

Miller, K. G., Kopp, R. E., Horton, B. P., Browning, J. V., and Kemp, A. C. (2013). A geological perspective on sea-level rise and its impacts along the U.S. mid-Atlantic coast. *Earth's Future* **1**: 3–18.

Millett, J. and Edmondson, S. (2013). The impact of 36 years of grazing management on vegetation dynamics in dune slacks. *Journal of Applied Ecology* **50**: 1367–1376.

Milton, S. L., Schulman, A. A., and Lutz, P. L. (1997). The effect of beach nourishment with aragonite versus silicate sand on beach temperature and loggerhead sea turtle nesting success. *Journal of Coastal Research* **13**: 904–915.

Minerals Management Service. (2001). *Development and Design of Biological and Physical Monitoring Protocols to Evaluate the Long-Term Impacts of Offshore Dredging Operations on the Marine Environment.* U.S. Department of the Interior, Minerals Management Service Final Report MMS 2001-089.

Minteer, B. A. and Manning, R. E. (1999). Pragmatism in environmental ethics: democracy, pluralism, and the management of nature. *Environmental Ethics* **21**: 191–207.

Mireille, E., Mendoza, E., and Silva, R. (2020). Micro sand engine beach stabilization strategy at Puerto Morelos, Mexico. *Journal of Marine Science and Engineering* **8**: 247–263.

Miselis, J. L., Andrews, B. D., Nicholson, R. S., Defne, Z., Ganju, N. K., and Navoy, A. (2016). Evolution of mid-Atlantic coastal and back-barrier estuary environments in response to a hurricane: implications for barrier-estuary connectivity. *Estuaries and Coasts* **39**: 916–934.

Mitsch, W. J. (1998). Ecological engineering – the 7-year itch. *Ecological Engineering* **10**: 119–130.

Mitteager, W. A., Burke, A., and Nordstrom, K. F. (2006). Landscape features and restoration potential on private shorefront lots in New Jersey, USA. *Journal of Coastal Research* **SI39**: 890–897.

Moore, R. and Davis, G. (2015). Cliff instability and erosion management in England and Wales. *Journal of Coastal Conservation* **19**: 771–784.

Morand, P. and Merceron, M. (2005). Macroalgal populations and sustainability. *Journal of Coastal Research* **21**: 1009–1020.

Moreno-Casasola, P. (1986). Sand movement as a factor in the distribution of plant communities in a coastal dune system. *Vegetatio* **65**: 67–76.

Moreno-Casasola, P. (2004). A case study of conservation and management of tropical sand dune systems: La Mancha-El Llano. In Martínez, M. L. and Psuty, N. P. (Eds.), *Coastal Dunes, Ecology and Conservation*. Berlin: Springer-Verlag, pp. 319–333.

Moreno-Casasola, P., Martínez, M. L., Castillo-Campos, G., and Campos, A. (2013). The impacts on natural vegetation following the establishment of exotic *Casuarina* plantations. In Martínez, M. L., Gallego-Fernández, J. B., and Hesp, P. A. (Eds.), *Restoration of Coastal Dunes*. New York: Springer, pp. 217–233.

Morgan, R. (1999). Preferences and priorities of recreational beach users in Wales, UK. *Journal of Coastal Research* **15**: 653–667.

Morgan, R. and Williams, A. T. (1999). Video panorama assessment of beach landscape aesthetics on the coast of Wales. *Journal of Coastal Conservation* **5**: 13–22.

Morris, R. K. A. (2012). Managed realignment: a sediment management paradigm. *Ocean & Coastal Management* **65**: 59–66.

Morris, R. L., Konlechner, T. M., Ghisalberti, M., and Swearer, S. E. (2019). From grey to green: efficacy of eco-engineering solutions for nature-based coastal defence. *Global Change Biology* **24**: 1827–1842.

Morton, R. A. (2002). Factors controlling storm impacts on coastal barriers and beaches – a preliminary basis for near real-time forcasting. *Journal of Coastal Research* **18**: 486–501.

Moschella, P. S., Abbiati, M., Aberg, P., et al. (2005). Low-crested coastal defence structures as artificial habitats for marine life: using ecological criteria in design. *Coastal Engineering* **52**: 1053–1071.

Moses, C. A. and Williams, R. B. G. (2008). Artificial beach recharge: the South East England experience. *Zeitschrift für Geomorphologie* **52**(Suppl. 3): 107–124.

Moss, D. and McPhee, D. P. (2006). The impacts of recreational four-wheel driving on the abundance of the ghost crab (*Ocypode cordimanus*) on a subtropical sandy beach in SE Queensland. *Coastal Management* **34**: 133–140.

Mossman, H. L., Davy, A. J., and Grant, A. (2012). Does managed realignment create saltmarshes with "equivalent biological characteristics" to natural reference sites? *Journal of Applied Ecology* **49**: 1446–1456.

Mulder, J. P. M., van de Kreeke, J., and van Vessem, P. (1994). Experimental shoreface nourishment, Terschelling (NL). In *Coastal Engineering: Proceedings of the Twenty-Fourth Coastal Engineering Conference*. New York: American Society of Civil Engineers, pp. 2886–2899.

Muñoz-Perez, J. J., Lopez de San Roman-Blanco, B., Gutierrez-Mas, J. M., Moreno, L., and Cuena, G. J. (2001). Cost of beach maintenance in the Gulf of Cadiz (SW Spain). *Coastal Engineering* **42**: 143–153.

Muñoz-Reinoso, J. C. (2003). *Juniperus oxycedrus* ssp. *macrocarpa* in SW Spain: ecology and conservation problems. *Journal of Coastal Conservation* **9**: 113–122.

Muñoz-Reinoso, J. C. (2004). Diversity of maritime juniper woodlands. *Forest Ecology and Management* **192**: 267–276.

Muñoz-Reinoso, J. C., Saavedra Azqueta, C., and Redondo Morales, I. (2013). Restoration of Andalusian coastal juniper woodlands. In Martínez, M. L., Gallego-Fernández, J. B., and Hesp, P. A. (Eds.), *Restoration of Coastal Dunes*. New York: Springer, pp. 145–158.

Murphy, A. L., Singers, N. J. D., and Rapson, G. L. (2019). Created dune slack wetlands effectively host rare early successional turf communities in a dynamic dunefield, New Zealand. *Journal of Coastal Conservation* **23**: 203–225.

Murray, A. B., Gopalakrishnan, S., McNamara, D., and Smith, M. D. (2013). Progress in coupling models of human and coastal landscape change. *Computers and Geosciences* **53**: 30–33.

Myatt, L. B., Scrimshaw, M. D., and Lester, J. N. (2003). Public perceptions and attitudes towards a current managed realignment scheme: Brancaster West Marsh, North Norfolk, U.K. *Journal of Coastal Research* **19**: 278–286.

Myers, M. R., Barnard, P. L., Beighley, E., et al. (2019). A multidisciplinary coastal vulnerability assessment for local government focused on ecosystems, Santa Barbara area, California. *Ocean & Coastal Management* **182**: 104921.

Nachite, D., Maziane, F., Anfuso, G., and Macias, A. (2018). Beach litter characteristics along the Moroccan Mediterranean coast: implications for coastal zone management. In Botero, C. M., Cervantes, O., and Finkl, C. W. (Eds.), *Beach Management Tools – Concepts, Methodologies and Case Studies*. Cham, Switzerland: Springer International Publishing, pp. 795–819.

Nairn, R., Johnson, J. A., Hardin, D., and Michel, J. (2004). A biological and physical monitoring program to evaluate long-term impacts from sand dredging operations in the United States outer continental shelf. *Journal of Coastal Research* **20**: 126–137.

Narayan, S., Pontee, N., Beck, M. W., and Hosking, A. H. (2016a). Nature based solutions: lessons from around the world. In Baptiste, A. (Ed.), *Coastal Management: Changing Coast, Changing Climate, Changing Minds*. International Coastal Management Conference. London: ICE Publishing, pp. 651–662.

Narayan, S., Beck, M. W., Reguero, B. G., et al. (2016b). The effectiveness, costs and coastal protection benefits of natural and nature-based defenses. *PLoS ONE* **11**: e0154735.

National Park Service. (2005). *Breezy Point District Adaptive Management Plan: Environmental Assessment*. Jamaica Bay Unit: NPS Gateway National Recreation Area.

National Park Service. (2011). *Assateague Island National Seashore General Management Plan/Environmental Impact Statement*. Newsletter 2, Summer 2011. Berlin: Assateague Island National Seashore.

National Research Council. (1995). *Beach Nourishment and Protection*. Washington, DC: National Academy Press.

National Research Council. (2007). *Mitigating Shore Erosion along Sheltered Coasts*. Washington, DC: National Academy Press.

National Research Council. (2014). *Reducing Coastal Risk on the East and Gulf Coasts*. Washington, DC: National Academy Press.

Nature Conservancy Council. (1991). *A Guide to the Selection of Appropriate Coast Protection Works for Geological SSSIs*. Peterborough: Nature Conservancy Council.

Naveh, Z. (1998). Ecological and cultural landscape restoration and the cultural evolution towards a post-industrial symbiosis between human society and nature. *Restoration Ecology* **6**: 135–143.

Naylor, L. A., Viles, H. A., and Carter, N. E. A. (2002). Biogeomorphology revisited: looking towards the future. *Geomorphology* **47**: 3–14.

Nelson, W. G. (1989). Beach nourishment and hardbottom habitats: the case for caution. In Tait, L. S. (Ed.), *Proceedings of the 1989 National Conference on Beach Preservation Technology*. Tallahassee: Florida Shore and Beach Preservation Association, pp. 109–116.

Nelson, W. G. (1993). Beach restoration in the southeastern US: environmental effects and biological monitoring. *Ocean & Coastal Management* **19**: 157–182.

Newell, R. C., Hitchcock, D. R., and Seiderer, L. J. (1999). Organic enrichment associated with outwash from marine aggregates dredging: a probable explanation for surface sheens and enhanced benthic production in the vicinity of dredging operations. *Marine Pollution Bulletin* **38**: 809–818.

Newell, R. C., Seiderer, L. J., Simpson, N. M., and Robinson, J. E. (2004). Impacts of marine aggregate dredging on benthic macrofauna of the south coast of the United Kingdom. *Journal of Coastal Research* **20**: 115–125.

Nicholls, R. J. and Branson, J. (1998). Coastal Resilience and planning for an uncertain future: an introduction. *The Geographical Journal* **164**: 255–258.

Nicholls, R. J. and Hoozemans, F. M. J. (1996). The Mediterranean: vulnerability to coastal implications of climate change. *Ocean & Coastal Management* **31**: 105–132.

Niven, R. J. and Bardsley, D. K. (2013). Planned retreat as a management response to coastal risk: a case study from the Fleurieu Peninsula, South Australia. *Regional Environmental Change* **13**: 193–209.

Nolet, C., and Riksen, M. J. P. M. (2019). Accommodation space indicates dune development potential along an urbanized and frequently nourished coastline. *Earth Surface Dynamics* **7**: 129–145.

Nordstrom, K. F. (1990). The concept of intrinsic value and depositional coastal landforms. *Geographical Review* **80**: 68–81.

Nordstrom, K. F. (1992). *Estuarine Beaches*. London: Elsevier Applied Science.

Nordstrom, K. F. (1994). Beaches and dunes of human-altered coasts. *Progress in Physical Geography* **18**: 497–516.

Nordstrom, K. F. (2000). *Beaches and Dunes of Developed Coasts*. Cambridge: Cambridge University Press.

Nordstrom, K. F. (2003). Restoring naturally functioning beaches and dunes on developed coasts using compromise management solutions: an agenda for action. In Dallmeyer, D. (Ed.), *Values at Sea: Ethics for the Marine Environment*. Athens: University of Georgia Press, pp. 204–229.

Nordstrom, K. F. (2005). Beach nourishment and coastal habitats: research needs for improving compatibility. *Restoration Ecology* **13**: 215–222.

Nordstrom, K. F. (2008). *Beach and Dune Restoration*. Cambridge: Cambridge University Press.

Nordstrom, K. F. (2014). Living with shore protection structures: a review. *Estuarine, Coastal and Shelf Science* **150**: 11–23.

Nordstrom, K. F. (2019). Coastal dunes with resistant cores. *Journal of Coastal Conservation* **23**: 227–237.

Nordstrom, K. F. and Arens, S. M. (1998). The role of human actions in evolution and management of foredunes in The Netherlands and New Jersey, USA. *Journal of Coastal Conservation* **4**: 169–180.

Nordstrom, K. F. and Jackson, N. L. (1995). Temporal scales of landscape change following storms on a human-altered coast, New Jersey, USA. *Journal of Coastal Conservation* **1**: 51–62.

Nordstrom, K. F. and Jackson, N. L. (2003). Alternative restoration outcomes for dunes on intensively developed coasts. In Özhan, E. (Ed.), *MEDCOAST '03*. Ankara: MEDCOAST Secretariat, pp. 1469–1478.

Nordstrom, K. F. and Jackson, N. L. (2013). Removing shore protection structures to facilitate migration of landforms and habitats on the bayside of a barrier spit. *Geomorphology* **199**: 179–191.

Nordstrom, K. F. and Jackson, N. L. (2018). Constraints on restoring landforms and habitats on storm-damaged shorefront lots in New Jersey. *Ocean & Coastal Management* **155**: 15–23.

Nordstrom, K. F. and Lotstein, E. L. (1989). Conflicting scientific, managerial, and societal perspectives on resource use of dynamic coastal dunes, *Geographical Review* **79**: 1–12.

Nordstrom, K. F. and Mauriello, M. N. (2001). Restoring and maintaining naturally-functioning landforms and biota on intensively developed barrier islands under a no-retreat scenario. *Shore and Beach* **69**(3): 19–28.

Nordstrom, K. F. and McCluskey, J. M. (1985). The effects of houses and sand fences on the eolian sediment budget at Fire Island, New York. *Journal of Coastal Research* *l*: 39–46.

Nordstrom, K. F., Lampe, R., and Vandemark, L. M. (2000). Re-establishing naturally-functioning dunes on developed coasts. *Environmental Management* **25**: 37–51.

Nordstrom, K. F., Jackson, N. L., Bruno, M. S., and de Butts, H. A. (2002). Municipal initiatives for managing dunes in coastal residential areas: a case study of Avalon, New Jersey, USA. *Geomorphology* **47**: 137–152.

Nordstrom, K. F., Jackson, N. L., and Pranzini, E. (2004). Beach sediment alteration by natural processes and human actions: Elba Island, Italy. *Annals of the Association of American Geographers* **94**: 794–806.

Nordstrom, K. F., Hartman, J. M., Freestone, A. L., Wong, M., and Jackson, N. L. (2007a). Changes in topography and vegetation near gaps in a protective foredune. *Ocean & Coastal Management* **50**: 945–959.

Nordstrom, K. F., Jackson, N. L., Hartman, J. M., and Wong, M. (2007b). Aeolian sediment transport on a human-altered foredune. *Earth Surface Processes and Landforms* **32**: 102–115.

Nordstrom, K. F., Lampe, R., and Jackson, N. L. (2007c). Increasing the dynamism of coastal landforms by modifying shore protection methods: examples from the eastern German Baltic Sea Coast. *Environmental Conservation* **34**: 205–214.

Nordstrom, K. F., Pranzini, E., Jackson, N. L., and Coli, M. (2008). The marble beaches of Tuscany. *Geographical Review* **98**: 280–300.

Nordstrom, K. F., Jackson, N. L., and de Butts, H. L. (2009). A proactive programme for managing beaches and dunes on a developed coast: a case study of Avalon, New

Jersey, USA. In Williams, A. and Micallef, A. (Eds.), *Beach Management: Principles and Practice*. London: Earth Scan, pp. 307–316.

Nordstrom, K. F., Jackson, N. L., Kraus, N. C., et al. (2011). Enhancing geomorphic and biologic functions and values on backshores and dunes of developed shores: a review of opportunities and constraints. *Environmental Conservation* **38**: 288–302.

Nordstrom, K. F., Jackson, N. L., Freestone, A. L., Korotky, K. H., and Puleo, J. A. (2012). Effects of beach raking and sand fences on dune dimensions and morphology. *Geomorphology* **179**: 106–115.

Nordstrom, K. F., Armaroli, C., Jackson, N. L., and Ciavola, P. (2015). Opportunities and constraints for managed retreat on exposed sandy shores: examples from Emilia-Romagna, Italy. *Ocean & Coastal Management* **104**: 11–21.

Nordstrom, K. F., Jackson, N. L., and Roman, C. T. (2016). Facilitating landform migration by removing shore protection structures: opportunities and constraints. *Environmental Science and Policy* **66**: 217–226.

Nordstrom, K. F., Liang, B., Garilao, E. S., and Jackson, N. L. (2018). Topography, vegetation cover and below ground biomass of spatially constrained and unconstrained foredunes in New Jersey, USA. *Ocean & Coastal Management* **156**: 117–126.

Norton, B. G. and Steinemann, A. C. (2001). Environmental values and adaptive management. *Environmental Values* **10**: 473–506.

Norton, D. A. (2000). Conservation biology and private land: shifting the focus. *Conservation Biology* **14**: 1221–1223.

Nourisson, D. H., Bessa, F., Scapini, F., and Marques, J. C. (2014). Macrofaunal community abundance and diversity and talitrid orientation as potential indicators of ecological long-term effects of a sand-dune recovery intervention. *Ecological Indicators* **36**: 356–366.

Novoa, A., González, L., Moravcová, L., and Pysek, P. (2013). Constraints to native plant species establishment in coastal dune communities invaded by *Carpobrotus edulis*: implications for restoration. *Biological Conservation* **164**: 1–9.

Nuryanti, W. (1996). Heritage and postmodern tourism. *Annals of Tourism Research* **23**: 249–260.

O'Brien, M. K., Valverde, H. R., Trembanis, A. C., and Haddad, T. C. (1999). Summary of beach nourishment activity along the Great Lakes' shoreline 1955–1996. *Journal of Coastal Research* **15**: 206–219.

O'Connell, T. O., Franze, C. D., Spalding, E. A., and Poirier, M. A. (2005). Biological resources of the Louisiana coast: Part 2. Coastal animals and habitat associations. *Journal of Coastal Research* **SI44**: 146–161.

O'Donnell, J. E. D. (2017). Living shorelines: a review of literature relevant to New England coasts. *Journal of Coastal Research* **33**: 435–451.

O'Donnell, T. (2019). Coastal management and the political-legal geographies of climate change adaptation in Australia. *Ocean & Coastal Management* **175**: 127–135.

O'Neil, K. M. and van Abs, D. J. (Eds.). (2016). *Taking Chances: The Coast after Hurricane Sandy*. New Brunswick, NJ: Rutgers University Press.

Ochieng, C. A. and Erftemeijer, P. L. A. (1999). Accumulation of seagrass beach cast along the Kenyan coast: a quantitative assessment. *Aquatic Botany* **65**: 221–238.

Odériz, I., Knöchelmann, N., Silva, R., Feagin, R. A., Martínez, M. L., and Mendoza, E. (2020). Reinforcement of vegetated and unvegetated dunes by a rocky core: a viable alternative for dissipating waves and providing protection. *Coastal Engineering* **158**: 103675.

Ofiara, D. D. and Brown, B. (1999). Assessment of economic losses to recreational activities from 1988 marine pollution events and assessment of economic losses from long-term contamination of fish within the New York Bight to New Jersey. *Marine Pollution Bulletin* **38**: 990–1004.

Ollerhead, J., Davidson-Arnott, R., Walker, I. J., and Mathew, S. (2013). Annual to decadal morphodynamics of the foredune system at Greenwich Dunes, Prince Edward Island, Canada. *Earth Surface Processes and Landforms* **38**: 284–298.

Orford, J. and Jennings, S. (1998). The importance of different time-scale controls on coastal management strategy: the problem of Porlock gravel barrier, Somerset, UK. In Hooke, J. (Ed.), *Coastal Defense and Earth Science Conservation*. Bath: The Geological Society, pp. 87–102.

Orford, J. D. and Pethick, J. (2006). Challenging assumptions of future coastal habitat development around the UK. *Earth Surface Processes and Landforms* **31**: 1625–1642.

Orr, M., Zimmer, M., Jelinski, D. E., and Mews, M. (2005). Wrack deposition on different beach types: spatial and temporal variation in the pattern of subsidy. *Ecology* **86**: 1496–1507.

Osswald, F., Dolch, T., and Reise, K. (2019). Remobilizing stabilized island dunes for keeping up with sea level rise? *Journal of Coastal Conservation* **23**: 675–687.

Owens, J. S. (1911). Royal Commission on Coast Erosion. *The Geographical Journal* **38**: 598–601.

Özgüner, H. and Kendle, A. D. (2006). Public attitudes towards naturalistic versus designed landscapes in the city of Sheffield (UK), *Landscape and Urban Planning* **74**: 139–157.

Pacini, M., Pranzini, E., and Sirito, G. (1997). Beach nourishment with angular gravel at Cala Gonone (eastern Sardinia, Italy). In *Proceedings of the Third International Conference on the Mediterranean Coastal Environment*. Ankara: MEDCOAST Secretariat, pp. 1043–1058.

Packham, J. R., Randall, R. E., Barnes, R. S. K., and Neal, A. (Eds.). (2001). *Ecology and Geomorphology of Coastal Shingle*. Seattle: Westbury Academic & Scientific Publishing.

Pagán, J. I., López, M., López, I., Tenza-Abril, A. J., and Aragonés, L. (2018). Study of the evolution of gravel beaches nourished with sand. *Science of the Total Environment* **626**: 87–95.

Palmer, M. A., Ambrose, R. F., and Poff, N. L. (1997). Ecological theory and community restoration ecology. *Restoration Ecology* **5**: 291–300.

Pardini, E. A., Vickstrom, K. E., and Knight, T. M. (2015). Early successional microhabitats allow the persistence of endangered plants in coastal sand dunes. *PLoS ONE* **10**(4): e0119567.

Parkinson, R. W. and Ogurcak, D. E. (2018). Beach nourishment is not a sustainable strategy to mitigate climate change. *Estuarine, Coastal and Shelf Science* **212**: 203–209.

Parrott, A. and Burningham, H. (2008). Opportunities of, and constraints to, the use of intertidal agri-environmental schemes for sustainable coastal defence: a case study of the Blackwater Estuary, southeast England. *Ocean & Coastal Management* **51**: 352–367.

Parsons, L. S., Sayre, J., Ender, C., Rodrigues, J. L., and Barberán, A. (2020). Soil microbial communities in restored and unrestored dune ecosystems in California. *Restoration Ecology* **28**(S4): 311–321.

Parsons, R. (1995). Conflict between ecological sustainability and environmental aesthetics: conundrum, canard or curiosity. *Landscape and Urban Planning* **32**: 227–244.

Parsons, R. and Daniel, T. C. (2002). Good looking: in defense of scenic landscape aesthetics. *Landscape and Urban Planning* **60**: 43–56.

Penland, S., Connor, P. F. Jr., Beall, A., Fearnley, S., and Williams, S. J. (2005). Changes in Louisiana's shoreline: 1855–2002. *Journal of Coastal Research* **SI44**: 7–39.

Pérez-Hernández, E., Ferrer-Valero, N., and Hernández-Calvento, L. (2020). Lost and preserved coastal landforms after urban growth. The case of Las Palmas de Gran Canaria city (Canary Islands, Spain). *Journal of Coastal Conservation* **24**: 26.

Pérez-Maqueo, O., Martínez, M. L., Lithgow, D., Mendoza-González, G., Feagin, R. A., and Gallego-Fernández, J. B. (2014). The coasts and their costs. In Martínez, M. L., Gallego-Fernández, J. B., and Hesp, P. A. (Eds.), *Restoration of Coastal Dunes*. New York: Springer, pp. 289–304.

Peterson, C. H. and Bishop, M. L. (2005). Assessing the environmental impacts of beach nourishment. *BioScience* **55**: 887–896.

Peterson, C. H. and Lipcius, R. N. (2003). Conceptual progress towards predicting quantitative ecosystem benefits of ecological restorations. *Marine Ecology Progress Series* **264**: 297–307.

Peterson, C. H., Hickerson, D. H. M., and Johnson, G. G. (2000). Short-term consequences of nourishment and bulldozing on the dominant large invertebrates of a sandy beach. *Journal of Coastal Research* **16**: 368–378.

Peterson, C. H., Kneib, R. T., and Manen, C.-A. (2003). Scaling restoration actions in the marine environment to meet quantitative targets of enhanced ecosystem services. *Marine Ecology Progress Series* **264**: 173–175.

Peterson, C. H., Bishop, M. J., D'Anna, L. M., and Johnson, G. A. (2014). Multi-year persistence of beach habitat degradation from nourishment using coarse shelly sediments. *Science of the Total Environment* **487**: 481–492.

Pethick, J. (1996). The sustainable use of coasts: monitoring, modelling and management. In Jones, P. S., Healy, M. G., and Williams, A. T. (Eds.), *Studies in European Coastal Management*. Cardigan: Samara Publishing Ltd., 83–92.

Pethick, J. (2001). The Anglian coast. In Bodungen, B. V. and Turner, R. K. (Eds.), *Science and Integrated Coastal Management*. Berlin: Dahlem University Press, pp. 121–133.

Pezzuto, P. R., Resgalla, C. Jr., Abreu, J. G. N., and Menezes, J. T. (2006). Environmental impacts of the nourishment of Balneário Camboriú beach, SC, Brazil. *Journal of Coastal Research* **SI39**: 863–868.

Phillips, M. R. (2009). Beach consequences of an industrial heritage. In Williams, A. and Micallef, A. (Eds.), *Beach Management: Principles & Practice*. London: Earthscan, pp. 359–368.

Pickart, A. J. (2013). Dune restoration over two decades at the Lanphere and Ma-le'l dunes in northern California. In Martínez, M. L., Gallego-Fernández, J. B., and Hesp, P. A. (Eds.), *Restoration of Coastal Dunes*. New York: Springer, pp. 159–185.

Pilkey, O. H. (1981). Geologists, engineers, and a rising sea level. *Northeastern Geology* **3/4**: 150–158.

Pilkey, O. H. (1992). Another view of beachfill performance. *Shore and Beach* **60**(2): 20–25.

Pilkey, O. H. Jr. and Cooper, J. A. G. (2012). "Alternative" shoreline erosion control devices: a review. In Cooper, J. A. G. and Pilkey, O. H. Jr. (Eds.), *Pitfalls of Shoreline Stabilization: Selected Case Studies*. Dordrecht: Springer Science +Business Media, pp. 187–214.

Pilkey, O. H. and Wright, H. L. III (1988). Seawalls versus beaches. *Journal of Coastal Research* **SI4**: 41–64.

Pinsky, M. L., Guannel, G., and Arkema, K. K. (2013). Quantifying wave attenuation to inform coastal habitat conservation. *Ecosphere* **4**: 95.

Pinto, C., Silovsky, E., Henley, F., Rich, L., Parcell, J., and Boyer, D. (1972). *The Oregon Dunes NPA Resource Inventory*. Portland, OR: U.S. Department of Agriculture, Forest Service, Pacific Northwest Region.

Pinto, C. A., Silveira, T. M., and Teixeira, S. B. (2020). Beach nourishment practice in mainland Portugal (1950–2017): overview and retrospective. *Ocean & Coastal Management* **192**: 105211.

Piotrowska, H. (1989). Natural and anthropogenic changes in sand dunes and their vegetation on the southern Baltic coast. In van der Meulen, F., Jungerius, P. D., and Visser, J. H. (Eds.), *Perspectives in Coastal Dune Management*. The Hague: SPB Academic Publishing, pp. 33–40.

Plassmann, K., Laurence, M., Jones, M., and Edwards-Jones, G. (2010). Effects of long-term grazing management on sand dune vegetation of high conservation interest. *Applied Vegetation Science* **13**: 100–112.

Platt, R. H., Salvesen, D., and Baldwin, G.H. II (2002). Rebuilding the North Carolina coast after Hurricane Fran: did public regulations matter. *Coastal Management* **30**: 249–269.

Pluis, J. L. A. and de Winder, B. (1990). Natural stabilization. *Catena* Supplement **18**: 195–208.

Polomé, P., Marzetti, S., and van der Veen, A. (2005). Economic and social demands for coastal protection. *Coastal Engineering* **52**: 819–840.

Pontee, N. and Tarrant, O. (2017). Promoting coastal resilience using green infrastructure. *Maritime Engineering* **170**: 37–38.

Pontee, N. I., Pye, K., and Blott, S. J. (2004). Morphodynamic behaviour and sedimentary variation of mixed sand and gravel beaches, Suffolk, U.K. *Journal of Coastal Research* **20**: 256–276.

Portz, L., Manzolli, R. P., Hermanns, L., and Alcántara-Carrió, J. (2015). Evaluation of the efficiency of dune reconstruction techniques in Xangri-lá (Rio Grande do Sul, Brazil). *Ocean & Coastal Management* **104**: 78–89.

Portz, L., Manzolli, R. P., and Alcántara-Carrió, J. (2018). Dune system restoration in Osório municipality (Rio Grande do Sul, Brazil): good practices based on coastal management legislation. In Botero, C. M., Cervantes, O., and Finkl, C. W. (Eds.), *Beach Management Tools – Concepts, Methodologies and Case Studies*. Cham, Switzerland: Springer International Publishing, pp. 41–58.

Posey, M. and Alphin, T. (2002). Resilience and stability in an offshore benthic community: responses to sediment borrow activities and hurricane disturbance. *Journal of Coastal Research* **18**: 685–697.

Powell, E. J., Tyrrell, M., Milliken, A., Tirpak, J. M., and Stoudinger, M. D. (2019). A review of coastal management approaches to support the integration of ecological and human community planning for climate change. *Journal of Coastal Conservation* **23**: 1–18.

Powell, K. A. (1992). Engineering with conservation issues in mind. In Barrett, M. G. (Ed.), *Coastal Zone Planning and Management*. London: Thomas Telford, pp. 237–249.

Pranzini, E. (2001). Updrift river mouth migration on cuspate deltas: two examples from the coast of Tuscany, Italy. *Geomorphology* **38**: 125–132.

Pranzini, E. (2009). Protection projects at Poetto and Cala Gonone beaches (Sardinia, Italy). In Williams, A. and Micallef, A. (Eds.), *Beach Management: Principles & Practice*. London: Earthscan, pp. 287–306.

Pranzini, E. (2013). Italy. In Pranzini, E. and Williams, A. (Eds.), *Coastal Erosion and Protection in Europe*. London, Routledge, pp. 294–323.

Pranzini, E. and Vitale, G. (2011). Beach sand colour: the need for a standardised assessment procedure. *Journal of Coastal Research* **SI61**: 66–69.

Pranzini, E., Simonetti, D., and Vitale, G. (2010). Sand colour rating and chromatic compatibility of borrow sediments. *Journal of Coastal Research* **26**: 798–808.

Pranzini, E., Anfuso, G., Boteo, C.-M., et al. (2016). Sand color at Cuba and its influence on beach nourishment and management. *Ocean & Coastal Management* **126**: 51–60.

Pranzini, E., Anfuso, G., and Boteo, C. M. (2018a). Nourishing tourist beaches. In Botero, C. M., Cervantes, O., and Finkl, C. W. (Eds.), *Beach Management Tools – Concepts, Methodologies and Case Studies*. Cham, Switzerland: Springer International Publishing, pp. 293–317.

Pranzini, E., Rossi, L., Lami, G., Jackson, N. L., and Nordstrom, K. F. (2018b). Reshaping beach morphology by modifying offshore breakwaters. *Ocean & Coastal Management* **154**: 168–177.

Prati, G., Albanesi, C., Pietrantoni, L., and Airoldi, L. (2016). Public perceptions of beach nourishment and conflict management strategies: a case study of Portonovo Bay in the Adriatic Italian coast. *Land Use Policy* **50**: 422–428.

Priskin, J. (2003). Physical impacts of four-wheel drive related tourism and recreation in a semi-arid, natural coastal environment. *Ocean & Coastal Management* **46**: 127–155.

Provoost, S., Laurence, M., Jones, M., and Edmondson, S. E. (2011). Changes in landscape and vegetation of coastal dunes in northwest Europe: a review. *Journal of Coastal Conservation* **15**: 207–226.

Psuty, N. P. and Moreira, M. E. S. A. (1992). Characteristics and longevity of beach nourishment at Praja da Rocha, Portugal. *Journal of Coastal Research* **SI8**: 660–676.

Psuty, N. P. and Silveira, T. M. (2010). Global climate change: an opportunity for coastal dunes? *Journal of Coastal Conservation* **14**: 153–160.

Purvis, K. G., Gramling, J. M., and Murren, C. J. (2015). Assessment of beach access paths on dune vegetation: diversity, abundance and cover. *Journal of Coastal Research* **31**: 1222–1228.

Pye, K. and Blott, S. J. (2015). Spatial and temporal variations in soft-cliff erosion along the Holderness coast, East Riding of Yorkshire, UK. *Journal of Coastal Conservation* **19**: 785–808.

Pye, K. and Blott, S. J. (2016). Dune rejuvenation Trials Overview Report. Natural Resources Wales Evidence Report No. 296, Natural Resources Wales, Bangor.

Pye, K. and Blott, S. J. (2017). Evolution of a sediment-starved, over-stabilized dunefield: Kenfig Burrows, South Wales, UK. *Journal of Coastal Conservation* **21**: 685–717.

Pye, K. and Blott, S. J. (2020). Is "re-mobilisation" nature restoration or nature destruction? A commentary. Discussion. *Journal of Coastal Conservation* **24**: 10.

Pye, K., Blott, S. J., and Howe, M. A. (2014). Coastal dune stabilization in Wales and requirements for rejuvenation. *Journal of Coastal Conservation* **18**: 27–54.

Rabenold, C. (2013). Coastal zone management: using no-build areas to protect the shorefront. *Coastal Management* **41**: 294–311.

Rai, P. K. and Kim, K.-H. (2020). Invasive alien plants and environmental remediation: a new paradigm for sustainable restoration ecology. *Restoration Ecology* **28**: 3–7.

Rakocinski, C. F., Heard, R. W., LeCroy, S. E., McLelland, J. A., and Simons, T. (1996). Responses by macrobenthic assemblages to extensive beach restoration at Perdido Key, Florida, U.S.A. *Journal of Coastal Research* **12**: 326–353.

Randall, R. E. (1996). The shingle survey of Great Britain and its implications for conservation management. In Salman, A. H. P. M., Langeveld, M. J., and Bonazountas, M. (Eds.), *Coastal Management and Habitat Conservation*. Leiden: EUCC, pp. 369–376.

Randall, R. E. (2004). Management of coastal vegetated shingle in the United Kingdom. *Journal of Coastal Conservation* **10**: 159–168.

Rankin, K. L., Bruno, M. S., and Herrington, T. O. (2004). Nearshore currents and sediment transport measured at notched groins. *Journal of Coastal Research* **SI33**: 237–254.

Ranwell, D. S. (1972). *Ecology of Salt Marshes and Sand Dunes*. London: Chapman and Hall.

Ranwell, D. S. and Boar, R. (1986). *Coast Dune Management Guide*. Institute of Terrestrial Ecology, NERC.

Ratnayake, N. P., Ranayake, A. S., Azoor, R. M., et al. (2019). Erosion processes driven by monsoon events after a beach nourishment and breakwater construction at Uswetakeiyawa Beach, Sri Lanka *SN Applied Sciences* **1**: 52.

Redi, B. H., van Aarde, R. J., and Wassenaar, T. D. (2005). Coastal dune forest development and the regeneration of millipede communities. *Restoration Ecology* **13**: 284–291.

Rees, S., Curson, J., and Evans, D. (2015). Conservation of soft cliffs in England 2002–2013. *Journal of Coastal Conservation* **19**: 761–769.

Reid, J., Santana, G. G., Klein, A. H. F., and Diehl, F. L. (2005). Perceived and realized social and economic impacts of sand nourishment at Piçarras Beach, Santa Catarina, Brazil. *Shore and Beach* **73**(4): 14–18.

Reilly, F. J. and Bellis, V. J. (1983). *The Ecological Impact of Beach Nourishment with Dredged Materials on the Intertidal Zone at Bogue Banks, North Carolina*. Miscellaneous Report 83-3. Ft. Belvoir, VA: U.S. Army Corps of Engineers, Coastal Engineering Research Center.

Reinicke, R. (2001). *Inseln der Ostsee: Landschaften und Naturschönheit*. Bremen: Giritz & Gottschalk.

Rhind, P. M. and Jones, P. S. (1999). The floristics and conservation status of sand-dune communities in Wales. *Journal of Coastal Conservation* **5**: 31–42.

Rhind, P. and Jones, R. (2009). A framework for the management of sand dune systems in Wales. *Journal of Coastal Conservation* **13**: 15–23.

Rhind, P., Jones, R., and Jones, L. (2013). The impact of dune stabilization on the conservation status of sand dune systems in Wales. In Martínez, M. L., Gallego-Fernández, J. B., and Hesp, P. A. (Eds.), *Restoration of Coastal Dunes*. New York: Springer, pp. 125–143.

Richards, E. G. and Burningham, H. (2011). Hippophae rhamnoides on a coastal dune system: a thorny issue? *Journal of Coastal Conservation* **15**: 73–85.

Riksen, M. J., Goossens, D., Huiskes, H. P., Krol, J., and Slim, P. A. (2016). Constructing notches in foredunes: effect on sediment dynamics in the dune hinterland, *Geomorphology* **253**: 340–352.

Risser, P. G. (1995). The status of the science examining ecotones. *Bioscience* **45**: 318–325.

Ritchie, W. and Gimingham, C. H. (1989). Restoration of coastal dunes breached by pipeline landfalls in northeast Scotland. *Proceedings of the Royal Society of Edinburgh* **96B**: 231–245.

Ritchie, W. and Penland, S. (1990). Aeolian sand bodies of the south Louisiana coast. In Nordstrom, K. F., Psuty, N. P., and Carter, R. W. G. (Eds.), *Coastal Dunes: Form and Process*. Chichester: John Wiley & Sons, pp. 105–127.

Roberts, C. M. and Hawkins, J. P. (1999). Extinction risk in the sea. *Trends in Ecology and Evolution* **14**: 241–246.

Roberts, N. (1989). *The Holocene: An Environmental History*. New York: Basil Blackwell.

Roberts, T. M. and Wang, P. (2012). Four-year performance and associated controlling factors of several beach nourishment projects along three adjacent barrier islands, west-central Florida, USA. *Coastal Engineering* **70**: 21–39.

Robley, A., Purdey, D., Johnston, M., Lindeman, M., Busana, F., and Long, K. (2007). Experimental trials to determine effective fence designs for feral cat and fox exclusion. *Ecological Management and Restoration* **8**: 193–198.

Roca, E. and Villares, M. (2012). Public perceptions of managed realignment strategies: the case study of the Ebro Delta in the Mediterranean basin. *Ocean & Coastal Management* **60**: 38–47.

Roca, E. and Villares, M. (2018). Integrating social perceptions in beach management. In Botero, C. M., Cervantes, O., and Finkl, C. W. (Eds.), *Beach Management Tools – Concepts, Methodologies and Case Studies*. Cham, Switzerland: Springer International Publishing, pp. 875–893.

Rodrigues, R. S., Mascarenhas, A., and Jagtap, T. G. (2011). An evaluation of flora from coastal sand dunes of India: Rationale for conservation and management. *Ocean & Coastal Management* **54**: 181–188.

Rodriguez, A. B., Fegley, S. R., Ridge, J. T., van Dusen, B. M., and Anderson, N. (2013). Contribution of aeolian sand to backbarrier marsh sedimentation. *Estuarine, Coastal and Shelf Science* **117**: 248–259.

Rogers, S. M. (1993). Relocating erosion-threatened buildings: a study of North Carolina housemoving. In *Coastal Zone 93*. New York: American Society of Civil Engineers, pp. 1392–1405.

Roman, C. T. and Nordstrom, K. F. (1988). The effect of erosion rate on vegetation patterns of an east coast barrier island. *Estuarine, Coastal and Shelf Science* **26**: 233–242.

Romano, B. and Zullo, F. (2014). The urban transformation of Italy's Adriatic coastal strip: fifty years of unsustainability. *Land Use Policy* **38**: 26–36.

Ruessink, B. G., Arens, S. M., Kuipers, M., and Donker, J. J. A. (2018). Coastal dune dynamics in response to excavated foredune notches. *Aeolian Research* **31A**: 3–17.

Rumbold, D. G., Davis, P. W., and Perretta, C. (2001). Estimating the effect of beach nourishment on *Caretta caretta* (loggerhead sea turtle) nesting. *Restoration Ecology* **9**: 304–310.

Runyan, K. and Griggs, G. B. (2003). The effects of armoring seacliffs on the natural sand supply to the beaches of California. *Journal of Coastal Research* **19**: 336–347.

Rupp-Armstrong, S. and Nicholls, R. J. (2007). Coastal and estuarine retreat: a comparison of the application of managed realignment in England and Germany. *Journal of Coastal Research* **23**: 1418–1430.

Ruz, M.-H. and Anthony, E. J. (2008). Sand trapping by brushwood fences on a beach-foreshore contact: the primacy of the local sediment budget. *Zeitschrift für Geomorphologie* **52**(Suppl. 3): 179–194.

Saathoff, F., Oumeraci, H., and Restall, S. (2007). Australian and German experiences on the use of geotextile containers. *Geotextiles and Geomembranes* **25**: 251–263.

Saffir, H. S. (1991). Hurricane Hugo and implications for design professionals and code-writing authorities. *Journal of Coastal Research* **SI8**: 25–32.

Salgado, K. and Martínez, M. L. (2017). Is ecosystem-based coastal defense a realistic alternative? Exploring the evidence. *Journal of Coastal Conservation* **21**: 837–848.

Salmon, J., Henningsen, D., and McAlpin, T. (1982). *Dune Restoration and Vegetation Manual*. SGR-48. Gainesville: Florida Sea Grant College Program.

Sancho-García, A., Guillén, J., and Ojeda, E. (2013). Storm-induced readjustment of an embayed beach after modification by protection works. *Geo-Marine Letters* **33**: 159–172.

Sanjaume, E. (1992). Valencia coast: human impact and dune conservation. *Coastline* **1**: 10–13.

Savard, J.-P. L., Clergeau, P., and Mennechez, G. (2000). Biodiversity concepts and urban ecosystems. *Landscape and Urban Planning* **43**: 131–142.

Schernewski, G., Schumacher, J., Weisner, E., and Donges, L. (2018). A combined coastal protection, realignment and wetland restoration scheme in the southern Baltic: planning process, public information and participation. *Journal of Coastal Conservation* **22**: 533–547.

Schlacher, T. A., Dugan, J., Schoeman, D. S., et al. (2006). Sandy beaches at the brink. *Diversity and Distributions* **13**: 556–560.

Schlacher, T. A., Noriega, R., Jones, A., and Dye, T. (2012). The effects of beach nourishment on benthic invertebrates in eastern Australia: impacts and variable recovery. *Science of the Total Environment* **435**: 411–417.

Schmahl, G. P. and Conklin, E. J. (1991). Beach erosion in Florida: a challenge for planning and management. In *Coastal Zone 91*. New York: American Society of Civil Engineers, pp. 261–271.

Schooler, N. K., Dugan, J. E., and Hubbard, D. M. (2019). No lines in the sand: impacts of intense mechanized maintenance regimes on sandy beach ecosystems span the intertidal zone on urban coasts. *Ecological Indicators* **106**: 105457.

Schreck Reis, C., Antunes do Carmo, J., and Freitas, H. (2008). Learning with nature: a sand dune system case study (Portugal). *Journal of Coastal Research* **26**: 1506–1515.

Schulze-Dieckhoff, M. (1992). Propagating dune grasses by cultivation for dune conservation. In Carter, R. W. G., Curtis, T. G. F., and Sheehy-Skeffington, M. J. (Eds.), *Coastal Dunes: Geomorphology, Ecology and Management for Conservation*. Rotterdam: A.A. Balkema, pp. 361–366.

Schupp, C. A., Winn, N. T., Pearl, T. L., Kumer, J. P., Carruthers, T. J. B., and Zimmerman, C. S. (2013). Restoration of overwash processes creates piping plover (*Charadrius melodus*) habitat on a barrier island (Assateague Island, Maryland). *Estuarine, Coastal and Shelf Science* **116**: 11–20.

Schwarzer, K., Crossland, C. J., De Luca Rebbello Wagener, A., et al. (2001). Group report: shoreline development. In Bodungen, B. V. and Turner, R. K. (Eds.), *Science and Integrated Coastal Management*. Berlin: Dahlem University Press, pp. 121–133.

Schwendiman, J. L. (1977). Coastal sand dune stabilization in the Pacific Northwest. *International Journal of Biometeorology* **21**: 281–289.

Scyphers, S. B, Powers, S. P., Heck, K. L. Jr., and Byron, D. (2011). Oyster reefs as natural breakwaters mitigate shoreline loss and facilitate fisheries. *PLoS ONE* **6**(8): e22396.

Scott, G. A. (1963). The ecology of shingle beach plants. *Journal of Ecology* **51**: 517–527.

Sea Isle City. (1982). *Sea Isle City Centennial 1882–1982*. Sea Isle City, NJ: City Hall.

Seabloom, E. W. and Wiedemann, A. M. (1994). Distribution and effects of *Ammophila breviligulata* Fern. (American beachgrass) on the foredunes of the Washington coast. *Journal of Coastal Research* **10**: 178–188.

Seltz, J. (1976). *The Dune Book: How to Plant Grasses for Dune Stabilization*. UNC-SG-76-16. Raleigh: North Carolina University Sea Grant.

Seymour, A. C., Ridge, J. T., Rodriquez, A. B., Newton, E., Dale, J., and Johnston, D. W. (2017). Deploying fixed wing unoccupied aerial systems (UAS) for coastal morphology assessment and management. *Journal of Coastal Research* **34**: 704–717.

Sharp, W. C. and Hawk, V. B. (1977). Establishment of woody plants for secondary and tertiary dune stabilization along the mid-Atlantic coast. *International Journal of Biometeorology* **21**: 245–255.

Sheik Mujabar, P. and Chandrasekar, N. (2013). Coastal erosion hazard and vulnerability assessment for southern coastal Tamil Nadu of India by using remote sensing and GIS. *Natural Hazards* **69**: 1295–1314.

Sherman, D. J. (1995). Problems of scale in the modeling and interpretation of coastal dunes. *Marine Geology* **124**: 339–349.

Sherman, D. J. and Nordstrom, K. F. (1994). Hazards of wind blown sand and sand drift. *Journal of Coastal Research* **SI12**: 263–275.

Shibutani, Y., Kuroiwa, M., and Matsubara, Y. (2016). Effect of coastal protection using the beach nourishment at Tottori sand dune coast. *Journal of Coastal Research* **SI75**: 695–699.

Shipman, H. (2001). Beach nourishment on Puget Sound: a review of existing projects and potential applications. In *Puget Sound Research 2001*. Olympia, WA: Puget Sound Water Quality Action Team, pp. 1–8.

Shipman, H., Stoops, K., and Hummel, P. (2000). *Seattle Waterfront Parks: Applications of Beach Nourishment*. Seattle: Washington Coastal Planner's Group.

Shipman, H., Dethier, M. N., Gelfenbaum, G., Fresh, K. L., and Dinicola, R. S. (Eds.). (2010). *Puget Sound Shorelines and the Impacts of Armoring*. Reston, VA: U.S. Geological Survey.

Shu, F., Cai, F., Qi, H., Liu, J., Lei, G., and Zheng, J. (2019). Morphodynamics of an artificial cobble beach in Tianquan Bay, Xiamen, China. *Journal of Ocean University of China* **18**: 868–882.

Shuisky, Y. D. and Schwartz, M. L. (1988). Human impact and rates of shoreline retreat along the Black Sea coast. *Journal of Coastal Research* **4**: 405–416.

Siddle, R. P., Rowe, S., and Moore, R. (2016). Adaptation to coastal cliff instability and erosion and property loss: case study into the Knipe Point cliff retreat Pathfinder Project. In Baptiste, A. (Ed.), *Coastal Management: Changing Coast, Changing Climate, Changing Minds*. International Coastal Management Conference. London: ICE Publishing, pp. 163–172.

Sigren, J. M., Figlus, J., Highfield, W., Feagin, R. A., and Armitage, A. R. (2018). The effects of coastal dune volume and vegetation on storm-induced property damage: analysis from Hurricane Ike. *Journal of Coastal Research* **34**: 164–173.

Silva, R., Martínez, M. L., Odériz, I., Mendoza, E., and Feagin, R. A. (2016). Response of vegetated dune-beach systems to storm conditions. *Coastal Engineering* **109**: 53–62.

Silveira, T. M., Franzão Santos, C., and Andrade, F. (2013). Beneficial use of dredged sand for beach nourishment and coastal landform enhancement – the case study of Tróia, Portugal. *Journal of Coastal Conservation* **17**: 825–832.

Simeoni, U., Calderoni, G., Tessari, U., and Mazzini, E. (1999). A new application of system theory to foredunes intervention strategy. *Journal of Coastal Research* **15**: 457–470.

Simpson, T. B. (2005). Ecological restoration and re-understanding ecological time. *Ecological Restoration* **23**: 46–51.

Skaradek, W., Miller, C., and Hocker, P. (2003). *Beachgrass Planting Guide for Municipalities and Volunteers*. Cape May Plant Materials Center, U.S. Department of Agriculture Natural Resources Conservation Service.

Skarregaard, P. (1989). Stabilisation of coastal dunes in Denmark. In van der Meulen, F., Jungerius, P. D., and Visser, J. H. (Eds.), *Perspectives in Coastal Dune Management*. The Hague: SPB Academic Publishing, pp. 151–161.

Slott, J. M., Murray, A. B., and Ashton, A. D. (2010). Large-scale responses of complex-shaped coastlines to local shoreline stabilization and climate change. *Journal of Geophysical Research* **115**: F03033.

Smallegan, S. M., Irish, J. L., van Dongeren, A. R., and Den Bieman, J. P. (2016). Morphological response of a sandy barrier island with a buried seawall during Hurricane Sandy. *Coastal Engineering* **110**: 102–110.

Smith, G. and Mocke, G. P. (1998). Coastline evolution in response to a major mine sediment discharge on the Namibian coastline. In *Coastal Engineering*. Reston, VA: American Society of Civil Engineers, pp. 2696–2709.

Smith, G., Mocke, G. P., van Ballegooyen, R., and Soltau, C. (2002). Consequences of sediment discharge from dune mining at Elizabeth Bay, Namibia. *Journal of Coastal Research* **18**: 776–791.

Smith, J. A. M., Niles, L. J., Hafner, S., Modjeski, A., and Dillingham, T. (2020). Beach restoration improves habitat quality for American horseshoe crabs and shorebirds in the Delawar Bay, USA. *Marine Ecology Progress Series* **645**: 91–107.

Smith, K. J. (1991). Beach politics – the importance of informed, local support for beach restoration projects. In *Coastal Zone 91*. New York: American Society of Civil Engineers, pp. 56–61.

Smith, R. A. (1992). Conflicting trends of beach resort development: a Malaysian case. *Coastal Management* **20**: 167–187.

Smyth, T. A. G. and Hesp, P. A. (2015). Aeolian dynamics of beach scraped ridge and dyke structures. *Coastal Engineering* **99**: 38–45.

Snyder, M. R. and Pinet, P. R. (1981). Dune construction using two multiple sand-fence configurations: implications regarding protection of eastern Long Island's south shore. *Northeastern Geology* **3**: 225–229.

Snyder, R. A. and Boss, C. L. (2002). Recovery and stability in barrier island plant communities. *Journal of Coastal Research* **18**: 530–536.

Soares, A. G., Scapini, F., Brown, A. C., and McLachlan, A. (1999). Phenotypic plasticity, genetic similarity and evolutionary inertia in changing environments. *The Malacological Society of London* **65**: 136–139.

Society for Ecological Restoration. (2002). *The SER Primer on Ecological Restoration.* www.ser.org/.

Somerville, S. E., Miller, K. L., and Mair, J. M. (2003). Assessment of the aesthetic quality of a selection of beaches in the Firth of Forth, Scotland. *Marine Pollution Bulletin* **46**: 1184–1190.

Soulsby, C., Hannah, D., Malcolm, R., Maizels, J. K., and Gard, R. (1997). Hydrogeology of a restored coastal dune system in northeastern Scotland. *Journal of Coastal Conservation* **3**: 143–154.

Sparks, P. R. (1991). Wind conditions in Hurricane Hugo and their effect on buildings in South Carolina. *Journal of Coastal Research* **SI8**: 13–24.

Speybroeck, J., Bonte, D., Courtens, W., et al. (2006). Beach nourishment: an ecologically sound coastal defence alternative? A review. *Aquatic Conservation: Marine and Freshwater Ecosystems* **16**: 419–435.

Spodar, A., Héquette, A., Ruz, M.-H., et al. (2018). Evolution of a beach nourishment project using dredged sand from navigation channel, Dunkirk, northern France. *Journal of Coastal Conservation* **22**: 457–474.

Stallins, J. A. and Corenblit, D. (2018). Interdependence of geomorphic and ecologic resilience properties in a geographic context. *Geomorphology* **305**: 76–93.

Starkes, J. (2001). *Reconnaissance Assessment of the State of the Nearshore Ecosystem: Eastern Shore of Central Puget Sound, Including Vashon and Maury Islands (WRIAS 8 and 9).* Seattle: King County Department of Natural Resources.

Stauble, D. K. and Nelson, W. G. (1985). Guidelines for beach nourishment: a necessity for project management. In *Coastal Zone 85*. New York: American Society of Civil Engineers, pp. 1002–1021.

Stein, E. D., Doughty, C. L., Lowe, J., Cooper, M., Sloane, E. B., and Liza, D. (2020). Establishing targets for regional coastal wetland restoration planning using historical ecology and future scenario analysis: the past, present, future approach. *Estuaties and Coasts* **43**: 207–222.

Steinitz, M. J., Salmon, M., and Wyneken, J. (1998). Beach renourishment and loggerhead turtle reproduction: a seven year study at Jupiter Island, Florida. *Journal of Coastal Research* **14**: 1000–1013.

Stive, M. J. F., de Schipper, M. A., Luijendijk, A. P., et al. (2013). A new alternative to saving our beaches from sea-level rise: the Sand Engine. *Journal of Coastal Research* **29**: 1001–1008.

Stocker, L. and Kennedy, D. (2009). Cultural models of the coast in Australia: toward sustainability. *Coastal Management* **37**: 387–404.

Stocker, T. F., Qin, D., Plattner, G.-K., et al. (Eds.). (2013). *IPCC, 2013: Climate Change 2013: The Physical Science Basis. Contribution of Working Group I to the Fifth Assessment Report of the Intergovernmental Panel on Climate Change.* Cambridge: Cambridge University Press.

Stratford, C. and Rooney, P. (2017). Special issue – coastal dune slack hydro-ecology. *Journal of Coastal Conservation* **21**: 573–576.

Stratford, C. J., Robins, N. S., Clarke, D., Jones, L., and Weaver, G. (2013). An ecohydrological review of dune slacks on the west coast of England and Wales. *Ecohydrology* **6**: 162–171.

Stull, K. J., Cahoon, L. B., and Lankford, T. E. (2016). Zooplankton abundance in the surf zones of nourished and unnourished beaches in southeastern North Carolina, U.S.A. *Journal of Coastal Research* **32**: 70–77.

Sturgess, P. (1992). Clear-felling dune plantations: studies in vegetation recovery. In Carter, R. W. G., Curtis, T. G. F., and Sheehy-Skeffington, M. J. (Eds.), *Coastal Dunes: Geomorphology, Ecology and Management for Conservation.* Rotterdam: A.A. Balkema, pp. 339–349.

Sturgess, P. and Atkinson, D. (1993). The clear-felling of sand dune plantations: soil and vegetational processes in habitat restoration. *Biological Conservation* **66**: 171–183.

Sutton-Grier, A. E., Wowk, K., and Bamford, H. (2015). Future of our coasts: the potential for natural and hybrid infrastructure to enhance the resilience of our coastal communities, economies and ecosystems. *Environmental Science and Policy* **51**: 137–148.

Swart, J. A. A., van der Windt, H. J., and Keulartz, J. (2001). Valuation of nature in conservation and restoration. *Restoration Ecology* **9**: 230–238.

Taylor, E. B., Gibeaut, J. C., Yoskowitz, D. W., and Starek, M. J. (2015). Assessment and monetary valuation of the storm protection function of beaches and foredunes on the Texas coast. *Journal of Coastal Research* **31**: 1205–1216.

Taylor, R. B. and Frobel, D. (1990). Approaches and results of a coastal dune restoration program on Sable Island, Nova Scotia. In Davidson-Arnott, R. G. D. (Ed.), *Proceedings of the Symposium on Coastal Sand Dunes.* Ottawa: National Research Council Canada, pp. 405–431.

Teixeira, L. H., Weisser, W., and Ganade, G. (2016). Facilitation and sand burial effect plant survival during restoration of a tropical coastal sand dune degraded by tourist cars. *Restoration Ecology* **24**: 390–397.

Télez-Duarte, M. A. (1993). Cultural resources as a criterion in coastal zone management: the case of northwestern Baja California, Mexico. In Fermán-Almada, J. L., Gómez-Morin, L., and Fischer, D. W. (Eds.), *Coastal Zone Management in Mexico: The Baja California Experience.* New York: American Society of Civil Engineers, pp. 137–147.

Thieler, R. E., Pilkey, O. H. Jr., Young, R. S., Bush, D. M., and Chai, F. (2000). The use of mathematical models to predict beach behavior for U.S. coastal engineering: a critical review. *Journal of Coastal Research* **16**: 48–70.

Throop, W. and Purdom, R. (2006). Wilderness restoration: the paradox of public participation. *Restoration Ecology* **14**: 493–499.

Tinley, K. L. (1995). *Coastal Dunes of South Africa*. South African National Scientific Programmes Report No. 109.

Titus, J. G. (1990). Greenhouse effect, sea level rise, and barrier islands: case study of Long Beach Island, New Jersey. *Coastal Management* **18**: 65–90.

Toft, J. D., Cordell, J. R., Heerhartz, S. M., Armbrust, E. A., and Simenstad, C. A. (2010). Fish and invertebrate response to shoreline armoring and restoration in Puget Sound. In Shipman, H., Dethier, M. N., Gelfenbaum, G., Fresh, K. L., and Dinicola, R. S. (Eds.), *Puget Sound Shorelines and the Impacts of Armoring*. Reston, VA: U.S. Geological Survey, pp. 161–170.

Toft, J. D., Ogston, A. S., Heerhartz, S. M., Cordell, J. R., and Flemer, E. E. (2013). Ecological response and physical stability of habitat enhancements along an urban armored shoreline. *Ecological Engineering* **57**: 97–108.

Toimil, A., Losada, I. J., Nicholls, R. J., Dalrymple, R. A., and Stive, M. J. F. (2020). Addressing the challenges of climate change risks and adaptation in coastal areas: a review. *Coastal Engineering* **156**: 103611.

Tomasicchio, U. (1996). Submerged breakwaters for the defence of the shoreline at Ostia: field experiences, comparison. In *Proceedings of the 25th International Conference on Coastal Engineering*. New York: American Society of Civil Engineers, pp. 2404–2417.

Tondeur, Y., Vining, B., Mace, K., Mills, W., and Hart, J. (2012). Environmental release of dioxins from reservoir sources during beach nourishment programs. *Chemosphere* **88**: 358–363.

Tonnen, P. K., Huisman, B. J. A., Stam, G. N., and van Rijn, L. C. (2018). Numerical modeling of erosion rates, life span and maintenance volumes of mega nourishments. *Coastal Engineering* **131**: 51–69.

Townend, I. H. and Fleming, C. A. (1991). Beach nourishment and socio-economic aspects. *Coastal Engineering* **16**: 115–127.

Trembanis, A. C. and Pilkey, O. H. (1998). Summary of beach nourishment along the U.S. Gulf of Mexico shoreline. *Journal of Coastal Research* **14**: 407–417.

Tresca, A., Ruz, M.-H., and Grégoire, P. (2014). Coastal dune development and sand drifting management along and artificial shoreline: the case of Dunkirk harbor, northern France. *Journal of Coastal Conservation* **18**: 495–504.

Tudor, D. T. and Willams, A. T. (2003). Public perception and opinion of visible beach aesthetic pollution: the utilisation of photography. *Journal of Coastal Research* **19**: 1104–1115.

Tukiainen, H., Kiuttu, M., Kalliola, R., Alahuhta, J., and Hjort, J. (2019). Landforms contribute to plant biodiversity at alpha, beta and gamma levels. *Journal of Biogeography* **46**: 1699–1710.

Tunstall, S. M. and Penning-Rowsell, E. C. (1998). The English beach: experiences and values. *The Geographical Journal* **164**: 319–332.

Turkenli, T. S. (2005). Human activity in landscape seasonality: the case of tourism in Crete. *Landscape Research* **30**: 221–239.

Turnhout, E., Hisschemöller, M., and Eijsackers, H. (2004). The role of views of nature in Dutch nature conservation: the case of the creation of a drift sand area in the Hoge Veluwe National Park. *Environmental Values* **13**: 187–198.

Tye, R. S. (1983). Impact of Hurricane David and mechanical dune restoration on Folly Beach, South Carolina. *Shore and Beach* **51**(2): 3–9.

U.S. Army Corps of Engineers. (1980). *Beach Erosion Control Colonial Beach, Virginia: Detailed Project Report*. Baltimore: U.S. Army Corps of Engineers.

U.S. Army Corps of Engineers. (1986). *Lincoln Park Shoreline Erosion Control Seattle Washington: Final Detailed Project Report and Final Environmental Assessment*. Seattle: U.S. Army Corps of Engineers.

US Army Corps of Engineers. (2001). *The New York District's Biological Monitoring Program for the Atlantic Coast of New Jersey, Asbury Park to Manasquan Section Beach Erosion Control Project*. Vicksburg, MS: Engineer Research and Development Center, Waterways Experiment Station.

U.S. Fish and Wildlife Service. (2002). *Draft Fish and Wildlife Coordination Act Report: Bogue Banks Shore Protection Project, Carteret County, NC*. Raleigh, NC: Raleigh Ecological Services Field Office, U.S. Fish and Wildlife Service.

Utizi, K., Corbau, C., Rodella, I., Nannini, S., and Simeoni, U. (2016). A mixed solution or a highly protected coast (Punta Marina, Northern Adriatic Sea, Italy). *Marine Geology* **381**: 114–127.

Valiela, I., Peckol, P., D'Avanzo, C., et al. (1998). Ecological effects of major storms on coastal watersheds and coastal waters: Hurricane Bob on Cape Cod. *Journal of Coastal Research* **14**: 218–238.

Valverde, H. R., Trembanis, A. C., and Pilkey, O. H. (1999). Summary of beach nourishment episodes on the U.S. east coast barrier islands. *Journal of Coastal Research* **15**: 1100–1118.

van Aarde, R. J., Wassenaar, T. D., Niemand, L., Knowles, T., and Ferreira, S. (2004). Coastal dune forest rehabilitation: a case study on rodent and bird assemblages in northern Kwazulu-Natal, South Africa. In Martínez, M. L. and Psuty, N. P. (Eds.), *Coastal Dunes, Ecology and Conservation*. Berlin: Springer-Verlag, pp. 103–115.

van Bohemen, H. D., and Meesters, H. J. N. (1992). Ecological engineering and coastal defense. In Carter, R. W. G., Curtis, T. G. F., and Sheehy-Skeffington, M. J. (Eds.), *Coastal Dunes: Geomorphology, Ecology and Management for Conservation*. Rotterdam: A.A. Balkema, pp. 369–378.

van Boxel, J. H., Jungerius, P. D., Kieffer, N., and Hampele, N. (1997). Ecological effects of reactivation of artificially stabilized blowouts in coastal dunes. *Journal of Coastal Conservation* **3**: 57–62.

van den Hoek, R. E., Brugnach, M., and Hoekstra, A. Y. (2012). Shifting to ecological engineering in flood management: introducing new uncertainties in the development of a Building with Nature pilot project. *Environmental Science and Policy* **22**: 85–99.

van der Biest, K., de Nocker, L., Povoost, S., Boerema, A., Staes, J., and Meire, P. (2017). Dune dynamics safeguard ecosystem services. *Ocean & Coastal Management* **149**: 148–158.

van der Biest, K., Meire, P., Schellekens, T., et al. (2020). Aligning biodiversity conservation and ecosystem services in spatial planning: focus on ecosystem processes. *Science of the Total Environment* **712**: 136350.

van der Hagen, H. G. J. M., Geelen, L. H. W. T., and de Vries, C. N. (2008). Dune slack restoration in Dutch mainland coastal dunes. *Journal for Nature Conservation* **16**: 1–11.

van der Laan, D., van Tongeren, O. F. R., van der Putten, W. H., and Veenbaas, G. (1997). Vegetation development in coastal foredunes in relation to methods of establishing marram grass (*Ammophila arenaria*). *Journal of Coastal Conservation* **3**: 179–190.

van der Maarel, E. (1979). Environmental management of coastal dunes in The Netherlands. In Jefferies, R. L. and Davy, A. J. (Eds.), *Ecological Processes in Coastal Environments*. Oxford: Blackwell, pp. 543–570.

van der Merwe, D., and McLachlan, A. (1991). The interstitial environment of coastal dune slacks. *Journal of Arid Environments* **21**: 151–163.

van der Meulen, F. and Salman, A. H. P. M. (1995). Management of Mediterranean coastal dunes. In Salman, A. H. P. M., Berends, H., and Bonazountas, M. (Eds.), *Coastal Management and Habitat Conservation*. Leiden: EUCC, pp. 261–277.

van der Meulen, F. and Salman, A. H. P. M. (1996). Management of Mediterranean coastal dunes. *Ocean & Coastal Management* **30**: 177–195.

van der Meulen, F., Bakker, T. W. M., and Houston, J. A. (2004). The costs of our coasts: examples of dynamic dune management from western Europe. In Martínez, M. L. and Psuty, N. P. (Eds.), *Coastal Dunes, Ecology and Conservation*. Berlin: Springer-Verlag, pp. 259–277.

van der Meulen, F., van der Valk, B., Baars, L., Schoor, E., and van Woerden, H. (2014). Development of new dunes in the Dutch Delta: nature compensation and "building with nature". *Journal of Coastal Conservation* **18**: 505–513.

van der Meulen, F., van der Valk, B., Vertegaal, K., and van Eerden, M. (2015). "Building with nature" at the Dutch dune coast: compensation target management in Spanjaard Duin at EU and regional policy levels. *Journal of Coastal Conservation* **19**: 707–714.

van der Putten, W. H. (1990). Establishment and management of *Ammophila arenaria* (marram grass) on artificial coastal foredunes in The Netherlands. In Davidson-Arnott, R. G. D. (Ed.), *Proceedings of the Symposium on Coastal Sand Dunes*. Ottawa: National Research Council Canada, pp. 367–387.

van der Putten, W. H. and Kloosterman, E. H. (1991). Large-scale establishment of *Ammophila arenaria* and quantitative assessment by remote sensing. *Journal of Coastal Research* **7**: 1181–1194.

van der Putten, W. H. and Peters, B. A. M. (1995). Possibilities for management of coastal foredunes with deteriorated stands of *Ammophila arenaria* (marram grass). *Journal of Coastal Conservation* **1**: 29–39.

Van der Salm, J. and Unal, O. (2003). Towards a common Mediterranean framework for beach nourishment projects *Journal of Coastal Conservation* **9**: 35–42.

van der Veen, A., Grootjans, A. P., de Jong, J., and Rozema, J. (1997). Reconstruction of an interrupted primary beach plain succession using a geographical information system. *Journal of Coastal Conservation* **3**: 71–78.

van der Wal, D. (1998). The impact of the grain-size distribution of nourishment sand on aeolian sand transport. *Journal of Coastal Research* **14**: 620–631.

van der Wal, D. (2000). Grain-size-selective aeolian sand transport on a nourished beach. *Journal of Coastal Research* **16**: 896–908.

van der Wal, D. (2004). Beach-dune interaction in nourishment areas along the Dutch coast. *Journal of Coastal Research* **20**: 317–325.

van der Windt, H. J., Swart, J. A. A., and Keulartz, J. (2007). Nature and landscape planning: exploring the dynamics of valuation, the case of The Netherlands. *Landscape and Urban Planning* **79**: 218–228.

van Duin, M. J. P., Wiersma, N. R., Walstra, D. J. R., van Rijn, L. C., and Stive, M. J. F. (2004). Nourishing the shoreface: observations and hindcasting of the Egmond case, The Netherlands. *Coastal Engineering* **51**: 813–837.

van Egmond, E. M., van Bodegom, P. M., Berg, M. P., et al. (2018). A mega-nourishment creates novel habitat for intertidal macroinvertebrates by enhancing habitat relief of the sandy beach. *Estuarine, Coastal and Shelf Science* **207**: 232–241.

van Koningsveld, M. and Lescinski, J. (2007). Decadal scale performance of coastal maintenance in The Netherlands. *Shore and Beach* **75**(1): 20–36.

van Koningsveld, M. and Mulder, J. P. M. (2004). Sustainable coastal policy developments in The Netherlands. A systematic approach. *Journal of Coastal Research* **20**: 375–385.

van Koningsveld, M., Stive, M. J. F., Mulder, J. P. M., de Vriend, H. J., Ruessink, B. G., and Dunsbergen, D. W. (2003). Usefulness and effectiveness of coastal research: a matter of perception. *Journal of Coastal Research* **19**: 441–461.

van Leeuwen, S., Dodd, N., Calvete, D., and Falqués, A. (2007). Linear evolution of a shoreface nourishment. *Coastal Engineering* **54**: 417–431.

van Puijenbroek, M. E. B., Limpens, J., de Groot, A. V., et al. (2017). Embryo dune development drivers: beach morphology, growing season precipitation, and storms. *Earth Surface Processes and Landforms* **42**: 1733–1744.

van Tomme, J., Vanden Eede, S., Speybroeck, J., Degraer, S., and Vincx, M. (2013). Macrofaunal sediment selectivity considerations for beach nourishment programmes. *Marine Environmental Research* **84**: 10–16.

Vandemark, L. M. (2000). Understanding opposition to dune restoration: attitudes and perceptions of beaches and dunes as natural coastal landforms. Unpublished Ph.D. dissertation. Department of Geography, Rutgers University, New Brunswick, NJ.

Vandenbohede, A., Lebbe, L., Adams, R., Cosyns, E., Durinck, P., and Zwaenepoel, A. (2010). Hydrogeological study for improved nature restoration in dune ecosystems-Kleyne Vlakte case study. *Journal of Environmental Management* **91**: 2385–2395.

Verstrael, T. J. and van Dijk, A. J. (1996). Trends in breeding birds in Dutch dune areas. In Salman, A. H. P. M., Langeveld, M. J., and Bonazountas, M. (Eds.), *Coastal Management and Habitat Conservation*. Leiden: EUCC, pp. 403–416.

Vestergaard, P. (2013). Natural plant diversity development on a man-made dune system. In Martínez, M. L., Gallego-Fernández, J. B., and Hesp, P. A. (Eds.), *Restoration of Coastal Dunes*. New York: Springer, pp. 49–66.

Vestergaard, P. and Hansen, K. (1992). Changes in morphology and vegetation of a man-made beach-dune system by natural processes. In Carter, R. W. G., Curtis, T. G. F., and Sheehy-Skeffington, M. J. (Eds.), *Coastal Dunes: Geomorphology, Ecology and Management for Conservation*. Rotterdam: A.A. Balkema, pp. 165–176.

Vieira da Silva, G., Hamilton, D., Murray, T., et al. (2020). Impacts of a multi-purpose artificial reef on hydrodynamics, waves and long-term beach morphology. *Journal of Coastal Research* **SI95**: 706–710.

Viola, S. M., Hubbard, D. M., Dugan, J. E., and Schooler, N. K. (2014). Burrowing inhibition by fine textured beach fill: implications for recovery of beach ecosystems. *Estuarine, Coastal and Shelf Science* **150**: 142–148.

Vousdoukas, M. I., Bouziotas, D., Giardino, A., Bouwer, L. M., Voukoucalas, E., and Feyen, L. (2018). Understanding epistemic uncertainty in large-scale coastal flood risk assessment for present and future climates. *Natural Hazards and Earth System Sciences* **18**: 2127–2142.

Waks, L. J. (1996). Environmental claims and citizen rights. *Environmental Ethics* **18**: 133–148.

Walker, I. J., Eamer, J. B. R., and Darke, I. B. (2013). Assessing significant geomorphic changes and effectiveness of dynamic restoration in a coastal dune ecosystem. *Geomorphology* **199**: 192–204.

Walmsley, C. A. and Davey, A. J. (1997a). The restoration of coastal shingle vegetation: effects of substrate composition on the establishment of seedlings. *Journal of Applied Ecology* **34**: 143–153.

Walmsley, C. A. and Davey, A. J. (1997b). The restoration of coastal shingle vegetation: effects of substrate composition on the establishment of container-grown plants. *Journal of Applied Ecology* **34**: 154–165.

Wamsley, T. V., Waters, J. P., and King, D. B. (2011). Performance of experimental low volume beach fill and clay core dune shore protection project. *Journal of Coastal Research* **SI59**: 202–210.

Wanders, E. (1989). Perspectives in coastal dune management. In van der Meulen, F., Jungerius, P. D., and Visser, J. H. (Eds.), *Perspectives in Coastal Dune Management*. The Hague: SPB Academic Publishing, pp. 141–148.

Warren, R. S., Fell, P. E., Rozsa, R., et al. (2002). Salt marsh restoration in Connecticut: 20 years of science and management. *Restoration Ecology* **10**: 497–513.

Warrick, J. A., Stevens, A. W., Miller, I. M., Harrison, S. R., Ritchie, A. C., and Gelfenbaum, G. (2019). World's largest dam removal reverses coastal erosion. *Scientific Reports* **9**: 13968.

Watson, J. J., Kerley, G. I. H., and McLachlan, A. (1997). Nesting habitat of birds breeding in a coastal dunefield, South Africa and management implications. *Journal of Coastal Research* **13**: 36–45.

Webb, C. E., Oliver, I., and Pik, A. J. (2000). Does coastal foredune stabilization with *Ammophila arenaria* restore plant and arthropod communities in Southeast Australia. *Restoration Ecology* **8**: 283–288.

Wells, J. T. and McNinch, J. (1991). Beach scraping in North Carolina with special reference to its effectiveness during Hurricane Hugo. *Journal of Coastal Research* **SI8**: 249–261.

Westhoff, V. (1985). Nature management in coastal areas of Western Europe. *Vegetatio* **62**: 523–532.

Westhoff, V. (1989). Dunes and dune management along the North Sea Coasts. In van der Meulen, F., Jungerius, P. D., and Visser, J. H. (Eds.), *Perspectives in Coastal Dune Management*. The Hague: SPB Academic Publishing, pp. 41–51.

Westman, W. E. (1991). Ecological restoration projects: measuring their performance. *The Environmental Professional* **13**: 207–215.

White, P. S. and Walker, J. L. (1997). Approximating nature's variation: selecting and using reference information in restoration ecology. *Restoration Ecology* **5**: 338–349.

Wiedemann, A. M. and Pickart, A. J. (1996). The Ammophila problem on the northwest coast of North America. *Landscape and Urban Planning* **34**: 287–299.

Wiedemann, A. M. and Pickart, A. J. (2004). Temperate zone coastal dunes. In Martínez, M. L. and Psuty, N. P. (Eds.), *Coastal Dunes, Ecology and Conservation*. Berlin: Springer-Verlag, pp. 53–65.

Wiegel, R. L. (1993). Artificial beach construction with sand/gravel made by crushing rock. *Shore and Beach* **61**(4): 28–29.

Wiens, J. A. and Hobbs, R. J. (2015). Integrating conservation and restoration in a changing world. *Bioscience* **65**: 302–312.

Williams, A. T. and Davies, P. (2001). Coastal dunes of Wales; vulnerability and protection. *Journal of Coastal Conservation* **7**: 145–154.

Williams, A. and Feagin, R. (2010). Sargassum as a natural solution to enhance dune plant growth. *Environmental Management* **46**: 738–747.

Williams, A. T. and Tudor, D. T. (2001). Temporal trends in litter dynamics at a pebble pocket beach. *Journal of Coastal Research* **17**: 137–145.

Williams, A. T., Davies, P., Curr, R., et al. (1993). A checklist assessment of dune vulnerability and protection in Devon and Cornwall, UK. In *Coastal Zone 1993*. New York: American Society of Civil Engineers, pp. 3394–3408.

Williams, A. T., Giardino, A., and Pranzini, E. (2016). Canons of coastal engineering in the United Kingdom: seawalls/groins, a century of change? *Journal of Coastal Research* **32**: 1196–1211.

Williams, G. D. and Thom, R. M. (2001). *Marine and Estuarine Shoreline Modification Issues*. White paper. Olympia, WA: Washington Department of Fish and Wildlife, Department of Ecology.

Willis, C. M. and Griggs, G. B. (2003). Reductions in fluvial sediment discharge by coastal dams in California and implications for beach sustainability. *The Journal of Geology* **111**: 167–182.

Wong, P. P. (Ed.). (1993). *Tourism vs Environment: The Case for Coastal Areas*. Dordrecht: Kluwer Academic Publishers.

Wood, D. W. and Bjorndal, K. A. (2000). Relation of temperature, moisture, salinity, and slope to nest site selection in loggerhead sea turtles. *Copeia* **1**: 119–128.

Woodhouse, W. W. (1974). *Stabilizing Coastal Dunes*. Reprint No. 70. Raleigh: North Carolina University Sea Grant.

Woodhouse, W. W. and Hanes, R. E. (1967). *Dune Stabilization with Vegetation on the Outer Banks of North Carolina*. TM-22. Washington, DC: U.S. Army Coastal Engineering Research Center.

Woodhouse, W. W. Jr., Seneca, E. D., and Broome, S. W. (1977). Effect of species on dune grass growth. *International Journal of Biometeorology* **21**: 256–266.

Woodruff, P. E. and Schmidt, D. V. (1999). Florida beach preservation – a review. *Shore and Beach* **67**(4): 7–13.

Wooldridge, T., Henter, H. J., and Kohn, J. R. (2016). Effects of beach replenishment on intertidal invertebrates: a 15-month, eight beach study. *Estuarine, Coastal and Shelf Science* **175**: 24–33.

Wootton, L. S., Halsey, S. D., Bevaart, K., McGough, A., Ondreicka, J., and Patel, P. (2005). When invasive species have benefits as well as costs: managing *Carex Kobomugi* (Asiatic sand sedge) in New Jersey's coastal dunes. *Biological Invasions* **7**: 1017–1027.

Wootton, L., Miller, J., Miller, C., Peek, M., Williams, A., and Rowe, P. (2016). *Dune Manual*. Sandy Hook: New Jersey Sea Grant Consortium.

Wortley, L., Hero, J.-M., and Howes, M. (2013). Evaluating ecological restoration success: a review of the literature. *Restoration Ecology* **21**: 537–543.

Wright, L. D. and Short, A. D. (1984). Morphodynamic variability of surf zones and beaches: a synthesis. *Marine Geology* **56**: 93–118.

Wright, S. and Butler, K. S. (1984). Land use and economic impacts of a beach nourishment project. In *Coastal Zone 83*, post conference volume. Sacramento: California State Lands Commission, pp. 1–18.

Zarkogiannis, S. D., Kontakiotis, G., Vousdoudas, M. I., Velegrakis, A. F., and Collins, M. B. (2018). Scarping of artificially-nourished mixed sand and gravel beaches: sedimentological characteristics of Hayling Island beach, Southern England. *Coastal Engineering* **133**: 1–12.

Zedler, J. B. (1991). The challenge of protecting endangered species habitat along the southern California coast. *Coastal Management* **19**: 35–53.

Zelo, I., Shipman, H., and Brennan, J. (2000). *Alternative Bank Protection Methods for Puget Sound Shorelines*. Ecology Publication #00-06-012. Olympia, WA: Washington Department of Ecology.

Zielinski, S., Botero, C. M., and Yanes, A. (2019). To clean or not to clean? A critical review of beach cleaning methods and impacts. *Marine Pollution Bulletin* **139**: 390–401.

Zinnert, J. C., Via, S. M., Nettleton, B. P., Tuley, P. A., Moore, L. J., and Stallins, J. A. (2020). Connectivity in coastal systems: Barrier island vegetation influences upland migration in a changing climate. *Global Change Biology* **25**: 2419–2430.

Zmyslony, J. and Gagnon, D. (2000). Path analysis of spatial predictors of front-yard landscapes: a random process? *Landscape and Urban Planning* **40**: 295–307.

Index

CPSIA information can be obtained
at www.ICGtesting.com
Printed in the USA
LVHW060822240222
711773LV00023B/102

9 781316 516157